Cosmology: Astronomy and Astrophysics

Cosmology: Astronomy and Astrophysics

Edited by Vivian Miles

SYRAWOOD
PUBLISHING HOUSE
New York

Published by Syrawood Publishing House,
750 Third Avenue, 9th Floor,
New York, NY 10017, USA
www.syrawoodpublishinghouse.com

Cosmology: Astronomy and Astrophysics
Edited by Vivian Miles

International Standard Book Number: 978-1-68286-841-6 (Hardback)

Cataloging-in-Publication Data

Cosmology : astronomy and astrophysics / edited by Vivian Miles.
    p. cm.
Includes bibliographical references and index.
ISBN 978-1-68286-841-6
1. Cosmology. 2. Astronomy. 3. Astrophysics. I. Miles, Vivian.
QB980 .C67 2019
523.1--dc23

TABLE OF CONTENTS

PREFACE

The study of the universe, its origin, evolution and future is under the scope of cosmology. Astronomy is the science concerned with celestial objects such as stars, galaxies, planets, gamma ray bursts, etc. and their associated phenomena, origin and evolution. Cosmology deals with the universe as a whole, while astronomy studies individual celestial objects of the universe. Astrophysics is a branch of astronomy, which integrates the principles of physics and chemistry for the study of the nature of astronomical objects. The objects studied in astrophysics are stars, extrasolar planets, galaxies, cosmic microwave background and the interstellar medium. The studies of these are approached from both theoretical and observational astrophysics. This book is a compilation of chapters that discuss the most vital concepts and emerging trends in the fields of cosmology, astrophysics and astronomy. Different approaches, evaluations, methodologies and advanced studies in these fields have been included in this book. It is meant for students who are looking for an elaborate reference text on these domains.

After months of intensive research and writing, this book is the end result of all who devoted their time and efforts in the initiation and progress of this book. It will surely be a source of reference in enhancing the required knowledge of the new developments in the area. During the course of developing this book, certain measures such as accuracy, authenticity and research focused analytical studies were given preference in order to produce a comprehensive book in the area of study.

This book would not have been possible without the efforts of the authors and the publisher. I extend my sincere thanks to them. Secondly, I express my gratitude to my family and well-wishers. And most importantly, I thank my students for constantly expressing their willingness and curiosity in enhancing their knowledge in the field, which encourages me to take up further research projects for the advancement of the area.

Editor

Stellar Populations in the Central Galaxies of Fossil Groups

Habib G. Khosroshahi[1] · Louisa A. Nolan[2]

[1] School of Astronomy, Institute for Research in Fundamental Sciences (IPM), PO Box 19395-5531, Tehran, Iran; email: habib@ipm.ir

[2] School of Physics and Astronomy, The University of Birmingham, Birmingham B15 2TT, UK

Abstract. It is inferred from the symmetrical and luminous X-ray emission of fossil groups that they are mature, relaxed galaxy systems. Cosmological simulations and observations focusing on their dark halo and inter-galactic medium properties confirm their early formation. Recent photometric observations suggest that, unlike the majority of non-fossil brightest group galaxies (BGGs), the central early-type galaxies in fossil groups do not have boxy isophotes, and are therefore likely to have formed in non-equal-mass, gas-rich galaxy-galaxy mergers.

Although the isophotal shapes of early-types can be used to infer the nature of the parent galaxies of their most recent mergers, detailed star-formation histories of the galaxies are also needed, in order to uncover the epoch of these mergers and the differences in the physical processes which produce fossil groups rather than 'normal' galaxy groups. In this study, we use a powerful long-baseline (UV-optical), multi-component, stellar population fitting technique to disentangle the star-formation history of the dominant central giant elliptical galaxy in a sample of five fossil groups, and compare this with a control sample of non-fossil BGGs, in an attempt to fully understand their merger histories. Our technique allows us to identify multiple epochs of star formation in a single galaxy, and constrain the metallicities of the constituent stellar populations. Resolving the populations in this way gives us a much clearer picture of galaxies' histories than simply using the mean age / metallicity of total stellar populations recovered by more conventional techniques using absorption line strengths.

We find that i) the dominant stellar components in both galaxy samples are old (> 10 Gyr), metal-rich ($\geq$ solar) and statistically indistinguishable in terms of their ages, metallicities and relative mass fractions; ii) the ages of the secondary, younger stellar components are also statistically indistinguishable between the two samples; iii) the central fossil galaxies have secondary stellar components which have significantly lower metallicities than the corresponding stellar populations in the BGG sample.

In a gas-rich merger, dissipation dilutes the previously-enriched gas in the central regions of the parent galaxies with gas which has not been metal-enriched, lowering the metallicity of the stars produced. Hence, we conclude that the central fossil galaxies are the products of early and exceptionally gas-rich mergers, which leaves them with lower metallicities in their most recent merger-induced stellar populations than their BGG counterparts. In contrast, although the last starburst-inducing merger of the 'normal' BGGs occurred at the same epoch as that of the fossil BGGs, they have undergone a subsequent dissipation-less merger, leaving them in the present day with boxy isophotes and more disturbed x-ray emission, but no further significant star formation.

Keywords: ISM: molecules, ISM: structure, instabilities

1   Introduction

Most galaxy groups and clusters have a luminous elliptical galaxy which resides at the centre of the gravitational potential of the system. It is generally accepted that these galaxies have acquired the majority of their mass via galaxy-galaxy mergers [1]. This is confirmed by later simulations [2, 3, 4], although the precise nature of such mergers is still debated, for example the relationship between the morphology of the merging galaxies and the merger product [5, 6].

An interesting class of galaxy groups known as 'fossil groups' have been at the centre of recent studies[7, 8, 9, 10]. It is believed that these are the end product of galaxy mergers within a group, and as such, they should not have undergone any recent major mergers. Hence, they represent simple laboratories in which to test merger scenarios. Fossils are defined as galaxy groups, or clusters, with an X-ray luminosity of at least $10^{42} erg/sec$ and a large luminosity gap (≥ 2 magnitude) between the two brightest galaxy in the system. This also means that typical luminosity galaxies ($L^\star$) galaxies are absent in the majority of the fossils, arguably as a result of multiple mergers of these galaxies within the group halo[11].

Fossils also show interesting properties in their inter-galactic medium (IGM). They have highly symmetric X-ray emission, which indicates the absence of recent group mergers. They are also outliers in some X-ray scaling relations. For instance, they are over luminous in X-ray for a given optical luminosity compared with 'normal' groups, and they have more concentrated dark matter halos [8]. These characteristics point to an early formation epoch for fossil groups. The study of fossils in the cosmological simulations [12] especially the recent detailed study of fossils in the Millennium Simulations by Dariush et al [13, 14] show that they assemble most of their mass earlier than non-fossils which have the same masses at the current epoch.

Galaxy groups usually have diverse properties [15, 16]. These are mainly attributed to their rapid evolution and recent formations [17] as the old groups usually don't survive in the hierarchical structure formation. Given the absence of recent group mergers in fossil groups and their early formation epoch, they are ideal environments to study the formation and evolution of galaxies, especially the formation of giant elliptical galaxies in mature halos. The central galaxies in fossils are at least as luminous as those in non-fossil groups and clusters [18, 19], however study of Khosroshahi et al [20] presented clues that isophotal shapes of those in fossil groups are non-boxy in contrast to the majority of massive ellipticals. This is later explored by Smith et al [9]. Numerical studies of galaxy mergers suggest that boxy isophotes are produced either in equal mass mergers, or mergers between bulge-dominated systems, regardless of mass ratio [6]. Central fossil galaxies are therefore likely to be the result of gas-rich mergers.

Probing the morphology of central fossil galaxies provides the first constraint on their merger and interaction history. Exploring their stellar populations allows us to further understand the nature of their merger history. Studies of the stellar populations of early-type galaxies have been carried out by, for example, Trager et al [21], Terlevich & Forbes [22] and van Zee et al [23]. These used absorption line strengths to extract luminosity-weighted *mean* ages and metallicities for the stellar populations of early-type galaxies. However, our new approach, fitting multi-component stellar population models to long-baseline (ultraviolet-optical) spectra, can uncover a more detailed star-formation history. The long-baseline spectra contain sufficient data to lift the age-metallicity degeneracy and also to disentangle multiple stellar components [24]. Hence, we can know the epoch of the last major merger-induced starburst more accurately than when fitting to individual spectral lines, which give us only an upper limit to the age of the last (presumably merger-induced) starburst in these galaxies.

In this spirit we study a sample of five central fossil galaxies, which although it is small, is still the largest sample available. We explore the nature and epoch of the last major interactions/mergers they have experienced, and compare with a control sample of seven brightest group galaxies. Section two describes the data and the observations. A short description of the method and the results are provided in section 3. Section 4 contains a summary and discussion of the results. We adopt a cosmology with $H_0 = 70$ km s^{-1} Mpc^{-1} and $\Omega_m = 0.3$ with cosmological constant $\Omega_\Lambda = 0.7$ throughout.

2   The Sample

This study uses a sample of five giant elliptical galaxies dominating fossil groups and seven BGGs in non-fossil groups, all with long-baseline optical spectra.

The spectroscopic observations of the fossil sample were performed using a multi-slit spectrograph on the KPNO 4-m telescope on the 2000 March 11th. The Ritchey-Cretien spectrograph and KPC-10 Å grating gave a dispersion of 2.75 Å pixel^{-1} over 3800-3850 Å and with 1.8-arcsec slitlets, a resolution of 6 Å [full width at half-maximum (FWHM)] was achieved. Risley prisms compensated for atmospheric dispersion. Spectra were obtained through three slitmasks, with typically an hour exposure on each. The spectroscopic data were reduced and analysed in the standard way using IRAF.

The brightest group galaxy (BGG) sample was selected from the well-studied groups in the Group Evolution Multi-wavelength Study, GEMS [25]. The groups in which the BGGs reside have a range of X-ray properties [16], and were selected for the archived availability of their ultraviolet-to-near-infrared spectra. Table 1 lists the sources for the spectra. The BGG names are the same as the names of their host groups.

For both the BGGs and the fossil BGGs, it is the central $\sim$ one fifth of the effective radius (r_e) of the galaxies which was observed.

We compared the properties of the BGG sample, namely the K_s-band luminosity, with a much larger sample of BGGs listed by Ellis & O'Sullivan [26] and noted that our sample has a mean solar K_s-luminosity of 11.1 with a standard deviation of 0.2 while the same quantities for the BGGs in the Ellis & O'Sullivan [26] sample are 11.3 and 0.3, respectively. Our fossil central galaxies are in general more luminous (mean K_s-luminosity of 11.7 and standard deviation of 0.4). As we argue below, this should strengthen our conclusions rather than the opposite.

3   The Analysis and Results

3.1   The Star-Formation History

To determine the star-formation history of the fossil and brightest group galaxies, we use the powerful long-baseline two-component stellar population fitting technique, as described in detail in Nolan et al [24]. We use the Bruzual & Charlot [27] single stellar population synthesis models, and note that our results in Nolan et al (2007) were robust to choice of model. The Bruzual & Charlot [27] models have ages ranging from $0.01 - 14$ Gyr. Solar metallicity ($Z_\odot$) is 0.02, and the available metallicities are: 0.02, 0.2, 0.4, 1.0 and 2.5 $Z_\odot$. The age, metallicity and relative stellar mass fraction of each component were allowed to vary as free parameters in the fitting process. The best-fitting parameters of the two stellar populations were recovered via χ^2 minimisation. As in Nolan et al [24], we fit the various spectral sections (ultraviolet, optical, near-infrared) for each galaxy simultaneously, but with the relative normalisations allowed to float independently, to compensate for any deviations

Table 1: Archive sources of the spectra for the non-fossil BGGs. The Bica & Alloin data are available at *ftp://cdsarc.u-strasbg.fr/cats/III/219/galaxy/*, and those for NGC 1052 and NGC 1407 are available at *http://www.stsci.edu/ftp/catalogs/nearby gal/sed.html*.

object	source	wavelength / Å
NGC 1052	IUE Newly Extracted Spectra	2500 − 3200
	Storchi-Bergmann, Calzetti & Kinney unpublished	3200 − 7500
NGC 1407	Storchi-Bergmann, Calzetti & Kinney unpublished	3200 − 7500
IC 1459	Bica & Alloin, unpublished	3100 − 5400
	Bica & Alloin, 1987a	3800 − 7500
NGC 3557	Bica & Alloin, 1987a	3800 − 7500
	Bica & Alloin, 1987b	6400 − 8500
NGC 3923	Bica & Alloin, unpublished	3100 − 5400
	Bica & Alloin, 1987a	3800 − 7500
	Bica & Alloin, unpublished	3100 − 5400
NGC 4697	Bica & Alloin, unpublished	3100 − 5400
	Bica & Alloin, 1987a	3800 − 7500
	Bica & Alloin, 1987b	6400 − 8500
NGC 7144	Bica & Alloin, unpublished	3100 − 5400
	Bica & Alloin, 1987a	3800 − 7500
	Bica & Alloin, 1987b	6400 − 8500

in the flux calibration. Some long single sections have been split into two to compensate for any potential flux deviations across the broad wavelength range. Sky lines and emission lines are masked out of the fit. We use two components rather than three or more as we expect the bulk of the stars in most early-type galaxies to be represented by two components [28], and increasing the number of components rapidly becomes computationally expensive for diminishing returns.

Figure 1 presents the best-fitting parameters recovered from the two-component fitting. The galaxy spectra, with the best-fitting two-component models overlaid, are shown in Figures 2 and 3. In Figure 1, it can be seen that, although the dominant populations in both samples are old and metal-rich ($\geq Z_\odot$), the younger, secondary, stellar populations are generally of a lower metallicity in the fossil population than for the BGGs. The statistics presented in Table 2 confirm this. The distributions of the ages, metallicities and relative mass fractions of the dominant stellar populations in the two samples are indistinguishable from each other, and the Kolmogorov-Smirnov probabilities suggest that it is highly likely that they are drawn from the same population. However, for the second components, the metallicity distributions are significantly different, although the ages are not. The Kolmogorov-Smirnov probability that the metallicities are drawn from the same distribution is less than 20%, and the BGG populations have a mean metallicity more than three times that of the fossils.

3.2   Photometric Analysis

The photometric properties of our sample of central fossil galaxies, i.e. their isophotal shapes and radial surface brightness distributions, were presented in Khosroshahi, Jones & Ponman [20]. The results were based deep R-band and K_s-band photometry of the sample observed

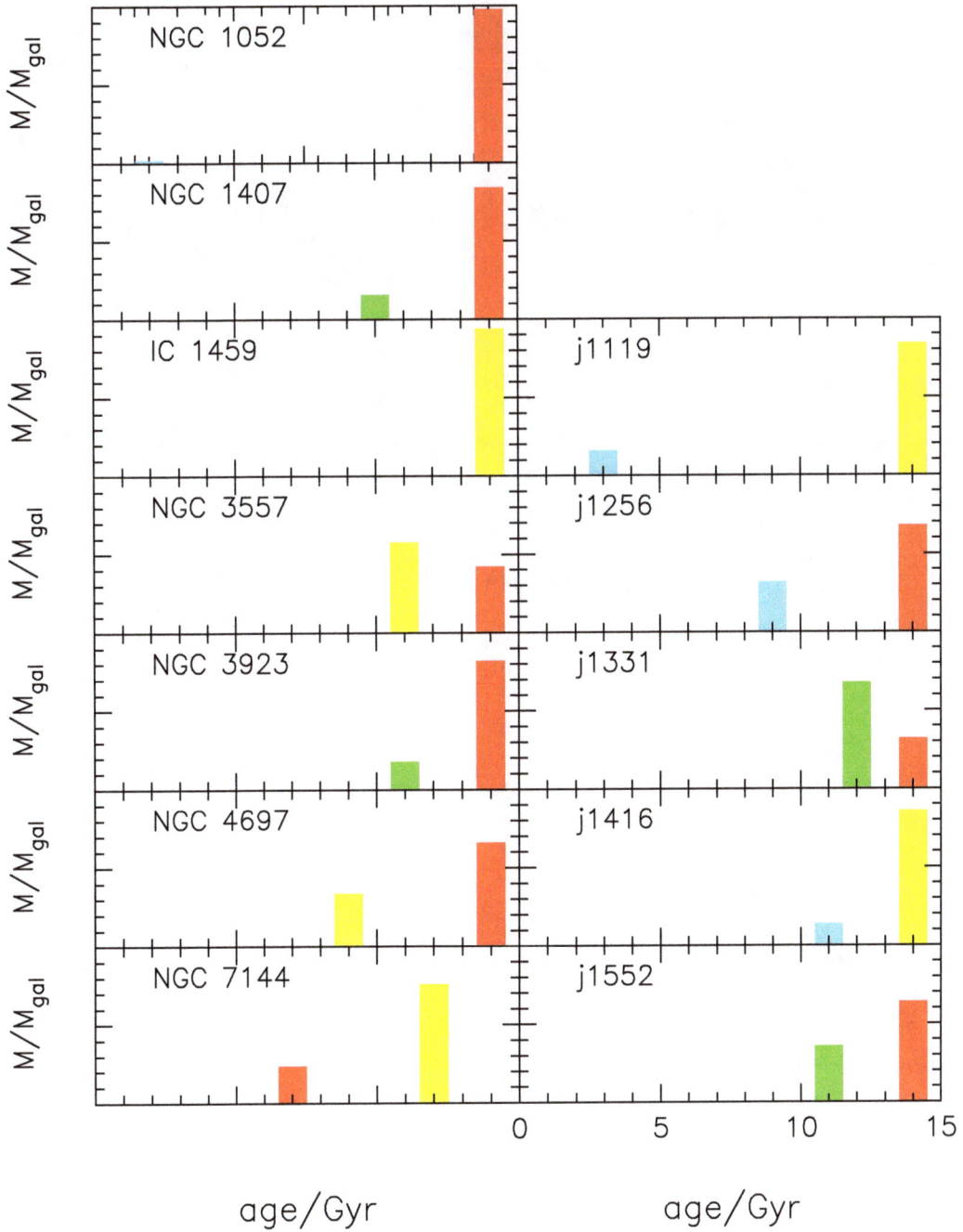

Figure 1: The best fitting parameters recovered from the two-component stellar population model fitting to the long-baseline spectra of the brightest group (**left**) and fossil galaxy (**right**) samples. The height of the histograms represent the fractional stellar mass of each component, and the colours represent the recovered metallicities: **red**: 5 $Z_\odot$; **yellow**: $Z_\odot$; **green**: 0.4 $Z_\odot$; **light blue**: 0.2 $Z_\odot$; **dark blue**: 0.02 $Z_\odot$.

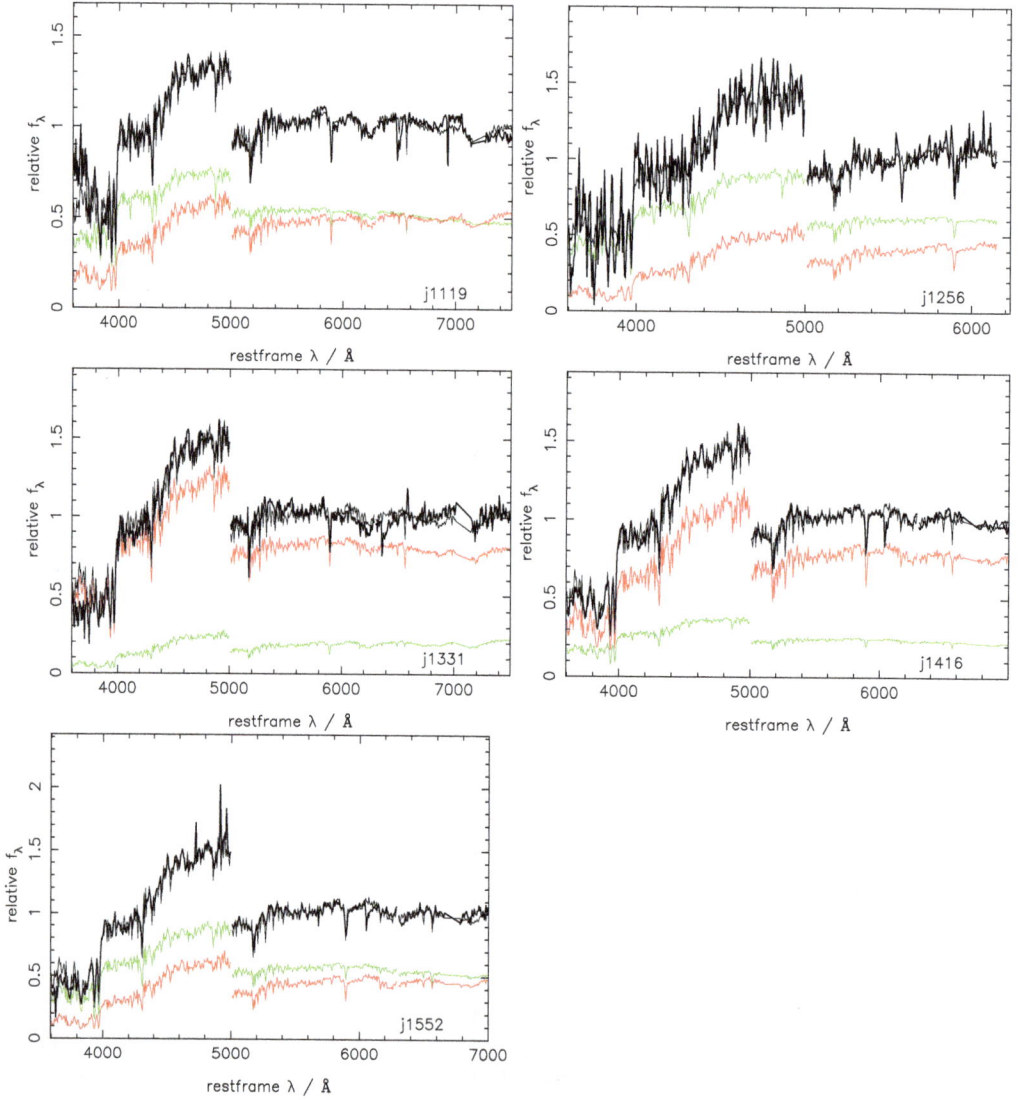

Figure 2: The best-fitting two-component model (thin black line) superimposed over the spectral sections of the fossil galaxies (thick black line). The two component populations are also shown; the dominant population is in red, and the secondary stellar population is in green. The flux normalisation has been arbitrarily adjusted for clarity.

Table 2: Mass-weighted mean values for the best-fitting star-formation history parameters for the fossil and brightest group galaxy samples. The older population is labelled 1, and the younger, 2. The uncertainties quoted are the standard deviations from the mass-weighted mean. Last column shows the mean stellar mass fraction of the older population . The Kolmogorov-Smirnov probabilities that the two samples are drawn from the same populations are also given.

	mean age 1 Gyr	mean age 2 Gyr	mean Z1 $Z_\odot$	mean Z2 $Z_\odot$	mass fraction
BGGs	13.7±0.7	9.8±3.6	2.03±0.68	1.10±0.91	0.79±0.18
fossils	14.0±0.5	10.3±3.4	1.74±0.75	0.33±0.11	0.67±0.19
KS probability	(1-2.4e-7)	(1-6.3e-5)	0.932	0.193	0.871

with wide field camera on Issac Newton Telescope and UIST on UK Infrared Telescope. We perform a similar analysis on the BGG comparison sample, using archival R-band data, observed by the WFI/ESO 2.2m telescope. Table 2 presents the results of the photometric analysis for the two samples.

Most of the non-fossil BGGs (5/6 for which the isophotes were fitted) have boxy isophotes. However, *none* of the fossil BGGs exhibit boxy isophotes [20]. The central fossil galaxies are on average more luminous than the non-fossil BGGs. Given that, in general, more luminous early-types are more likely to have boxy isophotes [29], one would expect that the fossil BGGs would have boxy isophotes if they had evolved in a similar manner to other brightest group or cluster galaxies, but this is not what we see.

The Sersic indices, n_S, are indistinguishable between the two samples. The mean for both samples is 4.1, and the Kolmogorov-Smirnov probability that the two samples are drawn from the same underlying population is 97 %. However, the fossil galaxies are much larger than the non-fossil BGGs, with a mean half-light radius in the sample $r_e = 23.1$ kpc, compared with mean $r_e = 4.6$ kpc for the non-fossil BGGs, and the fossils are on average $\sim$ 1 magnitude brighter than the non-fossils. The Kolmogorov-Smirnov probability that these two samples are drawn from the same underlying population is only 65 %.

4 Discussion and Conclusions

We have studied the stellar populations of five central fossil group galaxies using a two-component spectral decomposition of their long-slit optical spectra. Thus, we find the epoch of their last major starburst, and hence their last gas-rich major merger/interaction. We combine this with the photometric analysis, from which we infer the nature of the parent galaxies in the last major merger. We compare these results with those from a control group of seven BGGs, chosen on the basis of the availability of archival data.

We find that i) the dominant stellar components in both galaxy samples are old (> 10 Gyr), metal-rich ($\geq Z_\odot$) and statistically indistinguishable in terms of their ages, metallicities and relative mass fractions; ii) the ages of the secondary, younger stellar components are also statistically indistinguishable between the two samples; iii) the central fossil galaxies have secondary stellar components which have significantly lower metallicities than the corresponding stellar populations in the BGG sample; iv) the fossil galaxies have non-boxy isophotal shapes, whereas the non-fossil BGGs are predominantly boxy [20].

Our sample of brightest fossil galaxies is not luminosity-matched to the non-fossil BGG sample; our fossil galaxies are, on average, more luminous than the non-fossils. However,

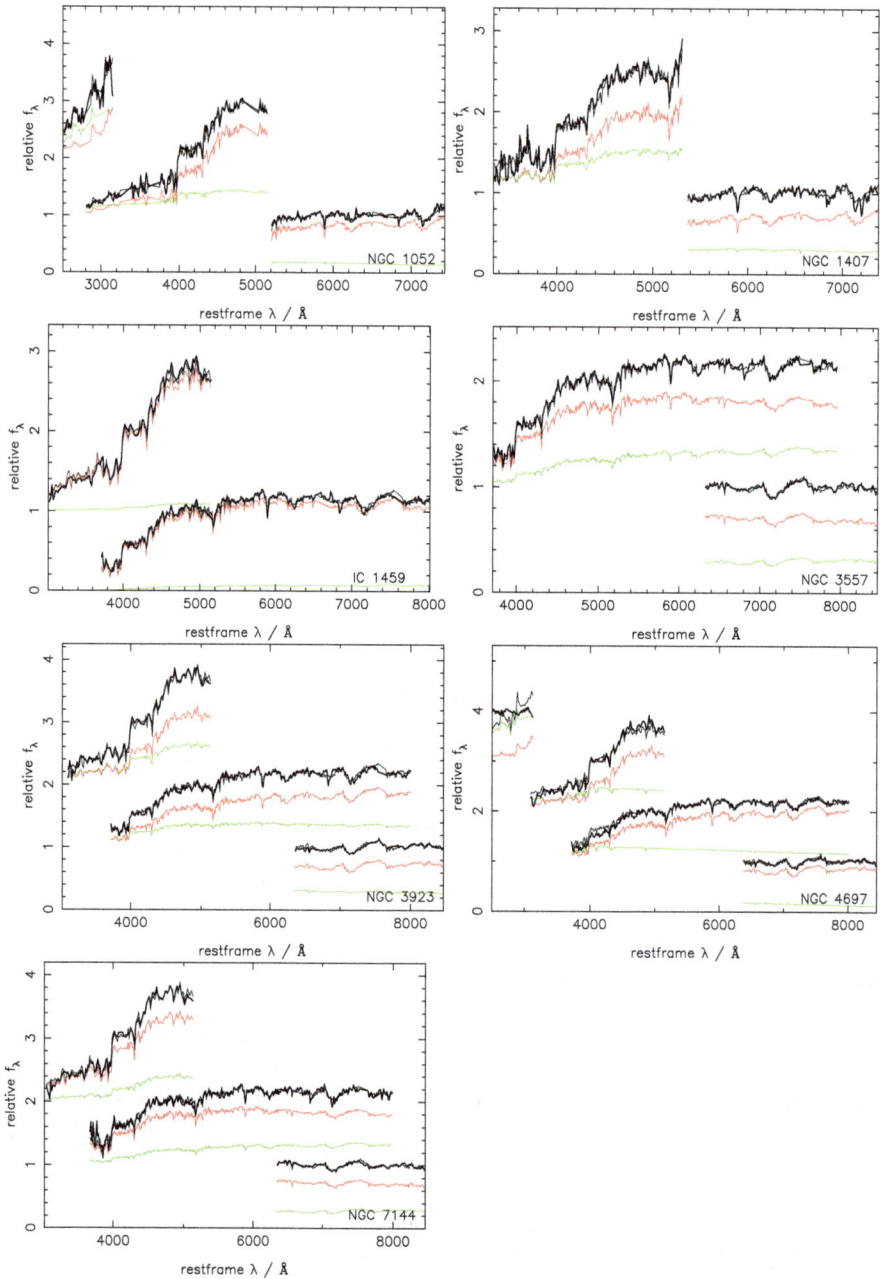

Figure 3: The best-fitting two-component model (thin black line) superimposed over the spectral sections of the brightest group galaxies (thick black line). The two component populations are also shown; the dominant population is in red, and the secondary stellar population is in green. The flux normalisation has been arbitrarily adjusted for clarity.

Table 3: Photometric properties of the sample.

Group	R.A. (J2000)	dec. (J2000)	z	M_R mag	a_4/a ×100	n_S	r_e kpc
fossil BGGs							
RX J1119.7+2126	11:19:43.6	+21:26:51	0.061	-22.1	0.1	5.1	10.1
RX J1256.0+2556	12:56:03.4	+25:56:48	0.232	-24.1	0.1	3.1	27.2
RX J1331.5+1108	13:31:30.2	+11:08:04	0.081	-22.9	0.3	4.3	11.0
RX J1416.4+2315	14:16:26.9	+23:15:32	0.137	-24.3	0.7	3.6	29.5
RX J1552.2+2013	15:52:12.5	+20:13:32	0.135	-24.0	0.5	4.6	37.9
non-fossil BGGs							
NGC 1052	02:41:04.8	-08:15:21	0.005	-21.6	0.0	4.2	2.12
NGC 1407	03:40:11.9	-18:34:49	0.006	-22.8	-0.3	3.5	4.62
IC 1459	22:57:10.6	-36:27:44	0.006	-22.7	–	3.7	2.95
NGC 3557	11:09:57.6	-37:32:21	0.010	-23.3	-0.2	4.6	10.53
NGC 3923	11:51:01:8	-28:48:22	0.006	-23.0	-0.4	–	–
NGC 4697	12:48:35.9	-05:48:03	0.004	-21.9	1.3	3.8	1.93
NGC 7144	21:52:42.4	-48:15:14	0.006	-21.7	-0.1	5.0	5.54

we expect more luminous galaxies to be *more* metal-rich than lower-luminosity galaxies [21]. Therefore, it seems unlikely that the lower metallicity of the more luminous fossil sample is a simple effect of our luminosity selection. Nor do we expect it to be an effect of the different aperture sizes probing different regions of the two classes of galaxy. Although the sizes of the apertures are different (in kpc), the region from which the spectra are sampled is $\sim r_e$ / 5 in both cases, so the metallicity gradient observed in the cores of early-type galaxies [30] should not be responsible for the significant differences between the metallicities in the young populations. Further, the well-known age-metallicity degeneracy [31] means that the secondary stellar populations in the central fossil galaxies would have to be younger if they were to match the higher metallicities of the corresponding stellar populations in the non-fossil galaxies, and we do not expect any such recent merger-induced major starbursts to be common in the relaxed, symmetrical fossil systems. In any case, the long-baseline spectra contain sufficient data to lift the age-metallicity degeneracy for the dominant populations in early-type galaxies [24].

As described above, we find no difference between the epochs of the last gas-rich merger in our two galaxy samples, but the younger stars in the central fossil galaxies have significantly lower metallicities than those in the non-fossil BGGS. The metallicity of a stellar population created in a starburst due to a major galaxy-galaxy merger or interaction can depend on a number of factors: i) the timescale of the starburst: the longer the duration of a starburst, the more time there is for metals to be created and made available for new stars; ii) how well the metals are retained in the potential well where the starburst is taking place; iii) the initial metallicity of the gas supplied by the parent galaxies; iv) how much un-enriched gas from outer regions is funnelled into the starburst to 'dilute' the metal-enrichment occurring in the starburst.

Conventional theory suggests that the later the epoch of formation, the more metal-rich a stellar population must be, as gas becomes increasingly enriched with metals released from earlier generations of stars. Whilst it is true that metals can only exist if earlier stellar generations have pre-existed, we do not in fact observe a simple metallicity-age correlation in the stellar populations of early-type galaxies. In fact, the most metal-rich stellar populations observed are generally associated with the oldest stellar populations, and we see relatively young stars with significantly lower metallicities [24, 32, 33]. This suggests that metallicity of stars produced in galaxy-galaxy mergers is dependent predominantly on local conditions, and in particular, on the amount of un-enriched gas available to a starburst event. In fact, Kewley et al [34], find evidence that the stronger a central, interaction-induced starburst is, the lower the metallicity of the resulting stellar population. This suggests that in a strong, and therefore gas-rich starburst, un-enriched gas from the outer regions of galaxies is funnelled into the central regions, diluting the pre-existing nuclear gas. As our fossil galaxies are more luminous than the non-fossil BGGs ($\sim$ 1 magnitude), a similar mass fraction in the younger stellar populations represents a more massive starburst in the fossils than in the non-fossil BGGs, consistent with this picture.

The isophotal shapes of early-type galaxies tell us something about the nature of their most recent merger. Recent numerical simulations show that boxy isophotes in early-type galaxies are formed in either equal-mass mergers or gas-poor mergers [6]. Hence, the last major merger / interaction of central fossil galaxies must have been gas-rich, in order to produce their non-boxy isophotes [20]. In this work, we can add to this our knowledge of the detailed star-formation history of the galaxies, which tells us when the last starburst-inducing merger occurred. Of course, in the case of the galaxies with boxy isophotes, the last starburst may not have occurred at the same time as the last major merger. If, as we expect, their last merger was gas-poor, there can have been no significant star-formation in that event. The younger population is therefore likely to be associated with an earlier merging event.

Combining our insights from the photometric and spectroscopic analyses detailed above, we tentatively arrive at the following picture for the formation history of non-fossil BGGs versus that of fossil BGGs, as sketched in Figure 4. Both classes of galaxy formed the bulk of their stars at an early epoch ($\geq$ 13 Gyr ago), and these stars have high metallicities ($>$ solar metallicity). They then underwent major, gas-rich, star-formation-inducing mergers at time t1, represented by the age of the younger stellar population, $\sim$ 10 Gyr. A lookback time of 10 Gyr is a redshift $\sim$ 2. Reassuringly, this is consistent with the epoch of peak star formation rate (e.g. Heavens et al. [35]), which is presumed to be the peak of merging-induced starbursts. In the fossil BGGs, the merging was very gas-rich, leading to low metallicities in the stars formed. The fossils completed this merging at an early epoch, and have evolved undisturbed since this epoch of merging. The non-fossil BGGs, however, have undergone another round of merging, at time t2 (which we do not know). As the parent galaxies are now the gas-poor early-type remnants of the previous round of merging, the final product has boxy isophotes, and no further significant star-formation. Hence, we can explain the stellar ages, isophotal shapes and X-ray emission of these two classes of galaxy in a fully consistent manner.

Although of course this study is based on a limited sample of galaxies, we have uncovered some tantalising evidence of differences between the photometric and stellar properties of BGGs and central fossil galaxies, which suggest different formation scenarios for these systems. We intend to follow up our study with a larger, statistically-selected sample, to further probe the statistical differences between star-formation history, luminosity, half-light radius etc. in fossil and non-fossil BGGs. We recognise that the N-body galaxy-galaxy merging simulations discussed here cover a limited parameter space, and we therefore intend to use the upcoming results from new simulations, which will sample a larger parameter space (e.g. mass ratio, orbital characteristics) than previously available, and include cooling, star-formation and black hole growth.

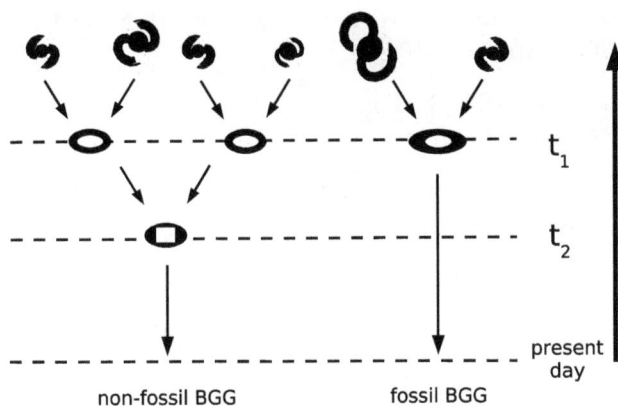

Figure 4: Merger history of fossil BGGs compared with non-fossil BGGs. See §4 for discussion.

Acknowledgements

This work uses INES data from the IUE satellite. We thank Laurence Jones for his earlier involvement in the spectroscopy of the fossil sample and Patricia Sanchez-Blazqez for discussion on the subject. We thank the anonymous referee for useful comments which improved the presentation of the paper.

References

[1] Toomre A., and Toomre J., 1972, ApJ, 178, 623

[2] Barnes J. E., 1988, ApJ, 331, 699

[3] Barnes J. E., Hernquist L., 1992, ARAA, 30, 705

[4] Dubinski J., 1998, ApJ, 502, 141

[5] Naab T., Burkert A., 2003, ApJ, 597, 893

[6] Khochfar S., Burkert A., 2005, MNRAS, 359, 1379

[7] Ponman T. J., Allan, D. J., Jones L. R., Merrifield M., MacHardy I. M., 1994, Nature, 369, 462

[8] Khosroshahi H. G., Ponman T. J., Jones L. R., 2007, MNRAS, 377, 595

[9] Smith G. P., Khosroshahi H. G., Dariush A., Sanderson A. J., Stott J. P., Haines C. P., Egami E., Stark D. P., 2010, MNRAS, 409, 169

[10] Alamo-Martínez K. A., West M. J., Blakeslee J. P., González-Lópezlira R. A., Jordán A., Gregg M., Côté P., Drinkwater M. J., van den Bergh S., 2012, A&A, 546A, 15

[11] Jones L. R., Ponman T. J., Horton A., Babul A., Ebeling H., Burke D. J., 2003, MNRAS, 343, 627

[12] D'Onghia E., Sommer-Larsen J., Romeo A. D., Burkert A., Pedersen K., Portinari L., Rasmussen J., 2005, ApJ, 630, 109

[13] Dariush A., Khosroshahi H. G., Ponman T. J., Pearce F., Raychaudhury S., Hartley W., 2007, MNRAS, 382, 433

[14] Dariush A., Raychaudhury S., Ponman T. J., Khosroshahi H. G., Benson, A. J., Bower R. G., Pearce F., 2010, MNRAS, 405, 187

[15] Mahdavi A., Böhringer H. Geller M. J., Ramella M., 2000, ApJ, 534, 114

[16] Osmond J. P. F., Ponman T. J., 2004, MNRAS, 350, 1511

[17] Rasmussen J., Ponman T. J., Mulchaey J. S., Miles T. A., Raychaudhury S., 2006, MNRAS, 373, 653

[18] Khosroshahi H. G., Maughan B., Ponman T. J., Jones L. R., 2006, MNRAS, 369, 1211

[19] Mendes de Oliveira C. L., Cypriano E. S., Sodré, L. Jr., 2006, AJ, 131, 158

[20] Khosroshahi H. G., Ponman T. J., Jones L. R., 2006, MNRAS Letters, 372, 68

[21] Trager S. C., Faber S. M., Worthey G., Gonzlez, J. Jesùs, 2000, AJ, 119,1645

[22] Terlevich A. I., Forbes D. A., 2002, MNRAS, 330, 547

[23] van Z. L., Barton E. J., Skillman E. D., 2004, AJ, 128, 279

[24] Nolan L. A., Dunlop J. S., Panter B., Jimenez R., Heavens A., Smith G., 2007, MNRAS, 375, 371

[25] Forbes D. A., et al., 2006, PASA, 23, 38

[26] Ellis S. C., O'Sullivan, E., 2006, MNRAS, 367, 627

[27] Bruzual G., Charlot S., 2003, MNRAS, 344, 1000

[28] Nolan L. A., Harva M. O., Kabán A., Raychaudhury S., 2006, MNRAS, 366, 321

[29] Bender R., Surma P., Doebereiner S., Moellenhoff C., Madejsky R., 1989, A&A, 217, 35

[30] Sánchez-Blázquez P., Forbes D. A., Strader J., Brodie J., Proctor R., 2007, MNRAS, 377, 759

[31] Worthey G., 1994, ApJS, 95, 107

[32] Maraston C., Thomas D., 2000, ApJ, 541, 126

[33] Kuntschner H., 2001, Ap&SS, 276, 885

[34] Kewley L. J., Geller M. J., Barton E. J., 2006, AJ, 131, 2004

[35] Heavens A., Panter B., Jimenez R. and Dunlop J., 2004, Nature, 428, 625

Segmentation of Photospheric Solar Images by using c-Means, k-Means, and FCM Algorithms

Mahdi Yousefzadeh[1] · Mohsen Javaherian[2] · Hossein Safari[2]

[1] Department of Physics, Institute for Advanced Studies in Basic Sciences (IASBS), Zanjan, P.O.Box 45195-1159, Iran

[2] Department of Physics, University of Zanjan, Zanjan, P.O.Box 45195313, Iran

Abstract. In this study, we use three kinds of clustering methods based on c-means, k-means, and fuzzy c-means (FCM) algorithms to segment solar ultra-violet (UV) images. The methods are applied on a sequence of quiet-Sun photospheric observations at 525 nm images taken by *Sunrise* on 9 June 2009. The comparison between these three algorithms represents a little bit differences in extraction of physical parameters (filling factors, brightness fluctuations, size distribution, etc.) from images. On the basis of FCM algorithm, the mean value of granule sizes is found to be about 1.8 arcsec^2 (0.85 Mm^2). Granules with sizes smaller than 2.8 arcsec^2 cover a wide range of brightness, while larger granules approaches a particular value. Granules may have lifetimes less than 10 minutes in this part of the Sun. Investigation of local fractal dimension of photospheric images shows that granulation pattern are approximately scale free in some resolutions.

Keywords: Sun: UV radiation, Sun: granulation, Techniques: image processing, Techniques: segmentation, Techniques: clustering

1 Introduction

Lots of events are stochastically appeared on the solar surface such as granules, bright points (BPs), etc. [1, 2] at times of both minimum and maximum solar activity [3]. Extracting physical properties of features happening on photosphere is important subjects of solar field. One of the features that has main role in solar activity is granulation pattern. It has been accepted the solar granulation is resulting from a convective turbulent process [4].

To optimize statistical analysis of granulation patterns, it is necessary to develop automatic detection techniques. The process of dividing data set into classes or clusters in the feature space is data clustering so that features (elements) in the same class have similarities but there is low resemblance between elements which are in various classes. One of the significant methods for granules segmentation and finding non-granular regions is using various kinds of clustering such as c-means and k-means. These algorithms are multichannel unsupervised and able to automatically segment regions fast [5]. Different kinds of clustering (e.g., spatial possibilistic clustering algorithm and fuzzy c-means clustering) have been applied on EUV solar images to segregate traditional regions into coronal hole (CH), quiet Sun (QS), and active regions (AR). The estimations of these areas characterizations are consistent with previous results [7]. We apply c-means, k-means, and fuzzy c-means (FCM) algorithms on the *Sunrise* photospheric UV images recorded on 9 June 2009 and break down this data set into regions that can be recognized as granule or non-granular region. The code is able to segment N-dimensional gray-scale images into c classes (more than two classes) by implementation of the c-means (k-means) clustering algorithm. The comparison between these two algorithms represents approximately the same results.

The paper is organized as follows: Data analysis is discussed in Section 2. Identification methods for granules and non-granular regions by using c-means, k-means, and FCM are explained in Section 3. The results are given in Section 4. The conclusions are discussed in Section 5.

2   Data Analysis

The *Sunrise* balloon-borne solar observatory was launched on 8 June 2009 [8, 9]. The high-resolution images in the UV and in the quiet Sun were begun to be recorded on 9 June 2009 [9]. *Sunrise* has a resolution of about 100 km (from a 1 m Gregory reflector telescope). This observatory is equipped with *Sunrise* filter imager [10], imaging magnetograph experiment [11], image stabilization and light distribution unit, and a correlating wave-front sensor [12]. The Imaging Magnetograph eXperiment (IMaX) data produces 936×936 pixel with resolution of 0.055 arcsec per pixel. IMaX uses a Zeeman triplet in the FeI 525.02 nm line. The *Sunrise* together with its instruments provides 4 levels of data. Level-0 is raw data. Level-1 data are fully reduced by phase-diversity reconstruction. Individual and averaged wave-front corrections are exerted on both level-2 and level-3 data, respectively. We used a sequence of IMaX (level-2) data (Figure 1A) recorded on 9 June 2009 (14:10-14:45 UT) at 525 nm to segregate granules and non-granular regions by three mentioned algorithms.

Using a subsonic filter, the global solar surface oscillations (e.g. 5-minute p-modes) are removed from data by modifying horizontal speeds above 5 km/s in momentum and frequency space.

3   Methods

3.1 c-Means Clustering

In c-mean clustering, called hard clustering, dataset is decomposed into certain classes where each feature belongs to exactly one cluster. For image clustering, pixels' brightness constructs our feature space [13]. Depending on the number of clusters, pixels with similar brightness set in the same class. As regions in this image set into two clusters, minimum and maximum brightness of image will be centers of our two clusters in feature space and the magnitude value of pixels brightness from centers will determine the class of each pixel. To achieve more computational efficiency, the histogram of the image brightness is calculated during the clustering process. When all of the pixels are clustered, mean of the each cluster is computed. The result of applying c-means clustering on *Sunrise* image is represented in Figure 1B.

3.2 k-Means Clustering

This algorithm, which is one of the unsupervised iterative non-deterministic learning algorithms, aims to divide n features into k clusters wherein each feature belongs to certain cluster with the nearest mean [13]. The purpose is minimizing an objective function, e.g. a squared error function.

Assume we have a finite collection of n features, $X = \{x_1, x_2, \ldots, x_n\}$,

1- k points are randomly chosen as a center of clusters and k clusters are generated.

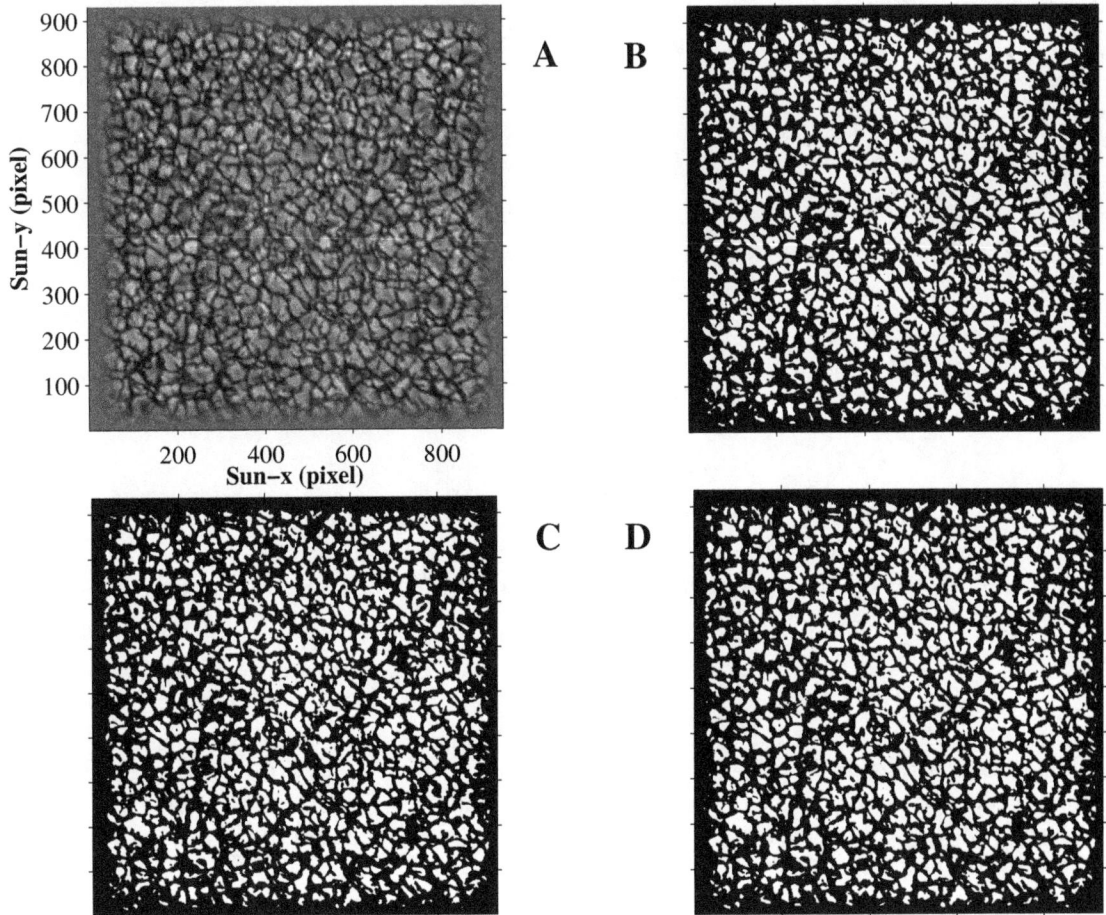

Figure 1: *Sunrise*/IMaX image (A) recorded on 9 June 2009 (14:16:00 UT). The segmented images based on *c*-means (B), *k*-means (C), and FCM procedure (D).

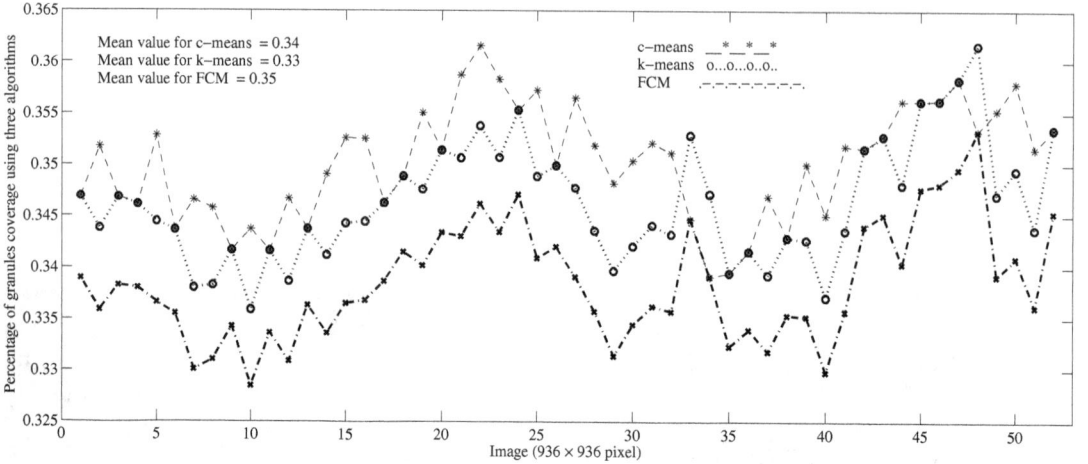

Figure 2: The filling factors of granules obtained by different clustering algorithms are presented. About one-third of each image is covered by granules.

2- Using association of each point to the nearest centroid, their belonging to a given data set is specified. Here, the objective function is introduced as

$$J = \sum_{j=1}^{k} \sum_{i=1}^{n} \parallel x_i^{(j)} - c_j \parallel^2 . \tag{1}$$

3- By averaging all of the pixels, centers of clusters are recomputed.

4- Step 2 and 3 are repeated until the centers of clusters achieve to some convergency (Figure 1C).

3.3   Fuzzy c-Means (FCM) Clustering

The FCM, which is rooted in fuzzy logic, aims to minimize an objective function such as k-means algorithm. In the other words, we could label this flexible algorithm as an extension of k-means clustering. Fuzzy c-means allows data points to be assigned into more than one cluster. Each data point has a degree of membership (or probability) of belonging to each cluster. We have an objective function with an extra parameter

$$J = \sum_{i=1}^{n} \sum_{j=1}^{c} w_{ij}^m \parallel x_i^{(j)} - c_j \parallel^2, \tag{2}$$

where w_{ij} is a fuzzy membership qualification representing the membership of x_i to j cluster and m determines the degree of fuzziness:

$$w_{ij}^m = \frac{1}{\sum_{k=1}^{c} \left(\frac{\parallel x_i - c_j \parallel}{\parallel x_i - c_k \parallel} \right)^{\frac{2}{m-1}}}. \tag{3}$$

Each element belongs to any fuzzy set with a degree of membership which takes a real value ranged from 0 to 1 [6]. Data points are given the partial degree of membership in multiple nearby clusters and final segmented image is obtained (Figure 1D).

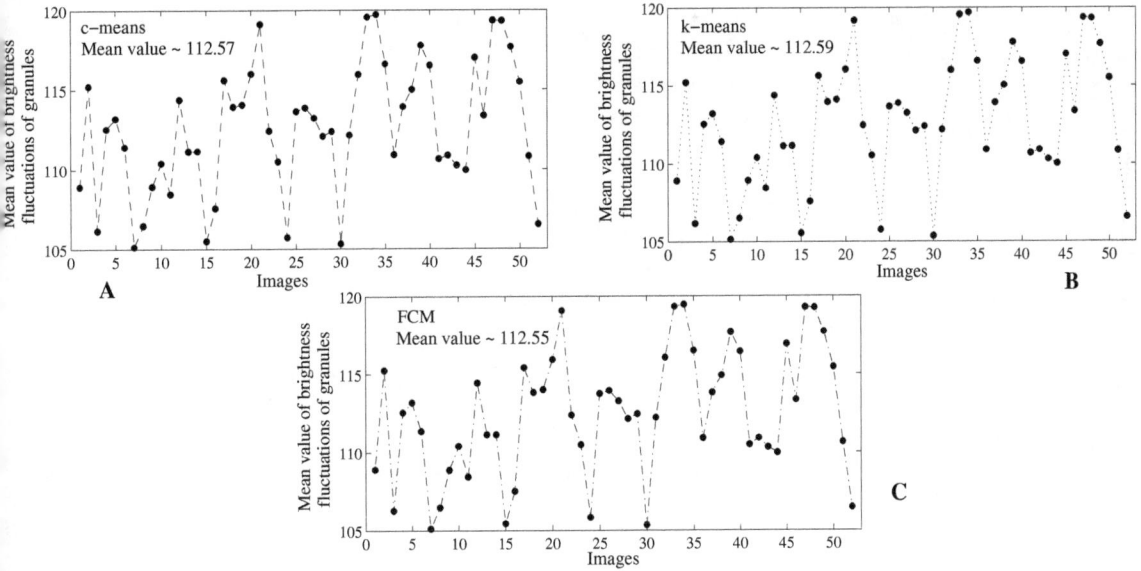

Figure 3: The brightness fluctuations of granules by using c-means (A), k-means (B), and FCM (C) algorithms are shown. The results approximately show the same behaviour of brightness fluctuations of granules extracted from algorithms.

4   Results

By applying the codes of c-means, k-means, and FCM algorithms, the sequence of photospheric images recorded by *Sunrise*/IMaX are segmented. Filling factors of granules (coverage area in each image) are calculated (Figure 2). The mean values of their filling factors derived from c-means, k-means, and FCM are about 0.34, 0.33 and 0.35, respectively. Fluctuations of filling factors obtained by algorithms are consistent and represent comparative results. Extracting the brightness fluctuations of granules from IMaX images show the same behaviour of these algorithms (Figure 3). The average of brightness fluctuations of granules is the same and equals with 112.5. The results represents that there is no preference in using these three methods to segment photospheric images. After applying FCM method on data (with membership degree of 0.5), the size distribution of granules is obtained (Figure 4). The power-law distribution is fitted with a power exponent $\alpha \approx 0.12$. The mean value of granule sizes is found to be about 1.8 arcsec^2 (0.85 Mm^2). The scatter plot of granules are shown in 5. It can be found granules follow two regimes that classifies them into two groups. Granules with sizes smaller than 2.8 arcsec^2 are more scattered in brightness, while larger ones are approaching mean value of brightness (1.3). A set of granules are followed from emergence up to disappearing in a series of images to attain lifetimes of granules with a normal size. The lifetimes are ranged from 6 to 10 minutes. The Number of granules is manually selected to compare with the results of these algorithms. It is found that false-positive detection appeared when granules are fragmented or dissolved.

The box-counting method, the technique to analyse the fractal dimension (FD) of an image, is employed to estimate changes between observations of detail and scale r by breaking our dataset. The purpose is finding the slope of the logarithmic regression line for both the number n and the size r of boxes. In this approach, by changing the resolution of box applied

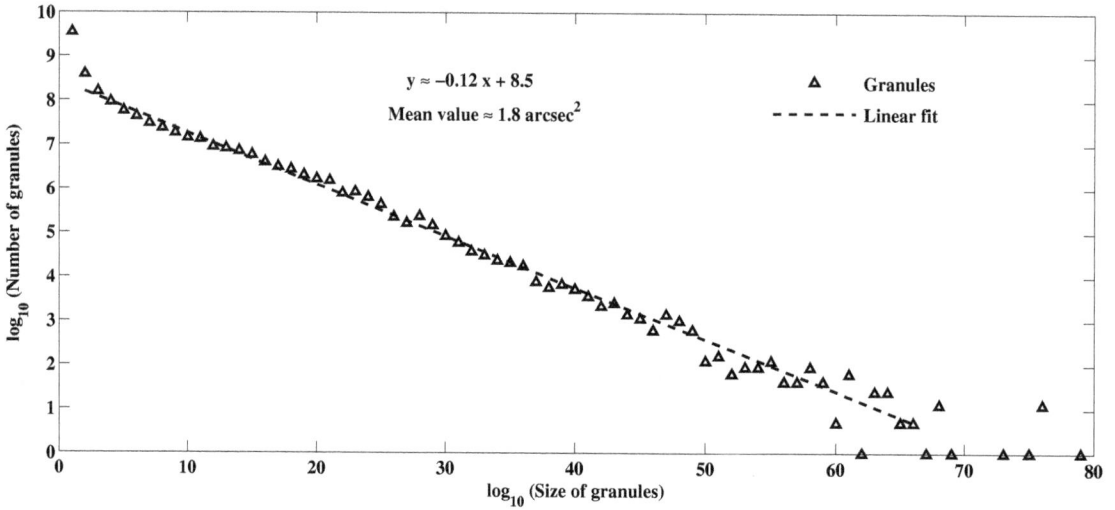

Figure 4: The size distribution of granules. The linear fit, $N \propto A^{-\alpha}$, is presented on a log scale with $\alpha \approx 0.12$.

on image pixels in each step (i.e., laying a series of grids), the size of elements that would be elongated within the box frequently varies in a form of power-low distribution [14]. For 2D (3D) images, if a power exponent is obtained with 2 (3), it shows random structure within the data set. The less values of power exponent demonstrate that how much the pattern of features captured in box is scale independence. In the case of local dimension box counting, the box for each r is centred on each pixel of interest. In box sizes ranged 10 and 100 the local FD goes to 1.8 (Figure 6). It can be said that granules appeared in grids with special box sizes represents their scale independence.

5   Conclusions

In this paper, an automated detection and characterization of photospheric granules were presented. Image processing methods and machine learning algorithms were applied to recognize and determine the physical properties of granules. The processes of three kinds of image segmentations (c-means, k-means, and FCM) were employed. The output images and the average of brightness fluctuation of granules derived from segmentations demonstrate that the detections of granules' locations are exactly correct. On the basis of Figure 2, it can be seen that one third of quiet-Sun is covered with granules. The mean value of granules size is consistent with previous studies [15, 16]. Two regimes governed on granules describe the turbulent eddies of smaller granules (< 2.8 arcsec^2) [15, 16]. The mean value of granular lifetimes is about 7 minutes in the quiet-Sun. Figure 6 indicates that it would be existed a scale range [50, 100] in which the FD estimate $D(r)$ shows the fractal characteristics of granulation. The entire computing time of procedures has taken about 5 minutes for data series in MATLAB environment. The top speed and high efficiency of unsupervised learning algorithms is the characteristics of these applications and make them easy to use.

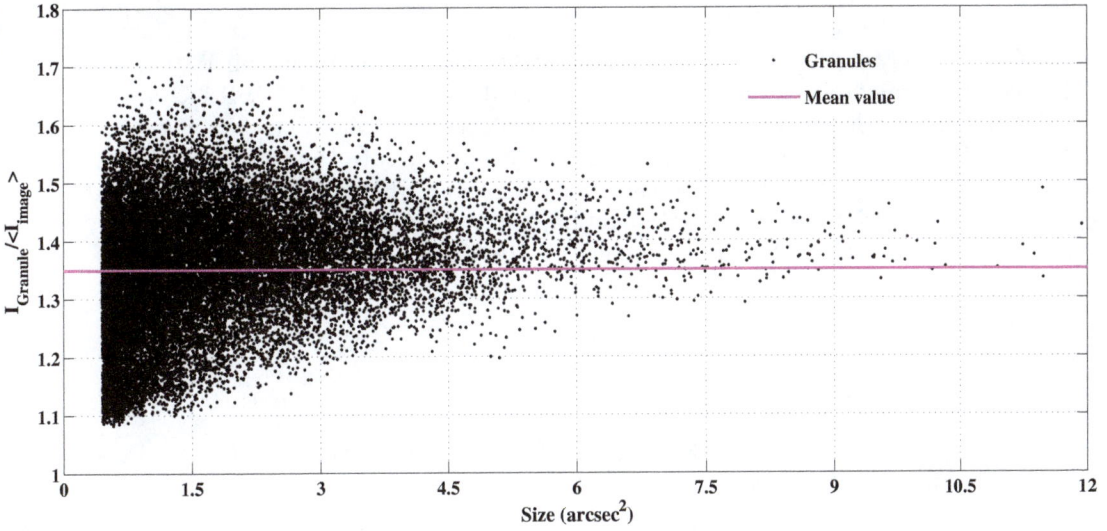

Figure 5: The scatter plot of granules brightness *vs.* size. The overall mean value of the scatter plot of the granule brightness is 1.35. For large granules, the brightness approaches a particular value (1.3).

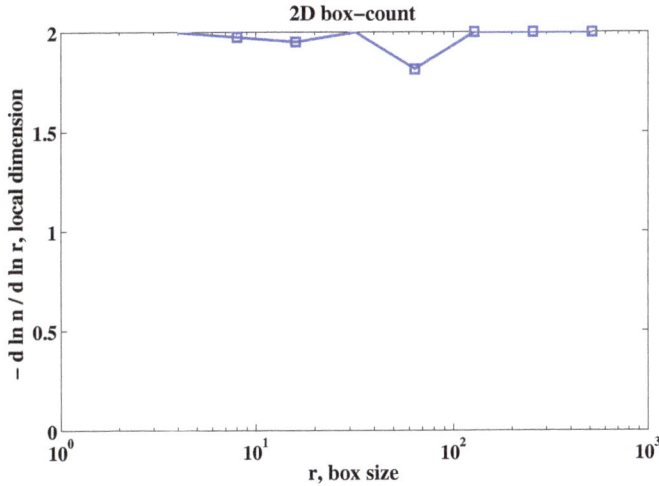

Figure 6: Local scaling exponent. If fractal characteristics are appeared over a limited range of box size r within the image, this can be demonstrated better by plotting the local exponent, $D(r) = -\frac{d \ln n}{d \ln r}$

Acknowledgment

The German contribution to *Sunrise* is funded by the Bundesministerium für Wirtschaft und Technologie through Deutsches Zentrum fr Luftund Raumfahrt e.V. (DLR), Grant No.50 OU 0401, and by the Innovationsfond of the President of the Max Planck Society (MPG). The Spanish contribution has been funded by the Spanish MICINN under projects ESP2006-13030-C06 and AYA2009-14105-C06 (including European FEDER funds). HAO/NCAR is sponsored by the National Science Foundation, and the HAO Contribution to *Sunrise* was partly funded through NASA grant number NNX08AH38G.

References

[1] de Wijn A.G., Stenflo J.O., Solanki S.K., Tsuneta S., 2008, Space Sci Rev 144, 275

[2] Riethmüller T.L., Solanki S.K., Martnez Pillet V., Hirzberger J., Feller A., Bonet J. A., et al., 2010, APJ Lett. 723, 169

[3] Berger T.E., Scherijver C.J., Shine R.A., Tarbell T.D., Title A.M., Scharmer G., 1995, APJ 454, 531

[4] Cloutman L.D., 1978 APJ 227, 614

[5] Barra V., Delouille V., Hochedez J.F., 2007, Adv. Spa. Res. 42, 917

[6] Bezdek J.C., Ehrlich R., Full W., 1984, Computers and Geosciences 10.2 191-203.

[7] Verbeeck C., Delouille V., Mampaey B., De Visscher R., 2013, A&A 561, A29 1

[8] Barthol P., Gandorfer A., Solanki S. K., Schüssler M., Chares B., Curdt W., et al., 2010, Solar Phys. 268, 1

[9] Solanki S.K., Barthol P., Danilovic S., Feller A., Gandorfer A., Hirzberger J., et al., 2010, APJ Lett. 723, 127

[10] Gandorfer A., Grauf B., Barthol P., Riethmüller T.L., Solanki S.K., Chares B., et al., 2010, Solar Phys. 268, 35

[11] Marínez Pillet V., del Toro Iniesta J.C., Álvarez-Herrero A., Domingo V., Bonet J.A., González Fernández L., et al., 2010, Solar Phys. 268, 57

[12] Berkefeld T., Schmidt W., Soltau D., Bell A., Doerr H.P., Feger B., et al., 2010, Solar Phys. 268, 103

[13] Jipkate B.R., Gohokar V.V., 2012, International Journal Of Computational Engineering 2 737

[14] Smith T.G., Lange G.D., Marks W.B., 1996, Journal of Neuroscience Methods 69 (2), 123

[15] Roudier T., Muller R., 1987, Solar Phys. 107, 11

[16] Javaherian M., Safari H., Amiri, A., Ziaei, S., 2014, Solar Phys. 289, 3969

3

Seismology of solar spicules based on Hinode/SOT observations

Vahid Abbasvand · Hossein Ebadi · Zahra Fazel

Astrophysics Departartment, Physics Faculty, University of Tabriz, Tabriz, Iran;
email: hosseinebadi@tabrizu.ac.ir

Abstract. We analyze the time series of Ca II H-line obtained from *Hinode*/SOT on the solar limb. The time-distance analysis shows that the axis of spicule undergos quasi-periodic transverse displacement. We determined the period of transverse displacement as $\sim 40 - 150 \pm 8$ s and the mean amplitude as $\sim 0.1 - 0.5$ arc sec. For the oscillation wavelength of $\lambda \sim 1/0.06$ arc sec ~ 11500 km, the estimated kink speed is ~ 13–83 km s^{-1}. We obtained the magnetic field strength in spicules as $B_0 = 2 - 12.5 \pm 1$ G and the energy flux as $7 - 227$ J m^{-2} s^{-1}.

Keywords: ISM: Sun, ISM: spicules, ISM: Dynamic parameters

1 Introduction

Spicules can be seen almost everywhere at the solar limb. They are highly-dynamic, thin, jet-like features in the solar chromosphere. Spicules have been observed for a very long time and have been the subject of numerous reviews [1, 2, 3, 4]. Perhaps the most interesting aspect about spicules is their potential to mediate the transfer of energy and mass from the photosphere to the corona. This potential has been recognized early [1, 5, 6], but the lack of high-quality observations prevented a better understanding of spicules and their link to the corona [7]. The advent of the Hinode Solar Optical Telescope(*SOT*) [8, 9, 10] provided a quantum leap in the understanding of spicules and their properties. With its seeing free high spatial and temporal resolution observations of the chromosphere, Hinode showed much more dynamic spicules than previously thought and re-ignited the discussion on their potential to heat the corona [11, 12, 13]. While there are several recent studies of the properties of spicules using Hinode/SOT [11, 14, 15], they are either preliminary results papers or cover only one or two data sets.

It is encouraging that the results from the various groups agree quite well, despite the difficulties in obtaining reliable magnetic field measurements at the limb. Lopez-Ariste & Casini [16] used full Stokes polarimetry and Hanle effect modelling of He I D3 spectra and find that the magnetic field strength is mostly 10 G at 3500 km height, with some spicules possibly showing fields up to 40 G. Trujillo et al. [17] used both the Hanle and Zeeman effects and full Stokes polarimetry of the He I 10830 line and find 10 G at 2000 km and again, some spicules with field strengths up to 40 G. Doppler velocities of solar limb spicules show oscillations with periods of $20 - 55$ and $75 - 110$ s. There is also clear evidence of 3-min oscillations at the observed heights. The observed upwards steady flows are $20 - 25$ km s^{-1} with short periods $(20 - 55$ s$)$ in type I spicules. Type I spicules have typical lifetimes of $150 - 400$ s and maximum ascending velocities of $15 - 40$ km s^{-1}. Type II Spicules in the quiet sun have lifetimes between $50 - 150$ s, and slightly less in coronal holes. They have a maximum velocity between $40 - 100$ km s^{-1}, with a very small amount with velocities greater than 150 km s^{-1} [1].

All spicule oscillations events are summarized in a recent review by Zaqarashvili & Erdélyi [18]. They suggested that the observed oscillation periods can be formally divided in two groups: those with shorter periods (<2 min) and those with longer periods (≥ 2 min)[18]. The most frequently observed oscillations lie in the period ranges of $3-7$ min and $50-110$ s. These spiky dynamic jets are propelled upwards at speeds of about $20-25$ km s^{-1} from photosphere into the magnetized low atmosphere of the sun. Their diameter varies from spicule to spicule having the values from 400 km to 1500 km. The mean length of classical spicules varies from 5000 km to 9000 km, and their typical life time is $5-15$ min.

The typical electron density at heights where the spicules are observed is $3.5 \times 10^{-10} - 2 \times 10^{-11}$ kg m^{-3}, and their temperatures are estimated $5000-8000$K [1]. Their periods are estimated $20-55$ and $75-110$ s. De Pontieu et al. [11], based on Hinode observations concluded that the most expected periods of transverse oscillations lay between $100-500$ s, which interpreted as signatures of Alfvén waves. Phase speed begins from ~ 40 km s^{-1} at lower heights and reaches to the maximum value of ~ 90 km s^{-1} at $\sim 2500-3000$ km. Then, it decreases to the minimum value of $\sim 20-25$ km s^{-1} at $\sim 3500-4500$ km [19]. Therefore, the observed quasi-periodic displacement of spicule axis can be caused due to fundamental standing mode of kink waves. The energy flux storied in the oscillation is estimated as 150 J m^{-2} s^{-1}, which is of the order of coronal energy losses in quiet Sun regions [19]. Spicule seismology, which means the determination of spicule properties from observed oscillations and was originally suggested by [18], has been significantly developed during last years [19, 20, 21].

In the present work, we study the observed oscillations in the solar spicules through the data obtained from Hinode. We trace the spicule axis oscillations via time slice diagrams for determining period, amplitude, phase speed, magnetic field, Alfvén speed and energy flux of the studied spicules. Observed waves can be used as a tool for spicule seismology. A description of the observations and image processing is made in Section 2 and in Section 3 we discuss spicules properties. Finally, our conclusions are summarized in Section 4.

2   Observations and data processing

We used a time series of Ca II H-line (396.86 nm) obtained on 22 January, 2007, during 23:26 to 23:42 UT by the Solar Optical Telescope onboard [9]. The spatial resolution reaches 0.2 arc sec (~ 150 km) and the pixel size is 0.109 arc sec (~ 80 km) in the Ca II H-line. The time series has a cadence of 20 seconds with an exposure time of 0.5 seconds. The position of $X-center$ and $Y-center$ of slot are 945 arc sec and 0 arc sec, while X-FOV and Y-FOV are 223 arc sec and 112 arc sec, respectively.

The "fgprep", "fgrigidalign"(available in solarsoft, http://www.lmsal.com/solarsoft) and "madmax" IDL algorithms are used to reduce the images spikes and jitter effect, to align time series, and to enhance the finest structures, respectively [22, 23].

3   Results and Discussions

In Figure 1, we presented the full view Ca II H-line image of the equator which contains the studied spicules. The processed spicules are indicated by red arrows on this image. This image was taken by Hinode/SOT telescope on 22 Jan, 2007, 23:26:04 UT. We used the time series images of this region and determined period, amplitude, phase speed, magnetic field, Alfvén speed and energy flux of the studied spicules. To this end, we selected 14 random spicules and used time-slice diagrams to illustrate their transversal oscillations.

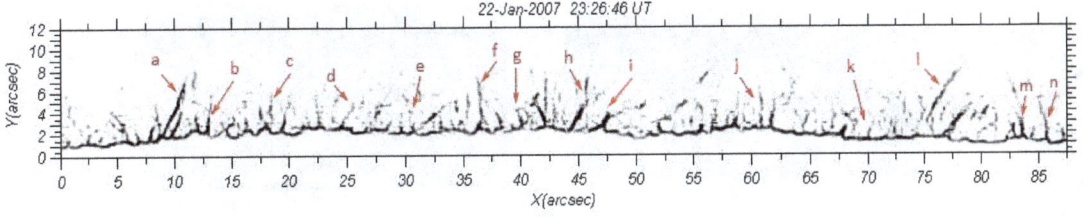

Figure 1: The full view Ca II H-line image of the solar equator which contains the processed spicules. The red arrows show the studied spicules. We used the "madmax" algorithm to enhance the finest structures.

Table 1: The amplitude (ξ), oscillation period(T_{obs}), phase speed(v_k), magnetic field(B_0) and energy flux(F) for the spicule oscillations.

No	T_{obs} (s)	$\xi(arcsec)$	$v_k(kms^{-1})$	B_0 (G)	$F(Jm^{-2}s^{-1})$
a	70	0.3	57	8.9	123
b	70	0.2	38	5.9	36
c	40	0.1	33	5	8
d	50	0.1	26	4	7
e	55	0.2	48	7.5	46
f	90	0.4	59	9.2	227
g	40	0.25	83	13	125
h	100	0.35	47	7.2	138
i	50	0.3	80	12.5	172
j	40	0.15	50	7.7	27
k	60	0.25	56	8.7	83
l	150	0.15	13	2	7
m	100	0.2	27	4.2	26
n	80	0.15	25	5.4	14

In Figure 2, we presented the time slice diagrams of the time series which consists of 30 consecutive images. The bright regions show the transversal oscillations of spicules axis. The clear quasi-periodic transverse motion of the spicules axis is seen on the figure. It should be noted that these diagrams are performed at the same height from the limb. In Figure 3, we presented the time slice diagrams and fitted polynomials for three spicules. As it is clear, the fitted functions are the best ones for our observed results. We estimated the oscillation period and amplitude at each diagram and presented them in Table 1. The density is almost homogeneous along the spicule axis [1]; therefore, the density scale height should be much longer than the spicule length. The variation of oscillation amplitude with height cannot be due to the decreased density. So, we may assume that the oscillation amplitude is proportional to sin(kz), where k is the oscillation wave number. This expression can be approximated in the long wavelength limit as $\sim$ kz. Hence, the oscillation wavelength is $\lambda \sim 1/0.06$ arc sec ~ 11500 km [24]. This allows the estimation of the kink speed as $v_k \sim 13\text{--}83$ km s^{-1}.

It is possible to estimate the Alfvén speed and consequently magnetic field strength through them. Kink waves are transverse oscillations of magnetic tubes and the phase

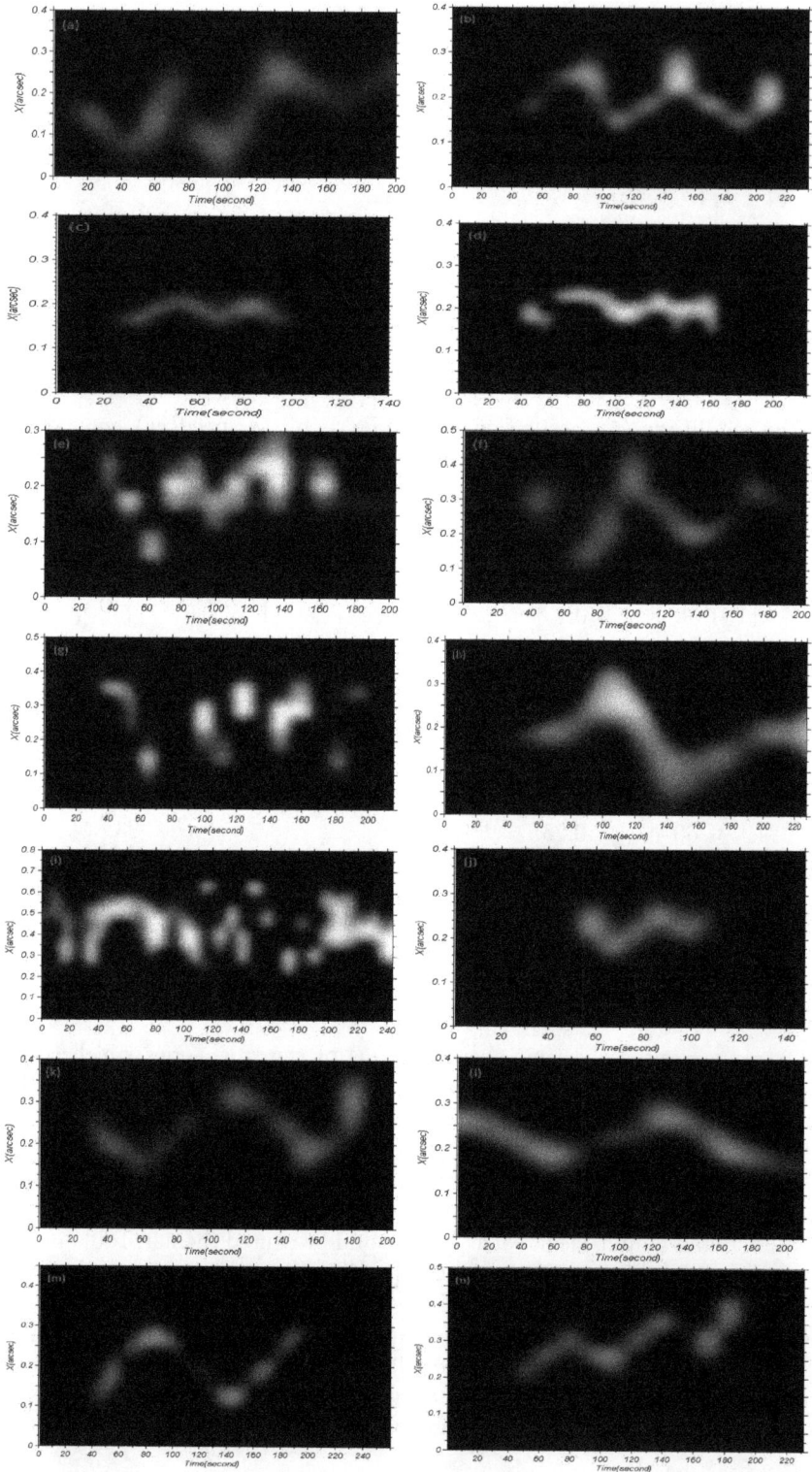

Figure 2: Time slice diagrams of the spicules in Ca II H-line. Axes are in heliocentric coordinates, where $1'' \approx 725$km.

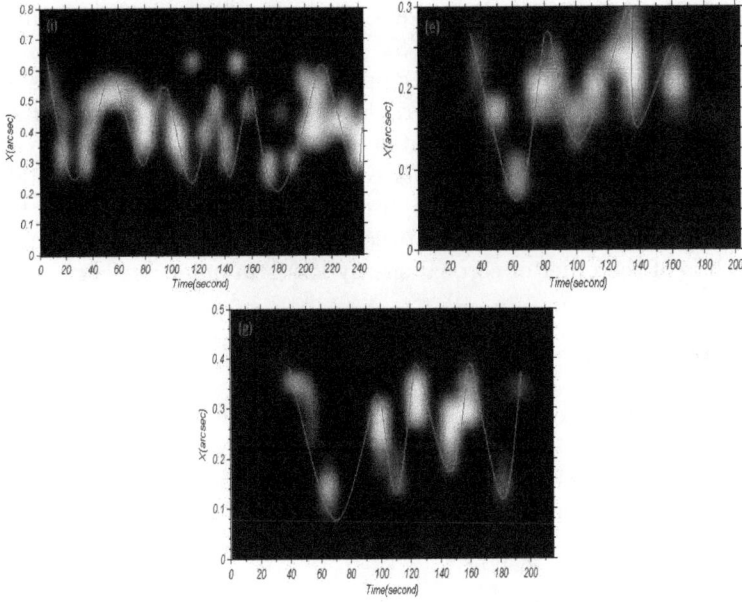

Figure 3: Time slice diagrams of spicule in Ca II H-line. The blue line is a fit to the oscillation of the spicule.

speed for a straight homogenous tube can be written as:

$$v_k = \frac{\lambda}{T_{obs}} = v_A \sqrt{\frac{\rho_0}{\rho_0 + \rho_e}}, \tag{1}$$

where $v_A \equiv B_0/\sqrt{\mu_0 \rho_0}$ is the Alfvén speed inside the tube, λ is the wavelength and T_{obs} is the observed oscillation period. ρ_e and ρ_0 are the plasma density outside and inside of tube, respectively. The density in spicules is 3×10^{-10} kg m^{-3} and $\rho_e/\rho_0 \simeq 0.02$ [18, 24].
The oscillation amplitude and phase speed allows to estimate the energy flux, F, storied in the oscillations:

$$F = \frac{1}{2}\rho \upsilon^2 v_k \tag{2}$$

where ρ, υ and v_k are density, wave amplitude and phase speed, respectively [25]. The wave velocity can be determined as $v = \xi/T_{obs}$, where ξ is the axis displacement estimated as $\sim 0.1 - 0.5$ arcsec and T_{obs} is the oscillation period estimated as $\sim 40 - 150 \pm 8$ s. With these parameters, the energy flux is estimated as $F = 7 - 227$ J m^{-2} s^{-1}.
Using the kink speed, the magnetic field strength in spicules is estimated as $B_0 = 2 - 12.5 \pm 1$ G. This is in good agreement with recently estimated magnetic field strengths in quiet-Sun spicules ($\sim$10G), which was obtained by spectropolarimetric observations of solar chromosphere in He I λ 10830 [17, 26, 24].

4   Conclusion

We performed the analysis of Ca II H-line time series at the solar limb obtained from *Hinode*/SOT in order to uncover the oscillations in the solar spicules. We studied 14 random spicules and found that their axis undergo quasi-periodic transverse displacements. The period of the transverse displacements are $\sim 40 - 150 \pm 8$ s and the amplitude are $\sim 0.1 - 0.5$ arc sec. The same periodicity was found in Doppler shift oscillation by Zaqarashvili & Erdélyi [18] and De Pontieu et al. [11], so the periodicity is probably common for spicules. The magnetic field strength in spicules is estimated as $B_0 = 2 - 12.5 \pm 1$ G. These magnetic field strengths are smaller than the values calculated by Zaqarashvili & Erdélyi [18].

We think that the observed quasi-periodic displacement of spicule axis can be caused due to standing kink waves. The energy flux storied in the oscillations is estimated as $7 - 227$ $J\,m^{-2}\,s^{-1}$, which is of the order of coronal energy losses in quiet Sun regions.

Acknowledgment

The authors are grateful to the Hinode Team for providing the observational data. Hinode is a Japanese mission developed and lunched by ISAS/JAXA, with NAOJ as domestic partner and NASA and STFC(UK) as international partners. Image processing Mad-Max program was provided by Prof. O. Koutchmy.

References

[1] Beckers, J.M., 1968, solphys, 3, 367

[2] Beckers, J.M., 1972, aap, 10, 73

[3] Suematsu, Y., Wang, H., Zirin, H., 1995, apj, 450, 411

[4] Sterling, A.C., 2000, solphys, 196, 79

[5] Pneuman, G. W., Kopp, R.A., 1997, solphys, 57, 49

[6] Athay, R.G., Holzer, T.E., 2005, apj, 255, 743

[7] Withbroe, G. L., 1983, apj, 267, 825

[8] Kosugi, T., Matsuzaki, K., Sakao, T., et al., 2007, solphys, 243, 3

[9] Tsuneta, S., Ichimoto, K., Katsukawa, Y., et al., 2008, solphys, 249, 167

[10] Suematsu, Y., Tsuneta, S., Ichimoto, K., et al., 2008, solphys, 249, 197

[11] De Pontieu, B., McIntosh, S.W., Carlsson, M., et al., 2007, Science, 318, 1574

[12] De Pontieu, B., McIntosh, S.W., Hansteen, V.H., Schrijver, C.J., 2009, apj, 701, L1

[13] De Pontieu, B., McIntosh, S.W., Carlsson, M., et al., 2011, Science, 331, 55

[14] Anan, T., Kitai, R., Kawate, T., et al., 2010, PASJ, 62, 871

[15] Zhang, Y.Z., Shibata, K., Wang, J.X., et al., 2012, apj, 750, 16

[16] Lopez-Ariste, A., Casini, R., 2005, aap, 436, 325

[17] Trujillo Bueno, J., Merenda, L., Centeno, R., Collados, M., Landi DeglInnocenti, E., 2005, apj, 619, L191

[18] Zaqarashvili, T.V., Erdélyi, R., 2009, ssr, 149, 335

[19] Ebadi, H., Khoshrang, M., 2014, apss, 352, 2

[20] Verth, G., Goossens, M., He, J.-S., 2011, apjl, 733, 15

[21] Ebadi, H., Ghiassi, M., 2014, apss, 353, 1

[22] Shimizu, T., Nagata, S., Tsuneta, S., et al., 2008, solphys, 249, 221

[23] Koutchmy, O., Koutchmy, S., In: O. von der Lühe (ed.): High Spatial Resolution Solar Observations: Proceedings of the Tenth Sacramento Peak Summer Workshop (Sunspot, NM, August 22–26, 1988), p. 217, National Solar Observatory/Sacramento Peak, Sunspot, NM 88349 (1989)

[24] Ebadi, H., Zaqarashvili, T.V., Zhelyazkov, I., 2012, apss, 337, 33

[25] Vranjes, J., Poedts, S., Pandey, B.P., De Pontieu, B., 2008, aap, 478, 553

[26] Singh, K.A.P., Dwivedi, B.N., 2007, New Astron., 12, 479

4

The study of Hydrodynamical wind on the observational properties of magnetized accretion flow with thermal conduction

Maryam Ghasemnezhad

Faculty of physics, Shahid Bahonar University of Kerman, Kerman, Iran

Abstract. In this paper, we examine the effect of a hydrodynamical wind on the observational properties of supercritical accretion discs (slim discs) with thermal conduction in the presence of magnetic field under a self similar treatment. The disk gas is assumed to be isothermal. In this flow, the mass-accretion rate $\dot{M}$ decreases with radius r as $\dot{M} \propto r^{(s+\frac{1}{2})}$, where s is an arbitrary constant and indication of the effect of wind. Cooling effects of outflows or winds are noticeable in luminosity and effective temperature of advection dominated accretion flows. We apply this model to black-hole X-ray binary LMC X-3, which is supposed to be under critical accretion rate. Increasing the effect of wind decreases the disc's temperature, luminosity and radiation flux of the disc, because of energy flux which is taken away by winds. The effect of thermal conduction is similar to the effect of wind in the disc's temperature, luminosity and radiation flux of the disc, but the influence of wind on the observational properties is bigger than the effect of thermal conduction.

Keywords: accretion, accretion flow, Thermal conduction, stars: winds, outflows

1 Introduction

Accretion disc, a disc-like flow of gas, or particles like free electrons and various types of ions around any gravitational object (like black hole) in which the material orbits in the gravitational field of the object, loses energy and angular momentum because of turbulence and viscosity or shear and magnetic fields as it spirals inward slowly. The gravitational energy of the gaseous matter is converted to heat. A fraction of heat is converted to radiation. The formation of stars and planets, the powerful emissions from quasars and X-ray binaries all involve accretion discs. Accretion disc physics is governed by a non-linear combination of many processes, including gravity, viscosity, radiation, magnetic fields and other parameters like thermal conduction or wind or outflow. Luminosity and observed flux are the observable physical quantity of radiation produced in accretion discs. There is a maximum possible luminosity at which gravity is able to balance the outward pressure of radiation. The limit for a steady, spherically symmetric accretion flow is given by the Eddington luminosity (Kato et al. 2008).

There are five models for accretion disc depending on their geometry (vertically thick versus thin), mass accretion rate (sub versus supper Eddington accretion rate) and optical depth (opaque versus transparent disc). These models named Shakura & Sunyaev disc, ADAF, slim disc, Polish doughnuts and ion tori (Abramowicze et al. 2010). Thin discs ($\frac{H}{r} < 1$) consist of Shakura & Sunyaev discs, ADAF, slim disc models and thick discs ($\frac{H}{r} > 1$) include Polish doughnuts and ion tori models. Each of these models have their own characteristics that have been studied in details by many researchers (e.g., Shakura, Sunyaev 1973, Narayan, Yi

1994 & 1995, Abramowicz et al. 1988 and Abramowicz et al. 1978). Following the article by Ghasemnezhad et al. 2013 (hear after GKA13), we are interested to study the effect of wind on the observational properties of LMC X-3 that is a black hole (BH) binary system $(7M_\odot \leq M_{BH} \leq 14M_\odot)$ in the large magelanic cloud at a distance of $48.1kpc$(Orosz et al. 2009, Cowley et al. 1983). LMC X-3 shows soft spectra and has a high luminosity. Also, this object has a low absorption column density along the line of sight. These properties make LMC X-3 an ideal laboratory for testing our understanding of accretion disc physics. Slim disc model describes accretion discs at high luminosity, while becoming to the Shakura & Sunyaev standard disc in the low luminosity limit (or low mass accretion rate). Slim disc model can explain the spectral behavior of several black hole binaries (like LMC X-3) which cannot be explained by the standard model (Straub et al. 2011). In the slim discs, the optical depth is very high and the radiation generated by viscosity can be trapped within the disc. Cooling processes in the slim disc model are Advection and blackbody radiation but in ADAF model are Advection, bremsstrahlung and Compton scattering. The structure and spectral properties of the slim disc model have been studied by many authors (e.g., Ghasemnezhad et al. 2013, Mineshige et al. 2000, Fukue 2000).

In this paper, we consider three important parameters in the slim disc model: magnetic field, thermal conduction and wind or outflow. The effect of magnetic field and thermal conduction on the surface temperature and observed flux of LMC X-3 were studied by GKA13. In this paper, we are interested to add the effect of wind parameter to evaluate the observational properties of slim discs.

Slim discs can explain the observed high temperature in the innermost regions (Watarai & Mineshige 2001). So, accreting materials are hot and ionized which will be affected by magnetic field. So, the magnetic field plays an crucial role in the dynamical structure and the observational properties of accretion disc. Kaburaki (2000), Abbassi et al. (2008, 2010)and Ghasemnezhad et al.(2012) studied the effect of magnetic field on optically thin advection dominated accretion flow (ADAF) in the recent years. But the magnetized slim disc model has been less studied. Numerical simulations have been developed to study the structure of slim discs (Ohsuga et al. 2005, Ohsuga & Mineshige 2011). GKA13 have studied the observational properties of the magnetized slim discs without wind parameter in two cases (LMC X-3 and narrow-line seyfert 1 galaxies). We also study the magnetized slim discs with wind parameter in LMC X-3.

Thermal conduction as a physical process has a great role in energy transport by ions or electrons in the accreting materials in a hot accretion disc where they are completely ionized. Recent observations of hot accretion disc around AGNa indicate that thermal conduction should be on collision-less regime. It seems that, it is only important for dilute (opacity is low, density is very low, mean free path of gases is very long, comparable to scale height) and accretion flows (e.g., Tanaka & Menou 2006)and is not important in the standard thin disc model (SSD) or Slim disc model. As we know, thermal conduction is the transfer of energy by ions or electrons;so, the electron thermal conduction is important at low accretion rate but ion heat conduction could be relevant at all accretion rates. We can study the importance of thermal conduction from two viewpoints: mean free pass of gases and thermal conduction time. As we know, the saturated conduction flux is $F \propto \rho c_s^3$. So, the ratio of the inflow time to the ion conduction time in a hot accretion flow (slim disc) may be as large as $\propto \frac{c_s}{V_r}$, where V_r and c_s are the radial flow velocity and the sound speed, respectively. Also, the dynamical properties of self similar solutions for the present slim disc case are the same as those for the ADAF. We used self-similar solutions for sound speed and radial velocity in section 3. We insert the numerical values for c_1 and α (used from Ghasemnezhad et et al. 2013). This ratio shows that the ion conduction time is comparable to the inflow time of the plasma. Therefore, thermal conduction can be important in slim discs. Also, Johnson &

Quataert (2007) has noted that Ion heat conduction may be important at higher accretion rates, but this is harder to assess because the ion conduction time is comparable to the inflow time.

The accretion flows lose their mass by the winds as they accrete onto the central body. Mass loss mechanism (in form of wind or outflow) is an important phenomenon in the structure and evolution of the accretion discs. Observational evidences confirm the existence of wind or outflow in various astronomical sites: in Active Galactic Nuclei(AGN), microquasars and Young Stellar Object (YSO) (Shadmehri 2009). An outflow emanating from an accretion disc can affect the energy dissipation rate and the effective temperature of the disc. As the result of the mass loss, the mass accretion rate is not constant and is dependent on radius as power law ($\dot{M} \sim r^s$ index s is an order of unity)(Blandford & Begelman 1999, Abbassi et al. 2010, 2013). Physical forces or sources that drive the discs wind, can be the thermal sources, the radiation field and the magnetic field. The name of winds in accretion disc depend on their driving forces. The wind mechanism has been investigated by many researchers (Meier 1979, Fukue 1989, Abbassi et al. 2010). We assume the wind in the disc causes the loss of angular momentum, mass and thermal energy. Recently radiation hydrodynamic (RHD) and radiation magnetohydrodynamic (RMHD) simulations, show that the strong radiation pressure force drives outflows above the slim discs (ohsuga et al. 2009, ohsuga et al. 2005). Also a mass losing accretion disc could produce spectra that will be different from that produced by standard accretion disc (Knigge 1999). Done & Davis (2008) studied the LMC X-3 spectra and concluded that the spectra of LMC X-3 could be affected by wind. Wilms et al. (2001) showed that the spectra of LMC X-3 consist of a disc black body with $KT \sim 0.8 - 1 Kev$ and a soft power law. In this paper, we are interested to know how the wind parameter affect the surface temperature and radiation flux of LMC X-3.

We improved the GKA13 by adding the wind parameter and then we compare the effect of wind and thermal conduction parameters in observational properties of LMC X-3. Knigge (1999) studied the effect of mass losing (wind) on the radial structure of the accretion disc by introducing two parameters (l & η). We have used these parameters for writing the angular momentum and energy equations. We describe the effects of (l & η) in the next section.

This paper is organized as follows. In section 2, we present the equations of magnetohydrodynamics as the basic equations. self-similar solutions are presented in section 3. In section 4, we consider the radiation properties of slim discs and the results and finally, we present the summary and conclusion in section 5.

2   The Basic Equations

We continue the paper presented by Abbassi et al. (2010) and investigate the effect of wind on the observational properties. Abbassi et al. (2010) studied the structure of magnetized accretion disc with thermal conduction and wind parameters. So, we use all the MHD equations and assumptions made by Abbassi et al. (2010). We use the cylindrical coordinates (r, φ, z) for steady state and axi-symmetric ($\frac{\partial}{\partial \phi} = \frac{\partial}{\partial t} = 0$) super critical accretion disc. We vertically integrate the flow equations; also, we suppose that all flow variables are only a function of r (radial direction). We ignore the relativistic effects and we use the Newtonian gravity in the radial direction. We adopt α -prescription for viscosity of rotating gas in accretion flow. We consider the magnetic field has just toroidal component.

The MHD equations are as the same as the equation made by Abbassi et al. (2010): The equation of continuity gives:

$$\frac{\partial}{\partial r}(r\Sigma V_r) + \frac{1}{2\pi}\frac{\partial \dot{M}_w}{\partial r} = 0 \qquad (1)$$

where V_r is the accretion velocity ($V_r < 0$) and $\Sigma = 2\rho H$ is the surface density at a cylindrical radius r. H is the disc half-thickness and ρ is the density. The mass-loss rate by wind is showed by $\dot{M}_w$. So

$$\dot{M}_w = \int (4\pi r' \dot{m}_w(r')dr'), \qquad (2)$$

where $\dot{m}_w(r)$ is the mass-loss per unit area from each disc face. On the other hand, we can rewrite the continuity equation as:

$$\frac{1}{r}\frac{\partial}{\partial r}(r\Sigma V_r) = 2\dot{\rho}H \qquad (3)$$

where $\dot{\rho}$ is the mass loss rate per unit volume. The equation of motion in the radial direction is:

$$V_r \frac{\partial V_r}{\partial r} = \frac{V_\varphi^2}{r} - \frac{GM_*}{r^2} - \frac{1}{\Sigma}\frac{d}{dr}(\Sigma c_s^2) - \frac{c_A^2}{r}$$

$$-\frac{1}{2\Sigma}\frac{d}{dr}(\Sigma c_A^2) \qquad (4)$$

where V_φ, G, c_s and c_A are the rotational velocity of the flow, the gravitational constant, sound speed and Alfven velocity of the gas respectively. The sound speed and the Alfven velocity are defined as $c_s^2 = \frac{p_{gas}}{\rho}$ and $c_A^2 = \frac{B_\varphi^2}{4\pi\rho} = \frac{2p_{mag}}{\rho}$, where B_φ, p_{gas} and p_{mag} are the toroidal component of magnetic field, the gas and magnetic pressure respectively.

By integration along z of the azimuthal equation of motion gives:

$$r\Sigma V_r \frac{d}{dr}(rV_\varphi) = \frac{d}{dr}(r^3\nu\Sigma\frac{d\Omega}{dr}) - \frac{\Omega(lr)^2}{2\pi}\frac{d\dot{M}_w}{dr} \qquad (5)$$

where ν is the kinematic viscosity coefficient. α-prescription (Shakura & Sunyaev 1973)for viscosity was assumed as:

$$\nu = \alpha c_s H \qquad (6)$$

where α is a constant less than unity. $\Omega(= \frac{V_\varphi}{r})$ is the angular speed and $\Omega_k$is the Keplerian angular speed. To write the angular momentum equation , we have considered the role of wind in transferring the angular momentum. The wind material moving along a steam line originating at radius r in the disc was assummed to co-rotate with the disc out to a radial distance lr. The wind material ejected at radius r on the disc and carries away specific angular momentum $(lr)^2\Omega$,where Ω related to a radial distance lr. Knigge (1999) define the l parameter as the lenght of the rotational lever arm that allows many types of accretoin disc winds models. The parameter $l = 0$ corresponds to a non-rotating wind, and the angular momentum is not extracted by the wind and the disc losses only mass because of the wind while $l = 1$ represents outflowing materials that carry away the angular momentum $(r^2\Omega)$, and radiation driven disc winds are corresponding to the $l = 1$ case (e.g. Murray & Chiang (1996)). $l > 1$ corresponds to wind material that can remove a lot of angular momentum from the disc. Centrifugally driven MHD disc winds are corresponding to $l > 1$ (e.g. Blandford & Payne (1982)). In this case, the lenght of the lever arm is $l = \frac{r_A}{r}$ where r_A is the Alfen radius. Also, thermally driven outflows belong to this class (Piram 1977).

By integrating along z of the hydrostatic balance, we have:

$$\frac{GM}{r^3}H^2 = c_s^2[1 + \frac{1}{2}(\frac{c_A}{c_s})^2] = (1 + \beta)c_s^2 \tag{7}$$

where $\beta = \frac{p_{mag}}{p_{gas}} = \frac{1}{2}(\frac{c_A}{c_s})^2$ which indicates the importance of magnetic field pressure as compare to gas pressure.

Now, we can write the energy equation considering cooling and heating processes in an ADAF. We assume the generated energy due to viscous dissipation and the thermal conducted are balanced by the advection cooling and energy loss of outflow/wind. Thus,

$$\frac{\Sigma V_r}{\gamma - 1}\frac{dc_s^2}{dr} - 2HV_rc_s^2\frac{d\rho}{dr} = \frac{f\alpha\Sigma c_s^2}{\Omega_k}r^2(\frac{d\Omega}{dr})^2 - \frac{2H}{r}\frac{d}{dr}(rF_s)$$

$$-\frac{1}{2}\eta\dot{m}_w(r)V_k^2(r) \tag{8}$$

where γ and f are adiabatic index and the advection parameter, respectively. In writing the energy equation, we have used the η parameter that defined by Knigge (1999). If a wind is driven from the surface of the optically thick disc, some of the dissipated accretion energy is converted to the wind energy. Strictly speaking, the fractions (η) of the wind 's binding energy and the wind's kinetic energy are provided by dissipation accretion energy. We have two models for the parameter (η) when $\eta = 0$ and $\eta = 1$. If $\eta = 0$, the effect of wind is minimized in this model, the wind is powered by the disc and the central star; but, in the maximal model $\eta = 1$, the effect of wind is maximized and the disc alone is responsible for powering the outflow (Knigge 1999).

So the last term on the right hand side of the energy equation is the energy loss due to the wind or outflow (Knigge 1999). So, η is a free and dimensionless parameter. In other words, the large η corresponds to more energy extraction from the disc because of the wind (Knigge 1999). Also, the second term on right hand side represents energy transfer due to the thermal conduction and $F_s = 5\Phi_s\rho c_s^3$ is the saturated conduction flux (Cowie & Makee 1977). Dimensionless coefficient Φ_s is less than unity. Finally, since we consider the toroidal component for the global magnetic field of central stars, the induction equation with field escape can be written as:

$$\frac{d}{dr}(V_rB_\varphi) = \dot{B}_\varphi \tag{9}$$

where $\dot{B}_\varphi$ is the field scaping/creating rate due to magnetic instability or dynamo effect.

3   Self-Similar Solutions

In the last section, we introduced the basic equations for an axi-symmetric, magnetized hot accretion flow in the presence of rotating wind. The basic equations of the model are a set of partial differential equations, which have a very complicated structure. The self-similar method is one of the most useful and powerful techniques to give an approximate solutions for differential MHD equations and has a wide range of applications in astrophysics. For the first time, this technique was applied by Narayan & Yi (1994) in order to solve ADAFs dynamical equations. By adopting Narayan & Yi (1994) self-similar scaling, in fact, the radial dependencies of all physical quantities are canceled out, and all of differential equations are transformed to algebraic equations. The properties of self similar solutions for the slim disc case are the same as those for the ADAF according to Fukue (2000, 2004). Following Abbassi et al. (2010)'s solutions, we introduce the physical quantities as follows:

$$V_r(r) = -c_1 \alpha V_k(r) \tag{10}$$

$$V_\varphi(r) = c_2 V_k(r) \tag{11}$$

$$c_s^2 = c_3 V_k^2 \tag{12}$$

$$c_A^2 \frac{B_\varphi^2}{4\pi\rho} = 2\beta c_3 \frac{GM}{r} \tag{13}$$

where

$$V_k(r) = \sqrt{\frac{GM}{r}} \tag{14}$$

and constant c_1, c_2 and c_3 are determined later from the magnetohydrodynamic equations. We will obtain the disc half-thickness H as:

$$\frac{H}{r} = \sqrt{c_3(1+\beta)} = \tan\sigma \tag{15}$$

If we assume a power law form for the surface density Σ as:

$$\Sigma = \Sigma_0 r^s \tag{16}$$

where s is constant. In order to have a valid solution for the self-similar treatment, the mass-loss rate per unit volume and the field escaping rate must have the following form:

$$\dot\rho = \dot\rho_0 r^{s-\frac{5}{2}} \tag{17}$$

$$\dot B_\varphi = \dot B_0 r^{\frac{s-5}{2}} \tag{18}$$

Substituting the above self-similar transformation in the MHD equations of the system, we 'll obtain the following system of dimensionless equations, which should be solve to having c_1, c_2 and c_3:

$$\dot\rho_0 = -(s+\frac{1}{2})\frac{c_1 \alpha \Sigma_0)}{2} \sqrt{\frac{GM_*}{(1+\beta)C_3}} \tag{19}$$

$$H = \sqrt{(1+\beta)c_3}r \tag{20}$$

$$-\frac{1}{2}c_1^2\alpha^2 = c_2^2 - 1 - [s - 1 + \beta(s+1)]c_3 \tag{21}$$

$$c_1 = 3(s+1)c_3 + (s+\frac{1}{2})l^2\dot m \tag{22}$$

$$(\frac{1}{\gamma-1} - \frac{1}{2})c_1 c_3 = \frac{9}{4}f c_3 c_2^2 - \frac{5\Phi_s}{\alpha}(s - \frac{3}{2})c_3^{\frac{3}{2}} - \frac{1}{8}\eta\dot m \tag{23}$$

$$\dot m = 2c_1 \tag{24}$$

After algebraic manipulations, we obtain a forth order algebraic equation for c_1:

$$D^2 c_1^4 + 2DB c_1^3 + (B^2 + 2D(E-1))c_1^2 + (2B(E-1) - A^2)c_1$$

$$+(E-1)^2 = 0 \tag{25}$$

where

$$D = \frac{1}{2}\alpha^2 \tag{26}$$

$$B = \frac{4}{9f}(\frac{1}{\gamma - 1} - \frac{1}{2}) - [s - 1 + \beta(s + 1)][\frac{(1 - 2(s + \frac{1}{2})l^2)}{3(s + 1)}] \tag{27}$$

$$A = \frac{20\Phi_s}{9f\alpha}(s - \frac{3}{2})[\frac{(1 - 2(s + \frac{1}{2})l^2)}{3(s + 1)}]^{\frac{1}{2}} \tag{28}$$

$$E = \frac{\eta}{3f}(\frac{s + 1}{1 - 2(s + \frac{1}{2})l^2}) \tag{29}$$

For the case $s = -\frac{1}{2}$, we have $\dot{\rho}_0 = 0$ which is correspond to no mass loss or wind in the hot magnetized flow (Abbassi et al. 2008). In this work, we focus on the wind case ($s > -\frac{1}{2}$). As we mentioned in introduction, the observation evidence shows that the outflow can exist in ADAF (like $sgrA^\star$) and slim disc (like LMC X-3). The outflow will affect the dynamics and the radiative flux of the accretion disc (Ohsuga et al. 2005,2009). As we know, c_1 determines the behaviour of the radial flow and it depends on the input parameter of the fluid such as α, Φ_s, β and f. Other flow's quantity such as c_2 and c_3 can be obtained easily from c_1:

$$c_2^2 = \frac{4c_1}{9f}[\frac{1}{\gamma - 1} - \frac{1}{2}] + \frac{20\Phi_s}{9f\alpha}(s - \frac{3}{2})[\frac{1 - 2(s + \frac{1}{2})l^2}{3(s + 1)}]^{\frac{1}{2}}c_1^{\frac{1}{2}}$$

$$+ \frac{\eta}{3f}(\frac{s + 1}{1 - 2(s + \frac{1}{2})l^2}) \tag{30}$$

$$c_3 = c_1(\frac{1 - 2(s + \frac{1}{2})l^2}{3(s + 1)}) \tag{31}$$

Abbassi et al. (2010) studied the structure of a magnetized ADAFs with thermal conduction and wind parameter by solving the above equations (c_1, c_2 and c_3). We use their results specially c_3 for studying the surface temperature and radiation flux of LMC X-3 as a slim disc in the next section.

4   radiation properties and results

As we know, ADAFs occur in two regimes depending on their mass accretion rate and optical depth. In optically thin ADAFs, the cooling time of accretion flow is longer than the accretion time scale. The generated heat by viscosity remains mostly in the accretion disc. The discs can not radiate their energy efficiently. So, the gas pressure dominates the optically thin ADAFs.

Also, it is not easy to calculate the radiative spectrum of optically thin ADAFs. This model of AFAFs do not radiate away like a black body radiation. The cooling process in the optically thin ADAFs are Bremsstrahlung, Synchrotron and Compton cooling. These process have possible roles to reproduce emission spectra.

In the optically thick ADAFs or slim disc models, the mass accretion rate and the optical depth is very high. So, the radiation generated by accretion disc can be trapped within the disc. The radiation pressure dominates the optically thick ADAFs and sound speed is related to radiation pressure. This model radiates away locally like a black body radiation (Remillard & McClintock 2006).

The averaged flux F is:

$$\Pi = \Pi_{rad} = \frac{1}{3}aT_c^4 2H = \frac{8H}{3c}\sigma T_c^4 \tag{32}$$

$$F = \sigma T_c^4 = \frac{3c}{8H}\Pi = \frac{3}{8}c\Sigma_0\sqrt{\frac{c_3}{1+\beta}}GMr^{s-2}, \tag{33}$$

where Π, T_c, c , σ is the height-integrated gas pressure, the disc central temperature, light speed and the Stefan-Boltzman constant. The optical thickness of the disc in the vertical direction is:

$$\tau = \frac{1}{2}\kappa\Sigma = \frac{1}{2}\kappa\Sigma_0 r^s \tag{34}$$

where κ is the electron-scattering opacity. Hence, the effective temperature of the disc surface becomes.

$$\sigma T_{eff}^4 = -\frac{16\sigma T^4}{3\kappa\rho}\frac{\partial T}{\partial z} \approx \frac{4\sigma}{3\tau}T^4 \tag{35}$$

$$\sigma T_{eff}^4 = \frac{\sigma T_c^4}{\tau} = \frac{3c}{4\kappa}\sqrt{\frac{c_3}{1+\beta}}\frac{GM}{r^2} = \frac{3}{4}\sqrt{\frac{c_3}{1+\beta}}\frac{L_E}{4\pi r^2} \tag{36}$$

$$T_{eff} = (\frac{3L_E}{16\pi\sigma}\sqrt{\frac{c_3}{1+\beta}})^{\frac{1}{4}}r^{\frac{-1}{2}} \tag{37}$$

where $L_E = 4\pi c\frac{GM}{\kappa}$ is the Eddington luminosity.

In figures 1,2 the surface temperature (T_{eff}) is plotted as a function of the dimensionless radius ($\frac{r}{r_g}$) for LMC X-3. It is obvious that the surface temperature decreases as $\frac{r}{r_g}$ increases. Figure 1 shows the effect of wind parameter on the surface temperature. The effect of wind or outflow decreases the temperature gradient. As we can see, the slim discs become colder for the case of strong wind ($s > -\frac{1}{2}$). This is because of energy flux which is taken away by winds. We have plotted together the effects of wind parameter and thermal conduction in figure 2 and then compared them together in reducing the surface temperature. We have showed that the wind parameter is more effective than the thermal conduction in reducing the slope of the temperature gradient and reducing the surface temperature.

If the assumption that the disc is optically thick in the z-direction holds, each element of the disc face radiates roughly as a black body with temperature $T(r)$. The temperature $T(r)$ plays a similar role as the effective temperature of a stars; so, we can approximate the intensity emitted by each element of area of the discs as $I_\nu = B_\nu(T_{eff}(r))$. As we have stated above, we assume that the disc surface radiates black body radiation B_ν with temperature $T_{eff}(r)$. Then, we can equate the disc radiative flux $F(r)$ to a black body flux $\sigma T_{eff}^4(r)$ where $T_{eff}(r)$ is the surface temperature. Since $\sigma T_{eff}^4(r) = \int_0^\infty \pi B_\nu(T_{eff}(r))d\nu$ we can equate the radiative flux $F_\nu(r) = \pi B_\nu(T_{eff}(r))$. Therefore, the radiative flux depends on the surface temperature. Then, the continuum spectrum (luminosity per frequency) L_ν can be calculated (see , e.g., Kato et al. 2008) by:

$$L_\nu = \int_{r_{in}}^{r_{cr}} \pi B_\nu(r)2\pi r dr \tag{38}$$

We will integrated the luminosity in a reasonable interval: ($r_{cr} = 5 - 50r_g$). We use the black-body function:

$$B_\nu(r) = \frac{2h}{c^2}\frac{\nu^3}{e^{\frac{h\nu}{k_B T_{eff}(r)}} - 1} \tag{39}$$

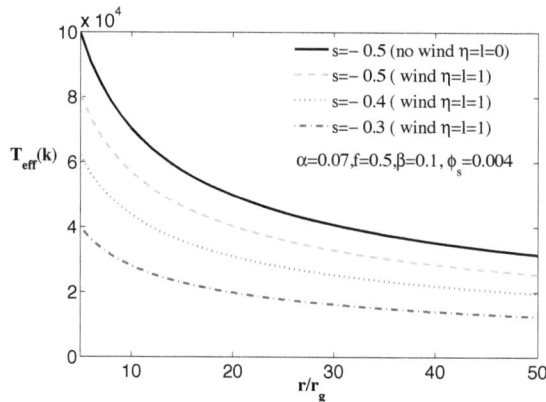

Figure 1: The surface temperature of LMC X-3 as a function of dimensionless radius $\left(\frac{r}{r_g}\right)$ for two cases: no wind solutions $s = -0.5$ and with wind solutions $s > -0.5$ for a fixed thermal conduction parameter $\phi_s = 0.004$.

When the disc is seen at an angle other than face-on (from above or below the plane exactly), the circular disc appears elliptical. We study the observed flux of LMC X-3 at inclination angle $i = 60$. Observed flux from an accretion disc depends on the distance $D(= 52kpc)$ and inclination angle i as:

$$F_\nu^{obs} = \frac{L^{obs}}{\pi D^2} = \frac{\cos i L_\nu}{\pi D^2} \tag{40}$$

We plotted the observed flux of LMC X-3 for inclination angle $i = 60$ for two cases using no wind and with wind for two values of thermal conduction in figure 3. We have shown the observed flux of disc decreases by increasing the effects of wind and thermal conduction. Davis et al. (2006) plotted the spectrum of LMC X-3 for $i = 60°$, $D = 52kpc$. The observed flux is approximately $1\frac{kev}{cm^2 s}$ and the peak of the spectrum is located in $E = 1kev$. As we can see in figure 3, by adding the effect of wind parameter, the observed flux is $2\frac{kev}{cm^2 s}$; Whereas, in our previous paper (GKA13) without wind parameter, the observed flux was $4\frac{kev}{cm^2 s}$. So the presence of wind parameter causes that the observed flux to be more compatible with paper presented by Davis et al. (2006).

5   Summary and Conclusion

In this paper, we have studied the magnetized slim discs (like LMC X-3) in the presence of the thermal conduction and wind parameter. We used the self-similar method for solving the equations in the cylindrical coordinates (r, φ, z). Although the self similar solutions are too simple, they improve our understanding of the physics of the accretion discs around black hole. For simplicity, we assume an axially symmetric and static state disc with α prescription of viscosity. Also, we ignore the relativistic effects and we use the newtonian gravity in the radial direction. We have studied the effect of wind on the observational properties of LMC X-3 by following GKA13. Our results reduce to their solutions when the effect of wind is neglected. In the slim disc models, the mass accretion rate and the optical depth are very high. So, the radiation generated by accretion disc can be trapped within the disc. In optically thick ADAFs, the radiation pressure dominates. This model

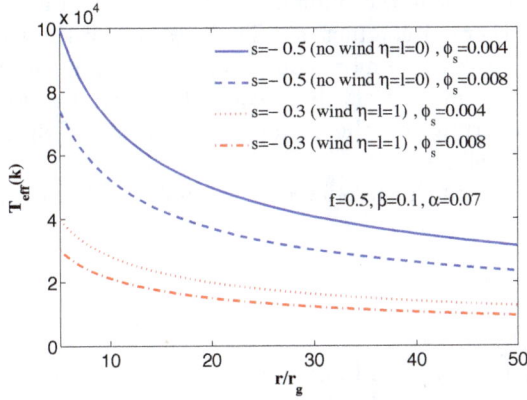

Figure 2: The comparison of the surface temperature of LMC X-3 as a function of dimensionless radius $\left(\frac{r}{r_g}\right)$ for two cases: no wind solutions $s = -0.5$ for two values of thermal conduction parameter $\phi_s = 0.004, \phi_s = 0.008$ and with wind solutions $s = -0.3$ for two values of thermal conduction parameter $\phi_s = 0.004, \phi_s = 0.008$.

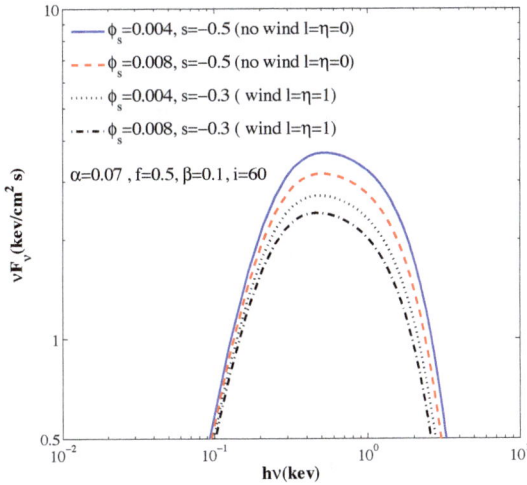

Figure 3: The observed flux of LMC X-3 with an inclination angle, $i = 60$ for two cases: no wind solutions $s = -0.5$ for two values of thermal conduction parameter $\phi_s = 0.004, \phi_s = 0.008$ and with wind solutions $s = -0.3$ for two values of thermal conduction parameter $\phi_s = 0.004, \phi_s = 0.008$.

radiates away locally like a black body radiation. We have shown the surface temperature and the observed flux of LMC X-3 decreases by increasing the effects of wind and thermal conduction. We have showed that the influence of wind on the observational properties is bigger than the effect of thermal conduction. The presence of the wind parameter causes the observed flux of LMC X-3 to be more compatible with paper presented by Davis et al. (2006). It is interesting to study the effects of convection, self gravity and global magnetic field on the radiation spectrum and the surface temperature of optically thick advection dominated accretion flows (slim discs).

The author would like to thank the referee for a number of his/her helpful and constructive comments.

References

[1] Abramowicz M. A., Jaroszynski M., Sikora M. 1978, AAP, 63, 221

[2] Abramowicz M. A., Czerny B., Lasota J. P., Szuszkiewicz E. 1988, APJ, 332, 646

[3] Abramowicz M. A., Jaroszynski M., Kato S., Lasota J. P., Rozanska A., Sadowski A. 2010, AAP, 512A, 15A

[4] Abbassi S., Ghanbari J., Najjar S. 2008, MNRAS, 388, 663

[5] Abbassi S., Ghanbari J., Ghasemnezhad M. 2010, MNRAS, 409, 1113

[6] Abbassi S., Nourbakhsh E., Shadmehri M. 2013, APJ, 765, 96

[7] Blandford R. D. & Begelman M. C. 1999, MNRAS, 303, L1

[8] Blandford R. D. & Payne D. G. 1982, MNRAS, 199, 883

[9] Cowie L. L. & Mackee C.F. 1977, APJ, 275, 641

[10] Cowley A.P., Crampton D., Hutchings J.B., Remillard R., Penfold J.E. 1983, APJ, 272, 118

[11] Davis S. W., Done C.,& Blaes O. M. 2006, APJ, 647, 525

[12] Done C. & Davis S.W. 2008, APJ, 683, 389

[13] Fukue J. 2000, PASJ, 52, 829

[14] Fukue J. 1989, PASJ, 41, 123

[15] Fukue J. 2004, PASJ, 56, 569-580

[16] Ghasemnezhad M., Khajavi M., Abbassi S. 2012, APJ,v750

[17] Ghasemnezhad M., Khajavi M., Abbassi S. 2013, APSS, 346, 341G

[18] Johnson B.M. & Quataert E. APJ, 660, 1273j

[19] Kato S., Fukue J., Mineshige S. 2008, Black hole accretion discs

[20] Kaburaki O. 2000, APJ, 660, 1273J

[21] Knigge C. 1999, MNRAS, 309, 409

[22] Meier D.L. 1979, APJ, 233, 664

[23] Mineshige S., Kawaguchi T., Takeuchi M., Hayashida K. 2000, PASJ, 52, 499

[24] Murray N. & Chiang J. 1996, Nat, 382, 789

[25] Narayan R., & Yi I. 1994,APJ, 428, L13

[26] Narayan R., & Yi I. 1995, APJ, 444, 238

[27] Orosz J.A., Steeghs D., MacClintock J.E. et al. 2009, APJ, 697, 573

[28] Ohsuga K., Mori M., Nakamoto T., Mineshige S. 2005, APJ, 628, 368

[29] Ohsuga K. & Mineshige S. 2011, astro-ph/1105.5474

[30] Ohsuga K., Mineshige S., Mori M., Kato Y. 2009, PASJ, 61, L7

[31] Piran T. 1977, MNRAS, 180, 45

[32] Remillard R.A. & McClintock J.E. 2006, ARAA, 44, 49

[33] Shadmehri M. 2009, MNRAS, 395, 877

[34] Shakura N. I., & Sunyaev R.A. 1973, AAP, 24, 337

[35] Straub O., Barsa M., Sadowski A., Steiner J.F., Abramowicz M.A., Kluzniak W., Mac-Clintock J.E., Narayan R.,Remillard R. 2011, AAP, 533A, 67S

[36] Tanaka T. & Menou K. 2006, ApJ, 649, 345

[37] Watarai K. & Mineshige S. 2001, PASJ, 53, 915

[38] Wilms J., Nowak M. A., Pottschmidt K., Heindl W. A., Dove J. B., Begelman M.C. 2001,MNRAS, 320, 327

5

Proper integration time of polarization signals of internetwork regions using SUNRISE/IMaX data

Akram Gheidi Shahran · Taghi Mirtorabi

Physics Department, Alzahra University, Vanak, 1993891176, Tehran, Iran

Abstract. Distribution of magnetic fields in the quiet-Sun internetwork areas has been affected by weak polarization (in particular Stokes Q and U) signals. To improve the signal-to-noise ratio (SNR) of the weak polarization signals, several approaches, including temporal integrations, have been proposed in the literature. In this study, we aim to investigate a proper temporal-integration time with which an optimum SNR maintained physical properties is obtained. We use magnetographs of Zeeman sensitive Fe I 5250.2 Å line recorded by SUNRISE/IMaX to determine fraction of areas with significant polarization signals after temporal integrations with different durations. We examine several thresholds for the noise level. We also perform simple numerical simulations to explore the effect of size and lifetime of the magnetic features in obtaining the proper integration time. We find that the maximum fraction of pixels with real detectable linear polarization signals in the quiet-Sun internetwork is achieved with a maximum integration time about 8 minutes. Variation of polarization signals with integration time is strongly dependent on lifetime and size of magnetic patches. The temporal integration should be performed with great caution since in the presence of relatively long-lived magnetic features (such as network patches) SNR increases monotonically by increasing the integration time. This monotonic increase does not necessarily correspond to the internetwork areas where the linear magnetic features are relatively short-lived.

Keywords: Sun: photosphere – Sun: magnetic fields – techniques: polarimetric

1 Introduction

Recent investigations have shown that the quiet-Sun internetwork (IN), which occupies a large fraction of the solar surface may contain most of the unsigned magnetic flux at any given time [26], with much of the magnetic flux at the solar surface being in the form of horizontal fields [20, 19, 9]. The majority of individual magnetic features in these regions have been found to be small and relatively short-lived [18]. Therefore, only recent high spatial and temporal resolution observations (i.e., from Hinode and SUNRISE) have allowed studying the distribution of magnetic fields in the IN regions in more detail [21, 19, 24, 25, 2, 28, 7, 17]. It has been shown that only 25% of the IN areas may contain significant polarization signals (i.e., signals larger than 4.5 times the noise level) of Hinode/SP measurements with an integration time of 67.2s [23]. Hence, due to the small signal-to-noise ratio (SNR) in Stokes Q, U, and/or V, the magnetic field strength and orientation in many parts of the IN are still uncertain. These measurements have often been based on Stokes inversion techniques that may fail to return correct parameters (such as magnetic field strength or inclination angle) when the SNR in the linear polarization Stokes profiles is too low ([7, 2, 6]; cf. [17]).

Most controversial is the inclination of magnetic features in the IN. Thus, a pervasive horizontal field has been reported [20, 24, 19, 14, 23], or an isotropic distribution [2], or even a predominantly vertical magnetic field [28]. Borrero et al. [7] have argued that contamination of Stokes Q, U, and V with photon noise leads to an overestimation of the horizontal component of the magnetic field. Because in the visible Q and U are generally weaker than Stokes V and hence are more likely to be swamped by noise. They performed several Monte Carlo simulations to retrieve through inversions the magnetic vector in pixels with predefined vertical field. They found that linear polarization signals that are not sufficiently above the noise caused the inversion code to return too horizontal fields. To achieve more reliable results, Borrero et al. [8] set up different selection criteria to single out pixels with high SNR, and found that due to the relative weakness of linear polarization signals compared to Stokes V, the selection based on only one of the Stokes parameters (i.e., where SNR of one of Stokes Q or U or V is larger than 4.5) may lead to a wrong result. They concluded that inversion codes retrieve the original (pre-defined) magnetic field when at least one of the linear components (Stokes Q or U) has an SNR of 4.5 or more. The latter criterion, however, restricts the number of invertible pixels to $5\% - 30\%$ of the entire area for the noise level of $\sigma = 1.0 \times 10^{-3}\ I_c$ and $\sigma = 2.8 \times 10^{-4}\ I_c$, respectively.

In their attempt to detect weak linear polarization signals, Lites et al. [19] temporally integrated the Hinode spectropolarimetric images over 67.2 s. Bellot et al. [4] extended the temporal integration of quiet-Sun Hinode/SP data to much longer integration times, and claimed to obtain the highest feasible SNR in the linear polarization signal. They reached an extremely low photon noise level of $7 \times 10^{-5}\ I_c$ after 25 min integration, which consequently led them to find that 69% of their quiet-Sun field-of-view (FOV) had at least SNR$\approx$ 4.5 in Stokes Q or U profiles. They also showed that even with an integration time of 10 min (with less image degradation compared to those with 25 min integration), about 60% of the internetwork had still a measurable linear polarization signal. In addition, they assumed that the linear polarization in the rest of the pixels with SNR$<$ 4.5 had to be real, since the Stokes Q and U in those pixels showed similar appearances in both Fe I 630 nm lines. From this, they concluded that *"the solar internetwork is pervaded by linear polarization signals"*.

This is an interesting and to a certain extent also surprising result, since the Sun has significantly evolved within 25 minutes, or even 10 minutes. Within these times multiple generations of granules are born, evolve and die, moving magnetic features around. Many magnetic features in the IN also live less long than these integration times. In particular the more horizontal magnetic elements, detected in the linear polarization, are short-lived, at least as seen in photospheric observations [20, 9]. This can conceivably limit the integration time beyond which the reduction in the noise level in polarization signals is more than offset by the decrease in polarization signal due to motion-smearing and, in particular, the finite lifetimes of the magnetic features.

Here we extend the work of Bellot et al. [4] by integrating data recorded by IMaX on the first flight of SUNRISE. These data complement those of Hinode /SP in that they were obtained by a filter instrument with a larger FOV and higher spatial resolution. The former property implies improved statistics, the latter a lower cut off in the size and hence possibly flux of magnetic features. In particular, we study the statistical behavior of noise and polarization signals from both observations and simple simulations. We investigate whether there is an integration time with which an optimum SNR can be achieved.

In Sect. 2 the data used in this study are described. Our analyses of the observational data and of the simulations are represented in Sect. 3. The last section is devoted to discussion and concluding remarks.

Table 1: Noise level (σ) of used dataset

σ_Q	σ_U	σ_V	σ_{LP}	σ_{CP}
8.3×10^{-4}	1.1×10^{-3}	1.0×10^{-3}	4.9×10^{-4}	5.2×10^{-4}

2 Data

We used a time series of 32 minutes duration acquired by IMaX magnetograph [22] on board the SUNRISE balloon-borne solar observatory [27, 3, 5, 10] on 9 June 2009. The FOV covers a 50×50 $arcsec^2$ of a quiet region near the center of the solar disk. Dataset consists of 58 consecutive frames of all four Stokes parameters of I, Q, U and V, with a cadence of 33 s and a pixel size of 0.055 arcsecs. Each polarization map was taken in five wavelength positions of ±40, ±80 and 227 mÅ from the center of line Fe I 5250.2 Å. The first four lie inside the absorption line while the last one located on a continuum point. The continuum images are used to estimate the noise level (i.e., the standard deviation of the continuum image).

The phase diversity reconstruction processing of data, increase the spatial resolution to 0.15 arcsecs that is two times larger than the resolution of non-reconstructed data, with the expense of increasing the noise by a factor of three [22]. Non-reconstructed (but flat fielded and corrected for instrumental effects) data with a lower noise are used in this work. To increase the SNR of polarization signals additionally, we average each of Stokes component over four wavelengths in the spectral line. It should be noted that total Q, U, V have a half of the noise of any individual wavelength summarised in Table 1. Also, we form the total linear polarization signal (LP) and circular polarization signal (CP) as follows:

$$LP = \frac{1}{4I_c} \sum_{i=1}^{4} \sqrt{Q_i^2 + U_i^2} \qquad (1)$$

$$CP = \frac{1}{4I_c} \sum_{i=1}^{4} a_i V_i, \qquad (2)$$

where i runs over the four wavelength positions, a is a vector equals to $[1, 1, -1, -1]$ [22], and I_c is the local continuum intensity taken from the fifth wavelength position. This definition of CP avoid any cancelation due to different signs in the blue and red wings of the absorption line.

3 Analysis

We expect that the SNR of circular and linear polarization signals should increase with increasing integration time of the data, if the magnetic patches in the quiet-Sun internetwork were to remain permanent and motionless, then we would expect an ideal increase in the SNR. For features with a finite lifetime, we expect there to be an 'ideal' integration time, i.e., one that leads to the largest increase in SNR. Integration times longer and shorter than this should lead to smaller enhancements in the SNR. For stationary magnetic features such

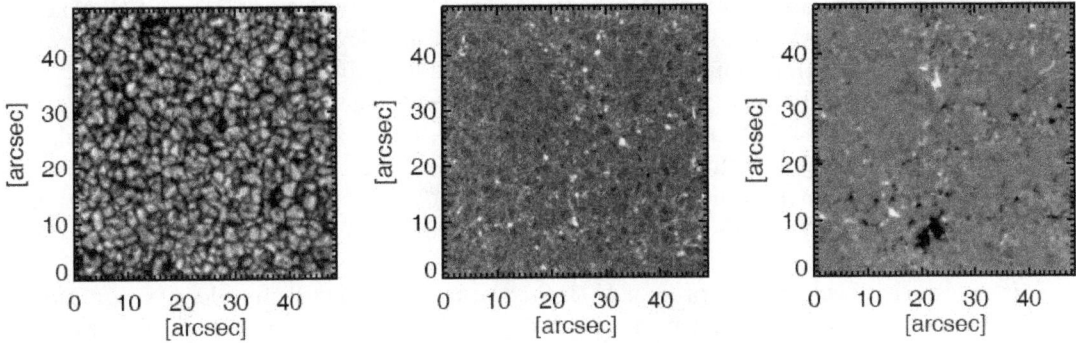

Figure 1: Continuum intensity map (left), linear polarization map (middle) and circular polarization map of SUNRISE/IMaX taken at quiet disk center on 9 June 2009.

an ideal integration time is expected to be comparable with the lifetimes of the magnetic features.

Horizontal magnetic features, located near the edges of granules, have been found to be smaller and shorter lived than more vertical magnetic elements that are concentrated in intergranular areas [19, 14, 9]. For the former a lifetime of 1-10 min is found, with a peak value at about 100 s [9, 14], while for the latter, a mean lifetime of 673 s [16] is found, assuming that vertical magnetic features are mostly associated with bright points. These limited lifetimes may be related to the lifetimes of the granules (of the order of 5-15 minutes; [1, 29, 13, 9]).

3.1　Temporal integration of SUNRISE/IMaX maps

In order to investigate the dependence of noise and the polarization signals on the integration time, we form the integrated images of IMaX as described in Sect. 2.

We also form the CP and the LP for each integrated map from their corresponding integrated images at each wavelength position, using Eqs. (2) and (1), respectively. Similarly, their noise levels are computed as the standard deviation of their integrated continuum positions. Figure 1 shows example maps of continuum intensity, linear polarization signals and circular polarization signals of IMaX data which we used in this work.

In Figure 2 variation of some parameters in terms of integration time are demonstrated. Plotted in Figure 2a, is the variation of rms contrast in the IMaX continuum intensity image. The values basically correspond to the granulation rms contrast (e.g. [12]). The contrast decreases through integration due to the proper motion, evolution and finite lifetime of the granules. It decreases from 8.2% for a single snapshot to 4.2% after integration time of 32 min.

It is important to note that all network patches have been excluded from the FOV prior to making these plots in the following, since we aim to investigate the polarization signals of internetwork areas only.

Reduction of the noise levels of Stokes Q, U, and V with the time integration is shown in Figure 2b (solid black line for Q, dotted blue line for U, and dashed red line for V). The noise levels are computed from the standard deviation of the continuum point of the corresponding Stokes images. After 30 minutes of integration, the noise level reduces to around $2 \times 10^{-4} - 3.5 \times 10^{-4}$.

Figures 2c and 2d are illustrated the fraction of the FOV covered by CP and LP signals larger than 4.5 times of noise level, versus integration time. From Figure 2c it can be seen that temporal integration amplifies the coverage of the FOV by significant CP signals for the first ≈ 5 minutes. For integration longer than this value, the CP coverage saturates at a relatively low level $\lesssim 14\%$ of the FOV. The surface coverage of pixels with detectable Stokes Q or U signals, i.e., these with SNR ≥ 4.5, do not increase monotonically with integration. An improving behavior is seen for surface coverage of Q or U signals up to an integration time of about 8 minutes then an overall gently decrease. Finally, it reaches to 2% of the FOV after 30 minutes integration time.

It is seen that the coverage of Q or U signals saturates at values an order of magnitude lower than that found by Bellot et al. [4]. Also, it should be noted that the increase in coverage of FOV by CP signals with temporal integration which not studied by them, is larger than that of Q or U coverage.

3.2   Temporal integration of synthesized maps

To test the results of the previous section from the observations and also investigate behavior of the variations of the polarization signals with integration time, we performed several simple numerical simulations.

3.2.1   Setup

We started by creating 60 blank frames of 500×500 $pixels^2$ in size. Each frame would represent a typical polarization map which could correspond to observational linear polarization image. At first a random noise with a normal distribution with a standard deviation of 4.9×10^{-4} (The same noise level as observational LP map) is added to all blank frames.

To simulate the linear polarization magnetic features, we created a set of magnetic patches each characterized by three quantities: size in pixels, lifetime in seconds (or number of consecutive frames they live) and amplitude. Danilovic et al. [9] found a rate of occurrence for the linear polarization magnetic patches equal to 7×10^{-4} s^{-1} $arcsec^{-2}$ for the IMaX images. To reproduce the occurrence rate in our synthetic frames, we added 1048 patches, in total, to the full set of 60 frames (i.e. about 17.5 patches in each frame, on average).

The distributions of the lifetime and size of the magnetic patches have also been taken from those obtained by Danilovic et al. [9] (Figures 2a and 2b in their paper). Exponentials with e-foldings of $\tau_T = 92$ s for the lifetime and $\tau_s = 0.24$ $arcsec^2$ for the size were fitted to the extended tail of their distributions. We use the correlations they obtained between size (S) and lifetime as well as between the maximum amplitude of linear polarization (LP_{max}) and the lifetime of linear patches (T), implying the following linear relationships:

$$S\,[pixel] = 0.66 \times T\,[second] + 11.2, \tag{3}$$

$$LP_{max}\,[\%] = 3.7 \times 10^{-4} \times T\,[second] + 0.15. \tag{4}$$

Also from Eq. 4, an exponential with e-folding of $\tau_{LP} = 7 \times 10^{-4}$ is fitted to the extended tail of the distribution of LP_{max}.

3.2.2   Assumptions and simplifications

We assumed a sudden birth and death for each patch with no growth rate, so that each magnetic patch displays a constant size and amplitude during its lifetime. Also, all the

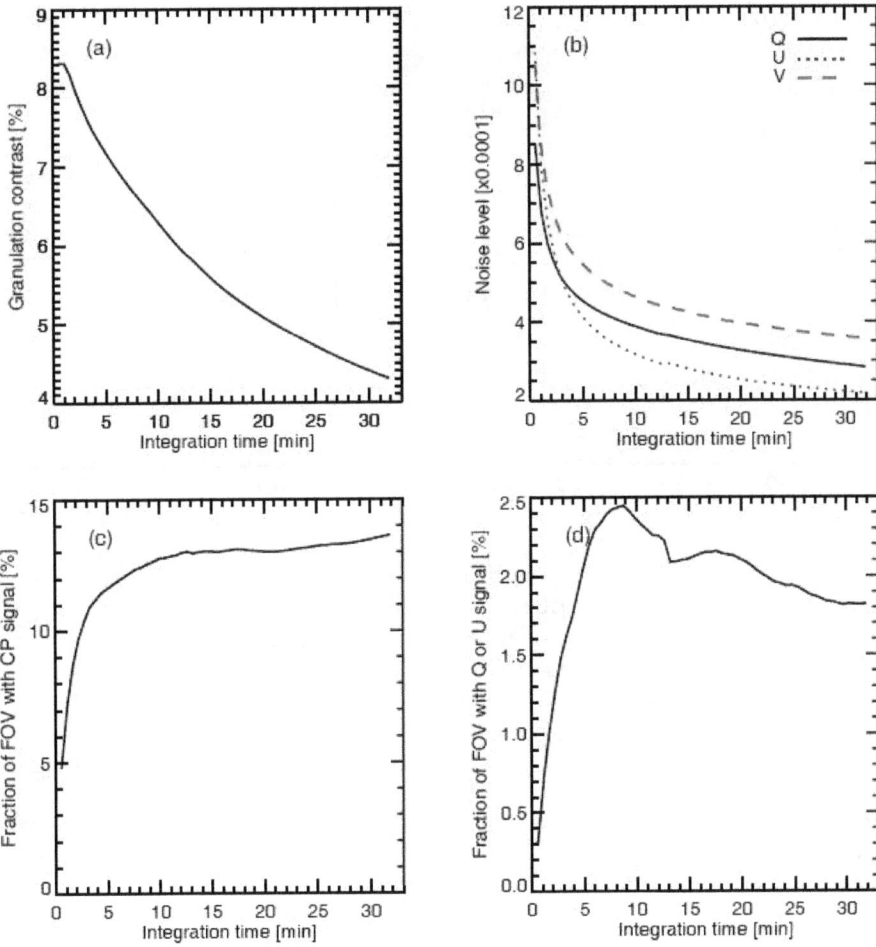

Figure 2: a) The rms continuum contrast versus integratin time. b) Noise levels of Stokes Q (solid black), U (dotted blue) and V (dashed red) as a function of integration time c) Fraction of field-of-view covered by circular polarization signals CP higher than 4.5 times of noise levels, versus integration time. d) Fraction of field-of-view covered by Q or U signals higher than 4.5 times of noise levels versus integration time.

pixels forming the patch surface, have the same amplitude of LP_{max}. This amplitude is determined according to the lifetime of the patch (Eq. 4)

In addition, the patches were assumed to be immobile. Thus, a patch appears in a random frame and stays alive and motionless in the next frames until its lifetime expires. The spatial motion of (horizontal and/or vertical) magnetic patches in the solar IN has been studied by several authors [9, 11, 15]. Horizontal motions reduce the statistical amplification of SNR by temporal integration. By assuming motionless magnetic patches, we determine an upper limit for SNR achievable by temporal integration of individual frames.

Furthermore, we combined these synthetic frames in the same way as we integrated the observed IMaX images in Sect 3.1 to explore the fraction al area covered by LP signals for different integration times.

In order to investigate the influence of individual quantities (i.e., lifetime, size, and LP

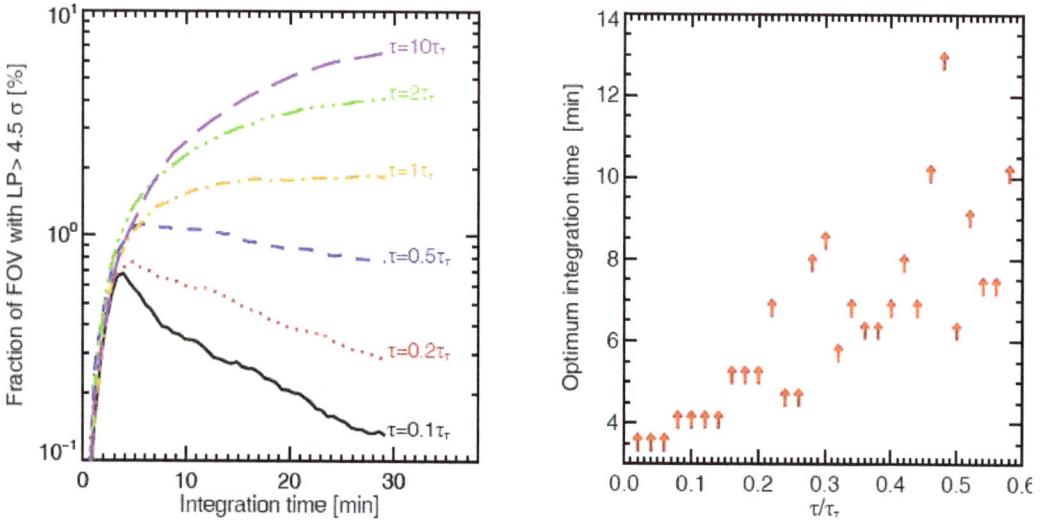

Figure 3: Area coverage of linear polarization signals larger than 4.5σ noise level versus integration time, for different distributions of lifetime of magnetic patches with different e-foldings (τ_T). All features are assumed to have the same size of 0.1 arcsec^2 and the same LP_{max} of 18.7×10^{-4} (left panel). Movement of peak position (i.e., the amount of optimum integration time which leads to the maximum area coverage of signals) with variable e-folding ($\tau_T = 92$ s) of lifetime distribution of patches (right panel).

amplitude) on the results, we simulated images for different distributions of these parameters (i.e., exponential distributions with different e-foldings). Thus the distribution of one of the three quantities is variable while the other parameters are left unchanged.

3.2.3 The effect of patch lifetime

To study the effect of lifetime of the magnetic patches on the optimum integration time of maps, we assume a variable e-folding ($\tau_T = 92$ s) ranging from $0.1\tau_T$ to $10\tau_T$ for the lifetime distributions. All patches are considered to have the same size of 0.1 arcsec^2 and same $LP_{max} = 18.7 \times 10^{-4}$.

Left panel of Figure 3 represents the fractional coverage of pixels with SNR$\geq$ 4.5 for different distributions of lifetime. For relatively short-lived patches, the graph mimics the behavior of the linear polarization component, from the IMaX data (appearance of a peak; Figure 2d). For these cases, it is also seen that position of the peak (the amount of optimum integration time which leads to a maximum area coverage of signals) shifts towards shorter integration times with smaller τ (right panel of Figure 3). Figure 3 (right) display dependence of peak position to the lifetime distribution of patches. For long-lived patches, it resembles the Figure 2c, i.e., the variation of significant signal coverage of CP with integration time.

3.2.4 The effect of patch size

Figure 4 (left panel) illustrates the effect of the size of the magnetic patches on the results. Variable e-foldings ($\tau_s = 0.24$ arcsec^2) between $0.1\tau_s$ to $10\tau_s$ are considered for the size of magnetic patches while a constant lifetime of 100 s and a constant $LP_{max} = 18.7 \times 10^{-4}$ are

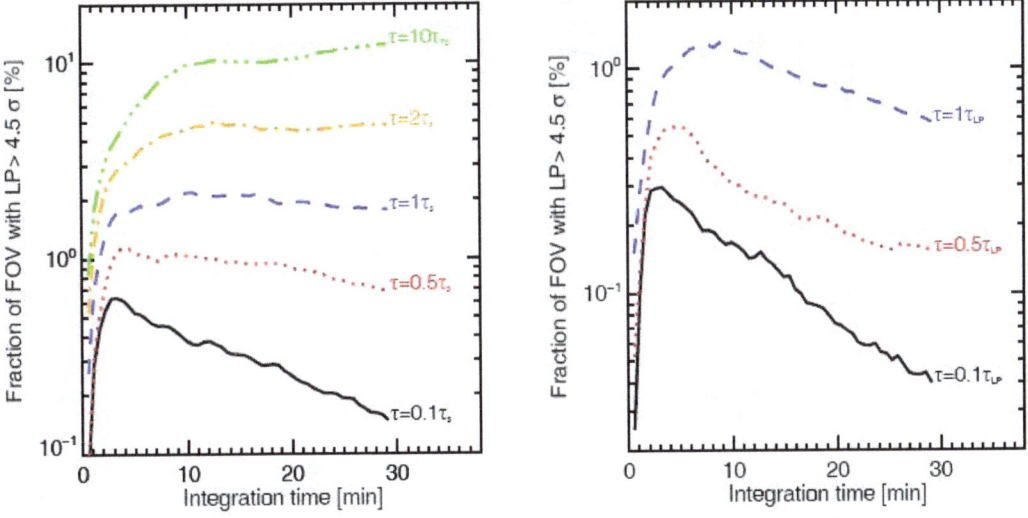

Figure 4: Same as Figure 3, but for different distribution of size of magnetic patches with different e-foldings of τ_S. Constant values for the lifetime and LP_{max} equal to 100 s and 18.7×10^{-4} are respectively assumed for all patches (left panel). Same as Figure 3, but for different distributions of LP_{max} of magnetic patches with different e-foldings of τ_{LP}. Both sizes and lifetimes of the patches are assumed unchanged equal to 100 s and 0.1 arcsec^2, respectively (right panel).

assumed. It is seen in Figure 4 (left panel) that curves get shallower on the left side of the peak for larger magnetic patches and drop slower, since larger patches more likely overlap each others. The larger the patches, the more monotonic the behavior of the variation of coverage of the FOV with detectable signal versus integration time are exhibited.

3.2.5   The effect of patch amplitude

In order to study the effect of different amplitudes of the magnetic patches, simulations for patches with constant size and lifetime of $0.1 arcsec^2$ and 100 s, respectively, but with variable e-foldings ($\tau_{LP} = 4.5 \times 10^{-4}$) ranging from $0.1\tau_{LP}$ to $10\tau_{LP}$, are performed (right panel of Figure 4).

From Figures 4, it is seen that finite lifetime of the patches, restricts the enough integration of the images to achieve the best SNR. Even for very large patches, integration seems to be effective till a finite time. Further integration leads to be seen an unchanged value or very slowly increasing behavior for the SNR.

Finally we provide a simulation for the case that all the three parameters (i.e., size, lifetime, and LP_{max}) differ for different patches (i.e., each has an exponential distribution with e-folding of τ_s, τ_T and τ_{LP} respectively.) and are coupled via equations 3 and 4 (see Figure 5).

4   Conclusions

Motivated by earlier work [19, 4], we have investigated a proper temporal-integration of the polarization signals with which an optimum SNR could be obtained. We used time-series

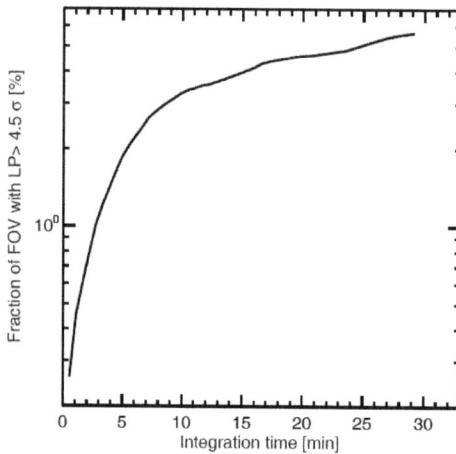

Figure 5: Fraction of the area covered by signals larger than 4.5σ noise level as a function of integration time, for synthesized polarization maps. Exponential with e-folding of τ_T, τ_S and τ_{LP} are assumed for distribution of lifetime, size, and amplitude of the patches. Each of these three parameters are coupled together via the Eqs. 3, 4.

of Stokes parameters from SUNRISE/IMaX to investigate the behavior of variation of SNR improvement with integration time.

We found that the fraction of area with a detectable linear polarization signal (i.e., when SNR$\geq$ 4.5) increases by increasing the temporal-integration time up to about 8 minutes (see Figure 2d). This variation is, however, different for the circular polarization signal. For CP, the fraction of FOV with SNR$\geq$ 4.5 shows a rapid increase up to 3 minutes integration, then it continues to increase with integration time up to about 8 minutes. After that, it basically saturates.

Danilovic et al. [9] have detected 4536 temporal magnetic features, appearing and disappearing in about 2000 arc second square in 30 minutes. The lifetime distribution of their linear magnetic features peaks at 100 s with an extended tail that can be fitted with exponential. From their Figure 2a, we estimated a mean lifetime of 92 s for the linearly polarized features. Visual inspection of the images of the linear and circular polarization shows that actually there are a few large and long life magnetic features which make the constant tail of the circular polarization plot in Figure 2, although the major contribution to those pixels with linear polarization above 4.5 σ coming from short-lived patches.

Statistically, a signal that is submerged in the noise can be extracted and detected after a long-enough temporal integration (as long as the signal exists). This is because that the signal amplification is proportional to integration time which is much faster than that of the noise which grows proportional to square root of integration time (for photon noise). For a static and long-lived (or permanent) magnetic patch, a very large SNR improvement can be achieved (i.e., with a long temporal integration comparable to the feature's lifetime). The SNR of a short-lived feature may, however, be improved up to a temporal integration whose length is given by the feature's lifetime. Further integration reduces the SNR again. (see Figure 2d).

We also explored the variations of coverage of detectable signals with integration time with the help of a simple numerical simulation. This clarified that the behavior of the variations strongly depended on lifetime and size of the magnetic features. As evidenced in Figures 3 and 4, the curves do not show any peak (corresponding to an integration time with optimum SNR) when relatively long-lived or relatively large magnetic patches exist in the FOV. Therefore, the existence of relatively large and long-lived magnetic patches in the FOV, that are mostly corresponded to the network elements (for the linear polarizations) results in an apparent monotonic improvement of the amount of detectable signals that do not necessarily represent the improvement of all signals in the FOV. In particular, such an improvement does not correspond to the relatively small and short-lived linear magnetic patches in internetwork areas. The maximum temporal-integration time of $\approx$ 8 minutes determined from the SUNRISE/IMaX observations agrees with those from our numerical simulations .

The results of our simulations may explain Figure 2 of the paper by Bellot et al. [4], where coverage of the significant linear signals monotonically increase with integration time. This could be due to the presence of large and long-lived magnetic patches in the FOV of their observations (see Figure 4 of their paper). In addition, we found a similar monotonic curve for the CP from the same dataset as they used in their study (but with much steeper increase and much larger SNR improvement, compared to the one for the LP in Figure 2 of their paper). Therefore, only based on such monotonic increases of LP and CP with integration time, the same statement as for the LP (i.e. pervaded horizontal field in the IN) could be also valid, even stronger, for the CP variation.

Summarizing, we conclude that temporal integration of polarization signals can indeed improve the SNR. However, the maximum allowed integration time is strongly limited to the lifetime and size of the magnetic features and also to the dynamical time-scale of the solar granules on the solar surface. Longer integration times would misinterpret the data, since the apparent SNR improvement could be due to, e.g. few longer-lived magnetic patches in the FOV, and not to smaller features.

References

[1] Alissandrakis, C. E., Dialetis, D., Tsiropoula, G. 1987, A&A, 174, 275

[2] Asensio Ramos, A. 2009, ApJ, 701, 1032

[3] Barthol, P., Gandorfer, A., Solanki, S. K., et al. 2011, Sol. Phys., 268, 1

[4] Bellot Rubio, L. R. , Orozco Surez, D. 2012, ApJ, 757, 19

[5] Berkefeld, T., Schmidt, W., Soltau, D., et al. 2011, Sol. Phys., 268, 103

[6] Bommier, V., Martnez Gonzlez, M., Bianda, M., et al. 2009, A&A, 506, 1415

[7] Borrero, J. M. , Kobel, P. 2011, A&A, 527, A29

[8] Borrero, J. M. , Kobel, P. 2012, A&A, 547, A89

[9] Danilovic, S., Beeck, B., Pietarila, A., et al. 2010, ApJ, 723, L149

[10] Gandorfer, A., Grauf, B., Barthol, P., et al. 2011, Sol. Phys., 268, 35

[11] Giannattasio, F., Del Moro, D., Berrilli, F., et al. 2013, ApJ, 770, L36

[12] Hirzberger, J., , Feller, A., Riethmller, T. L., et al. 2010, ApJ, 723, L154

[13] Hirzberger, J., Bonet, J. A., Vzquez, M., Hanslmeier, A. 1999, ApJ, 515, 441

[14] Ishikawa, R. , Tsuneta, S. 2009, A&A, 495, 607

[15] Jafarzadeh, S., Cameron, R. H., Solanki, S. K., et al. 2014a, A&A, 563, A101

[16] Jafarzadeh, S., Solanki, S. K., Feller, A., et al. 2013, A&A, 549, A116

[17] Jafarzadeh, S., Solanki, S. K., Lagg, A., et al. 2014b, A&A, 569, A105

[18] Lamb, D. A., Howard, T. A., DeForest, C. E. 2014, ApJ, 788, 7L

[19] Lites, B. W., Kubo, M., Socas-Navarro, H., et al. 2008, ApJ, 672, 1237

[20] Lites, B. W., Leka, K. D., Skumanich, A., Martnez Pillet, V., Shimizu, T. 1996,ApJ, 460, 1019

[21] Lites, B. W., Socas-Navarro, H., Kubo, M., et al. 2007, PASJ, 59, 571

[22] Martnez Pillet, V., Del Toro Iniesta, J. C., lvarez-Herrero, A., et al. 2011, Sol. Phys., 268, 57

[23] Orozco Surez, D. Bellot Rubio, L. R. 2012, ApJ, 751

[24] Orozco Surez, D., Bellot Rubio, L. R., del Toro Iniesta, J. C., et al. 2007a, ApJ, 670, L61

[25] Orozco Surez, D., Bellot Rubio, L. R., del Toro Iniesta, J. C., et al. 2007b, PASJ, 59, 837

[26] Snchez Almeida, J. 2004, Astrophys. Space Phys. Res., 325, 115

[27] Solanki, S. K., Barthol, P., Danilovic, S., et al. 2010, ApJ, 723, L127

[28] Stenflo, J. O. 2010, A&A, 517, A37

[29] Title, A. M., Tarbell, T. D., Topka, K. P., et al. 1989, ApJ, 336, 475

6

Gott-Kaiser-Stebbins (GKS) effect in an accelerated expanding universe

S. Y. Rokni[1] · H. Razmi[2] · M. R. Bordbar[3]

[1] Department of Physics, University of Qom, Qom, I. R. Iran;
email: s.y.rokni@gmail.com
[2] Department of Physics, University of Qom, Qom, I. R. Iran;
email: razmi@qom.ac.ir
[3] Department of Physics, University of Qom, Qom, I. R. Iran;
email: mbordbar@qom.ac.ir

Abstract. We want to find the cosmological constant influence on cosmic microwave background (CMB) temperature due to moving linear cosmic strings. Using the space-time metric of a linear cosmic string in an accelerated expanding universe, the Gott-Kaiser-Stebbins (GKS) effect, as an important mechanism in producing temperature discontinuity in the (CMB), is considered; then, its modification due to the effect of the cosmological constant is calculated. The result shows that a positive cosmological constant (i.e. the presence of cosmic strings in an accelerated expanding universe) weakens the discontinuity in temperature so that a stronger resolution is needed to detect the corresponding influences on the CMB power spectrum and anisotropy.

Keywords: ISM: Cosmic Strings; Cosmological Constant; Cosmic Microwave Background (CMB)

1 Introduction

Topological defects have formed during symmetry breaking phase transitions in the early universe [1]. Among different known defects, scientists have paid most attention to cosmic strings [1-2]. These objects, in addition to their importance in cosmology, have been recently under consideration because of their similarities to fundamental strings [3]. Although cosmic strings were under study as the necessary seeds for the large scale structure we see today [4], the data from $COBE$ and $BOOMERANG$ have shown that they cannot be considered as the main candidate for the early density fluctuations in our universe [5]. Indeed, precise measurements in cosmic microwave power spectrum have shown that the most possible contribution of cosmic strings in the early fluctuations can be at most up to 10 percentage; this value constrains the mass density of strings (G_μ) up to the limit of $\sim 10^{-7}$ [6]. The recent results from Planck also confirm this constraint ($G_\mu < 10^{-7}$) [7].
There are some observational effects for detecting cosmic strings. Loop cosmic strings may be observed via their gravitational radiation [8]; they may be considered as the origin of high-energy cosmic sources [9]. The long (linear) cosmic strings can have gravitational lensing effects [10-11]; they may have some effects on CMB [12-14]. There are possible mechanisms for the contribution of linear cosmic strings in producing temperature discontinuity in CMB [15]; among them, the Gott Kaiser Stebbins (GKS) effect is one of the most important ones [12]. According to this effect, the light ray reaching an observer in front of a linear cosmic string is blue shifted while the ray behind the string remains unchanged; so, the observer sees a small temperature discontinuity due to an angular separation at the order of the string

deficit angle. Although for a single cosmic string, it is very hard to observe such an effect, for a network of strings, at small angular scales (high multiples), the GKS effect can have a dominant contribution in angular power spectrum [13].

Considering today observational evidence for an accelerated expanding universe [16], it is natural to study possible modifications of different already known gravitational and cosmological phenomena under the influence of the positive cosmological term Λ which is usually considered as the driving force for this acceleration. Here, we want to find the cosmological constant influence on the CMB temperature discontinuity due to cosmic strings (the modified GKS effect).

2   A short review of GKS effect [16]

Considering the line element of a linear long cosmic string of mass density $\mu(G\mu << 1)$ [10]

$$ds^2 = dt^2 - dz^2 - d\rho^2 - (1 - 8G\mu)\rho^2 d\varphi^2 \qquad (1)$$

with the deficit angle $\Delta = 2\pi - \frac{\int_0^{2\pi} g_{\phi\phi}^{1/2} d\phi}{\int_0^\rho g_{\rho\rho}^{1/2} d\rho} = 8\pi G\mu$, for two particles moving with the same velocity $\vec{v}$ relative to the string, the observed temperature T_{obs} in terms of the background temperature T_0 is

$$T_{obs}(\Theta) = T_0 \frac{(1 - u^2)^{\frac{1}{2}}}{(1 + u\cos(\Theta)} \qquad (2)$$

where $u = 2v\sin\frac{\Delta}{2}$ and $\cos\Theta = +1(\cos\Theta = -1)$ when the source of emission of the observed photons and the observer are moving away from (toward) each other.

The temperature discontinuity is simply found as:

$$\frac{\delta T(\Theta)}{T_0} = \frac{T_{obs} - T_0}{T_0} \approx -u\cos\Theta + \frac{u^2}{2}(\cos(2\Theta)) + O(u^3) \qquad (3)$$

Or:

$$\frac{\delta T}{T_0} \approx 8\pi\gamma(v)vG\mu \qquad (4)$$

where $\gamma(v)$ is the Lorentz relativistic gamma factor which is about 1 ($\gamma(v) \approx 1$) when $v \ll c$.

The relation (4) is known as the GKS effect.

3   Cosmological constant influence on the CMB anisotropy due to cosmic strings

Using the following already known line element around a linear long cosmic string under influence of a positive cosmological constant Λ [18]

$$ds^2 = \cos^{\frac{3}{4}}[\frac{\sqrt{3\Lambda}}{2}\rho](dt^2 - dz^2) - d\rho^2 - \frac{4(1 - 4G\mu)^2}{3\Lambda}\cos^{\frac{3}{4}}[\frac{\sqrt{3\Lambda}}{2}\rho]tan^2[\frac{\sqrt{3\Lambda}}{2}\rho]d\varphi^2 \qquad (5)$$

the deficit angle, the relative velocity, and the temperature discontinuity are modified as:

$$\Delta = 2\pi - 2\pi \left[\frac{2(1 - 4G\mu)}{\sqrt{3\Lambda}\rho} \cos^{\frac{-1}{3}} \left[\frac{\sqrt{3\Lambda}}{2}\rho \right] \sin\left[\frac{\sqrt{3\Lambda}}{2}\rho \right] \right] \approx 8\pi G\mu - \frac{1}{40}\pi\Lambda^2\rho^4 + \frac{1}{40}\pi\Lambda^2\rho^4 G\mu \qquad (6)$$

$$u = \gamma(v)\left(\sin\frac{\Delta_{sl}}{2} + v\sin\frac{\Delta_{so}}{2}\right) \approx \gamma(v)v(8\pi G\mu) - \frac{1}{80}\pi\Lambda^2\rho_{sl}^4 - -\frac{1}{80}\pi\Lambda^2\rho_{so}^4) \qquad (7)$$

and

$$\frac{\delta T}{T_0} = 8\pi\gamma(v)vG\mu\left(1 - \frac{1}{640}\frac{1}{G\mu}\Lambda^2(\rho_{sl}^4 + \rho_{so}^4)\right) \qquad (8)$$

where ρ_{sl} and ρ_{so} are the distances from the string to the last scattering surface and the observer respectively.

The relation (8) can be considered as the modified GKS effect.

4   Conclusion

Although Λ has an absolute very small value ($\sim 10^{-52}m^{-2}$), for cosmic scales values of ρ_{sl} and ρ_{so} ($\leq R_H \sim 10^{25}m$) and because of a small value of $G\mu(< 10^{-7})$, the modification term $\frac{1}{640}\frac{1}{G\mu}\Lambda^2(\rho_{sl}^4 + \rho_{so}^4))$ in (8) not only isn't negligible but also may be comparable to 1. This means the modification of the GKS effect can weaken the standard GKS effect considerably. Among other things, an important consideration is that one needs a stronger resolution to detect the discontinuity in the CMB temperature due to cosmic strings. Therefore, considering current observational apparatuses, it may take a long time to be able to detect cosmic strings. As we know, a cosmic string alone cannot affect on the CMB; but, it is a set of these topological defects (e.g. a network) which can dominantly affect the CMB anisotropy power spectrum for larger values of the orders of the spherical harmonics multiple expansion ($l > 3000$) [19]. It seems it is necessary to reconsider the CMB anisotropy power spectrum based on the relation (8).

Finally, it is good to checking the special limiting case $G_\mu \ll \Lambda^2\rho_{sl,so}^4$ which may occur for $G_\mu \ll 10^{-7}$:

$$\frac{\delta T(\Theta)}{T_0} \approx -\frac{\pi\gamma(v)v}{80}\Lambda^2(\rho_{sl}^4 + \rho_{so}^4) \qquad (9)$$

The negative definite value of this result is justified based on this fact that Λ acts as an antigravity force.

References

[1] A. Vilenkin , E. P. S. Shellard, Cosmic Strings and Other Topological Defects (Cambridge University Press, Cambridge, England, 1994); T. W. Kibble, J. Phys. A 9, 1387 (1976).

[2] E. J. Copeland, T. W. B. Kibble, Proc. R. Soc. A 466, 623 (2010); A. Achucarro, C. J. A. P. Martins, (2008), arXiv: 0811.1277; M. B. Hindmarsh, T. W. B. Kibble, Rept. Prog. Phys. 58, 477 (1995); H. B. Nielsen, P.Olesen, Nucl. Phys. B61, 45, (1973).

[3] A.-C. Davis, T. Kibble, Contemp. Phys. 46, 313322 (2005) [hep-th/0505050]; G. Dvali , A. Vilenkin, JCAP 0403, 010 (2004) [hep-th/0312007]; E. J. Copeland, R. C. Myers, and J. Polchinski, JHEP 0406, 013 (2004) [hep-th/0312067]; S. Sarangi, S. H. H. Tye, Phys.Lett. B536, 185-192 (2002) [hep-th/0204074].

[4] Y. F. R. Bouchet, D. Bennett, Ap. J. 354, L41 (1990); N. Turok, R. H. Brandenberger, Phys. Rev. D33, 2175 (1986); H. Sato, Prog. Theor. Phys. 75, 1342 (1986); A. Stebbins, Astrophys. J. Lett. 303, L21 (1986); J. Silk , A. Vilenkin, Phys. Rev. Lett. 53, 1700(1984); N. Turok, Nuc. Phys. B 242, 520 (1984); A. Vilenkin, Phys. Rev. Lett. 46, 1169 (1981); B. Zeldovich, Mon. Not. Roy. Astron. Soc. 192, 663 (1980).

[5] F.R. Bouchet, P. Peter, A. Riazuelo and M. Sakellariadou, Phys. Rev. D65, 021301, (2002); R. Durrer, M. Kunz ,and A. Melchiorri, Phys. Rep. 364, 181, (2002), astro-ph/0110348; A. Albrecht, R. Battye, and J. Robinson, Phys. Rev. D69, 023508 (1999); A. Albrecht, R. Battye, and J. Robinson, Phys. Rev. Lett.79, 4736 (1997).

[6] C. Dvorkin, M. Wyman, and W. Hu, Phys. Rev. D 84, 123519 (2011); M. Landriau, E. P. S. Shellard, Phys. Rev. D 83,043516 (2011); R. Battye, A. Moss, Phys. Rev. D 82,023521 (2010); L. Pogosian, S. H. H. Tye, and I. Wasserman and M. Wyman, JCAP 0902, 013 (2009); N. Bevis, M. Hindmarsh, M. Kunz, and J. Urrestilla, Phys. Rev. Lett. 100, 021301 (2008); A.A. Fraisse, C. Ringeval, D.N. Spergel, and F.R. Bouchet, Phys. Rev. D78, 043535,)2008(;N. Bevis, M. Hindmarsh, M. Kunz, and J. Urrestilla, Phys. Rev. D75, 065015 (2007); A. A. Fraisse, JCAP 0703, 008 (2007); U. Seljak, A. Slosar, and P. McDonald, JCAP 0610, 014 (2006); M. Wyman, L. Pogosian, and I. Wasserman, Phys. Rev. D72, 023513 (2005); L. Pogosian, S. H. H. Tye, I. Wasserman, and M. Wyman, Phys. Rev. D68, 023506 (2003).

[7] P. A. R. Ade, et al. (Planck Collaboration), AA 571, A25 (2014).

[8] D. P. Bennett, F. R. Bouchet, Phys. Rev. D 41, 2408 (1990); T. Vachaspati, A. Vilenkin, Phys. Rev. D35, 1131 (1987); T.Vachaspati, A.Vilenkin, Phys. Rev. D 31, 3052(1985); C. J. Hogan, M. J. Rees, Nature 311, 109 (1984); A. Vilenkin, Phys. Lett. 107B, 47 (1981); A. Vilenkin, Phys. Rev. Lett. 46, 1169 (1981).

[9] T. Vachaspati, Phys. Rev. D 81, 043531(2010); P. Bhattacharjee and G. Sigl, Phys. Rep. 327, 109 (2000); P. Bhattacharjee, Phys. Rev. D 40, 3968 (1989).

[10] A. Vilenkin, Phys. Rev. D 23, 4, 852 (1981).

[11] R. Gregory, Phys. Rev. Lett. 59, 740 (1987); J. Gott, Ap. J. 288, 422 (1985); W. Hiscock, Phys. Rev. D31, 3288 (1985); B. Linet, Gen. Rel. Grav. 17, 1109 (1985); D. Garfinkle, Phys. Rev. D32, 1323 (1985); A. Vilenkin, Ap. J. 282, L51 (1984); A. Vilenkin, Phys. Rev. D23, 852 (1981).

[12] A. Stebbins, Ap. J. 327, 584. (1988); J. R. Gott, Ap. J. 288, 422 (1985); N. Kaiser, A. Stebbins, Nature 310, 391(1984).

[13] A. A. Fraisse, C. Ringeval, D. Spergel, and F. Bouchet, Phys. Rev. D 78, 043535 (2008); L. Pogosian, M. Wyman, Phys. Rev. D 77, 083509(2008).

[14] K. Takahashi, A. Nakuro, Y. Sendouda, D. Yamauchi, C. Yoo, and M. Sasaki, JCAP 10, 003 (2009); N. Bevis, M. Hindmarsh, M. Kunz, and J. Urrestilla, Phys. Rev. D 76, 043005 (2007); N. Turok, U.-L. Pen, U. Seljak, Phys. Rev. D 58,023506 (1998).

[15] L. Perivolaropoulos, (1994), arXiv: astro-ph/9410097v1.

[16] S. Perlmutter et al., Ap.J. 517, 565 (1999) [astro-ph/9812133]; A. G. Riess, A. V. Filippenko, and W. Li and B. P. Schmidt, Astron. Journ. 118, 2668 (1999) [astro-ph/9907038]; B. P. Schmidt et al., Ap. J.. 507, 46 (1998) [astro-ph/9805200]; A. G. Riess et al., Ap. J. 116, 1009 (1998) [astro-ph/9805201]; S. Perlmutter et al., Nature 391, 51(1998) [astro-ph/9712212]; S. Perlmutter et al., Ap. J.. 483, 565 (1997) [astro-ph/9608192].

[17] B. Shlaer, A. Vilenkin, and A. Loeb, arXiv: 1202.1346 (2012).

[18] A. H. Abbassi, A. M. Abbassi, and H. Razmi, Phys. Rev. D 67, 103504 (2003); Q. Tian, Phys. Rev. D 33, 3549 (1986); B. Linet, J. Math. Phys. 27, 1817 (1986).

[19] A. A. Fraisse, et al., Phys. Rev. D 78, 043535 (2008); L. Pogosian and S. H. Henry Tye, JCAP 0902, 013 (2009).

7

Effects of shear and bulk viscosity on head-on collision of localized waves in high density compact stars

Azam Rafiei · Kurosh Javidan · Mohammad Ebrahim Zomorrodian

Department of Physics, Ferdowsi University of Mashhad, 91775-1436, Mashhad, Iran

Abstract. Head on collision of localized waves in cold and dense hadronic matter with and without shear and bulk viscosities is investigated. Non-relativistic dynamics of propagating waves is studied using the hydrodynamics description of the system and suitable equation of state. It will be shown that the localized waves are described by solutions of the Burgers equation. Simulations show that the propagating waves in viscous media travel longer distances in comparison with inviscid similar fluids. In this way, the traveling distance of localized waves is a suitable criterion for evaluating the viscosity of hadronic fluids.

I. Introduction

It is believed that nuclear matters are created in the core of compact stars and in heavy ion collisions [1, 2]. There are some similarities between these two situations, like high density and also some contrasts such as difference in the medium temperature. Since the temperature in the core of compact stars is approximately a few mega electron volts, cold nuclear matter exists in such these places; on the contrary high temperature (a few tens of mega electron volts) nuclear matter can be produced after the relativistic heavy ion collisions. Phase transition from confined hadronic matter to deconfined quark gluon plasma (QGP) has been observed in both situations [1]. In the case of compact star, formation of the QGP regularly is started at the center of a compact star by increasing of its density above the critical density. Afterward the new phase will spread to the environs.

Because of inaccessibility of the interior of compact stars as a unique natural sample at very dense ($5 - 10$ times larger than normal saturation density, $\rho_0 = 0.16 fm^{-3}$) and very low temperature in terrestrial laboratory [2, 3], analysis of dynamic behavior of hadronic matter is investigated by heavy ion collisions [1, 4, 5]. Neutron stars and white dwarfs are two kinds of compact stars [6].Studying the propagation and collision of localized waves in such these media, produces valuable information about the stars formation and the dynamics of compact objects [7]. Propagation and interaction of small amplitude localized waves in the form of shock profilesare investigated in this paper.

A neutron star is characterized by the following conditions: the mass is nearly $1.5M_\odot$ ($M_\odot$ is the mass of sun), the radius is approximately $10Km$ and its initial temperature is about $10^{11}K$ [8]. At the center of the star the energy density upon c^2 changes from 10^{14} up to $10^{15}g.cm^{-3}$ and it reaches to zero at the surface of the star. For a typical neutron star, five different regions can be distinguished. *I*) Outer crust by $10^4 \leq \rho \leq 10^{11}g.cm^{-3}$, where contains lattice of neutron-rich nuclei in a free gas of (relativistic) electrons. *II*) The neutron drip line with the density $\rho \approx 10^{11}g.cm^{-3}$. Because of weakly bounding in this region, the neutrons are able to drip out of the nuclei and they become free increasingly. *III*) Inner crust (free neutron phase) with $10^{11} \leq \rho \leq 10^{14}g.cm^{-3}$, where neutron-rich nuclei are situated in a free gas of electrons and neutrons. IV) The outer core with $\rho \approx 5 \times 10^{14} g.cm^{-3}$ which contains neutrons, electrons, protons and muons in the form of a homogeneous liquid. V) Finally unfamiliar inner core with ρ several times $10^{14} g.cm^{-3}$. It is expected that the Fermi energies of the constituent particles could exceed the rest masses of heavier particles and therefore hyperons can be produced in this region [3]. Transition to deconfined quark matter most likely occurs in this situation [8].

Equation of state (EoS) plays an important role to investigate the behavior of matter in different phases. The EoS in hadronic phase is obtained from the famous nonlinear Walecka model. In the QGP phase the Bag model is utilized [9, 10, 11].

There are two different approaches to describe the hadronic matter. In the "microscopic approach" particle trajectories are pursued, while in the "macroscopic view" the hydrodynamic variables like temperature, pressure and velocity of the fluid are specified during the evolution of the system [12]. Generally there is not any precise knowledge on the microscopic details of reactions [13], especially where the nucleon mean free path is shorter than the dimension of the system [12]. In this situation the system behaves like a perfect fluid with a low viscosity (in QGP phase) or a viscous fluid (for hadron gas) [14, 15]. Hence the relativistic hydrodynamics model is constructed based on the local equilibrium [13, 16], however non-equilibrium degrees of freedom can be added to the problem in different ways. In this framework the variables of the model are the energy-momentum tensor, $T_{\mu\nu}$, net particle density, ρ, and the entropy density, S_μ. The relativistic fluid equation of state is obtained using the local conservation of energy-momentum, the relativistic continuity equation and considering the first law of thermodynamics [12, 15, 16, 17, 18, 19, 20] .

The investigation of localized anisotropies and perturbations help us for finding valuable information about the nature of the hydrodynamics medium, where the waves are propagated. There exist four different sources of density fluctuations which create localized waves propagating in the medium. These are: 1)initial state fluctuations, 2)hydrodynamic fluctuations, 3)fluctuations induced by hard processes and 4)freeze-out fluctuations. Quantum fluctuations in the densities of two colliding nuclei supplemented with energy fluctuations are called initial state fluctuations. Local thermal fluctuations of the energy density and flow velocity produce hydrodynamic fluctuations. Energy loss due to propagation of energetic partons causes hard process fluctuations. Finally there are event-by-event fluctuations during and after the freeze-out stage which are called freeze-out fluctuations [15, 21, 22, 23]. Such these perturbations are able to create nonlinear localized waves in the medium which can be detected and also studied during the evolution of the system. Therefore, the propagation of nonlinear waves and their collisions are very interesting subjects.

Nonlinear solitary waves utilize in various branches of physics [24, 25]. They are unique solutions which travel a long distance in nonlinear medium while save their shapes. Behavior of solitary waves during their collisions provides interesting information about the medium such as phase shift and traveling distance of the waves after the collision. Indeed, after the collision of two solitons, they emerge out with almost the same shapes and velocities that they entered in, but with different relative phase shifts. The phase shifts are functions of the soliton characters and also the medium properties [26, 27, 28]. Propagation of localized waves in super dense hadronic matters has been investigated in non-relativistic and viscose medium. But interaction of localized waves and especially head-on collision of solitary waves have not been studied before. Motivated by these situations, propagation of localized waves due to fluctuations and their head on collisions in hadronic gas are investigated in this paper. In the next section the equations which govern the dynamics of hadron gas are obtained from the Lagranginan density of the medium particles. Thermodynamic relations for these medium will be introduced in this section too. The standard perturbation method in head-on collision will be presented in the third section. The Burgers equations are derived for propagation of localized perturbations in hadron gas and the phase shifts of traveling waves, after the collision will be obtained in this section. The effects of viscosity on the behavior of localized waves are studied using numerical stimulations in the forth section. Finally, conclusions and some remarks are presented in the last section.

II. Hadronic matter

Dynamics of particles in a hadronic matter can be described using the nonlinear mean field theory (NMFL) approximation [29]. According to the chiral power counting, the famous Walecka model characterizes the properties of cold and high density nuclear matter which exist in the super dense nuclear matter, neutron stars and supernovas. This model is also recognized as (σ, ω) model or quantum hydrodynamics model. Based on the principal specifications of head-on collisions, interaction between two nucleons happens via the exchange of virtual σ and ω mesons. These mesons prepare the intermediate range attraction and short range repulsion respectively. So the Walecka model [29, 30, 31] is defined by the following Lagrangian density (1)

$$
\begin{aligned}
\mathcal{L} = \ & \overline{\psi} \left[\gamma_\mu \left(i \partial^\mu - g_\omega \omega^\mu \right) - \left(M - g_\sigma \sigma \right) \right] \psi + \frac{1}{2} \left(\partial_\mu \sigma \partial^\mu \sigma - m_\sigma^2 \sigma^2 \right) \\
& - \frac{1}{4} F_{\mu\nu} F^{\mu\nu} + \frac{1}{2} m_\omega^2 \omega_\mu \omega^\mu - \frac{\kappa}{3} \sigma^3 - \frac{\lambda}{4} \sigma^4
\end{aligned}
\tag{1}
$$

where nucleons (baryon fields), ψ, neutral Lorentz scalar field, σ, and neutral vector meson field, ω_μ, with their couplings and masses are the degrees of freedom for the theory. The expression $M^* = M - g_\sigma \sigma$ is the nucleon effective mass and the weights of the nonlinear scalar terms are shown by the couplings κ and λ while $F_{\mu\nu} = \partial_\mu \omega_\nu - \partial_\nu \omega_\mu$. According to the NMFT the equation of state are driven considering the meson fields act classically [15, 32] as follow

$$
\omega_\mu \to \langle \omega_\mu \rangle \equiv \delta_{\mu 0} \, \omega_0 \quad , \quad \sigma \to \langle \sigma \rangle \equiv \sigma_0
\tag{2}
$$

In which σ_0 and ω_0 are constant. Assuming there is spatially unlimited nuclear matter in statistical, homogeneous and isotropic state at zero temperature where the intense baryonic sources couple to meson fields strongly. In this case the above mentioned classical assumption is an acceptable approach. Thus the equation of motion are driven from [15, 32]

$$
m_\omega^2 \omega_0 = g_\omega \psi^\dagger \psi
\tag{3}
$$

$$
m_\sigma^2 \sigma_0 = g_\sigma \overline{\psi} \psi - \kappa \sigma_0^2 - \lambda \sigma_0^3
\tag{4}
$$

$$
\left[i \, \gamma_\mu \partial^\mu - g_\omega \gamma_0 \omega_0 - \left(M - g_\sigma \sigma_0 \right) \right] \psi = 0
\tag{5}
$$

The baryon density, ρ_B, is introduced by $\psi^\dagger \psi \equiv \rho_B = \frac{\gamma}{6\pi^2} k_F^3$ where k_F is the Fermi momentum. Therefore, the vector Meson ω_0 will be derived by using equation (3) in terms of baryon density as $\omega_0 = g_\omega \rho_B / m_\omega$. The Dirac Eq. which is performed through the equation (5) couples the nucleons to the vector mesons. From the calculations implemented in [15] the corresponding energy density could be derived using the average of the energy-momentum tensor [15, 32] as follows

$$
\begin{aligned}
\varepsilon = \ & \frac{g_\omega^2}{2 m_\omega^2} \rho_B^2 + \frac{m_\sigma^2}{g_\sigma^2} \left(M - M^* \right) + \kappa \frac{\left(M - M^* \right)^3}{3 g_\sigma^3} \\
& + \lambda \frac{\left(M - M^* \right)^4}{4 g_\sigma^4} + \frac{\gamma}{(2\pi)^3} \int_0^{k_F} d^3 k \sqrt{\vec{k}^2 + M^{*2}}
\end{aligned}
\tag{6}
$$

where the nucleon degeneracy factor is shown by γ that equals to 4 . In the above equation, the integral term takes into account the fermion contribution. The self-consistency relation obtained from the minimization of $\varepsilon(M^*)$ with respect to M^* determines the nucleon effective mass as follow

$$
\begin{aligned}
M^* = {} & M - \frac{g_\sigma^2}{m_\sigma^2} \frac{\gamma}{(2\pi)^3} \int_0^{k_F} d^3k \sqrt{\vec{k}^2 + M^{*2}} \\
& + \frac{g_\sigma^2}{m_\sigma^2} \left[\frac{\kappa}{g_\sigma^3}(M - M^*)^2 + \frac{\lambda}{g_\sigma^4}(M - M^*)^3 \right]
\end{aligned}
\tag{7}
$$

The following numerical values for masses and couplings are used for calculation from [15, 32] $M = 939 \; MeV$, $m_\omega = 783 \; MeV$, $m_\sigma = 550 \; MeV$, $\kappa = 13.47 fm^{-1}$, $g_\omega = 9.197$, $g_\sigma = 8.81$, and $\lambda = 43.127$. The baryon density ρ_B varies in the range of $\rho_0 \le \rho_B \le 2\rho_0$ in which $\rho_0 = 0.17 fm^{-3}$ is the nuclear baryon density. If the equation (7) is numerically solved, the nucleon effective mass in term of baryon density will be obtained. In this way, the energy density as a function of baryon density is achieved in [15, 32]

$$
\begin{aligned}
\varepsilon = {} & \left(0.1 \frac{m_\sigma^2}{g_\sigma^2} + 0.04 \frac{\kappa}{g_\sigma^3} + 0.01 \frac{\lambda}{g_\sigma^4} \right) + \left(4 + 2\frac{m_\sigma^2}{g_\sigma^2} + \frac{\kappa}{g_\sigma^3} + 0.43\frac{\lambda}{g_\sigma^4} \right) \rho_B \\
& + \left(-3.75 + \frac{g_\omega^2}{2m_\omega^2} + 8\frac{m_\sigma^2}{g_\sigma^2} + 7.6\frac{\kappa}{g_\sigma^3} + 5.42\frac{\lambda}{g_\sigma^4} \right) \rho_B^2 \\
& + \left(21.26\frac{\kappa}{g_\sigma^3} + 30.35\frac{\lambda}{g_\sigma^4} \right) \rho_B^3 + \left(63.73\frac{\lambda}{g_\sigma^4} \right) \rho_B^4 \\
& - 1.22\rho_B^{\frac{8}{3}} + 2.61 \; \rho_B^{\frac{5}{3}} - 1.4\rho_B^{2/3}
\end{aligned}
\tag{8}
$$

Infinitely high density hadronic plasmas with shear viscosity, v, and bulk viscosity, ζ, are characterized by the continuity and non-relativistic Navier-Stokes equations

$$
\frac{\partial \rho}{\partial t} + \nabla.(\rho \vec{v}) = 0
\tag{9}
$$

and

$$
\frac{\partial v^i}{\partial t} + v^k \frac{\partial v^i}{\partial x^k} = -\frac{1}{\rho}\frac{\partial p}{\partial x^k} - \frac{1}{\rho}\frac{\partial \Pi^{ki}}{\partial x^k}
\tag{10}
$$

with

$$
\Pi^{ki} = -v\left(\frac{\partial v^i}{\partial x^k} + \frac{\partial v^k}{\partial x^i} - \frac{2}{3}\delta^{ki}\frac{\partial v^l}{\partial x^l} \right) - \zeta\delta^{ki}\frac{\partial v^l}{\partial x^l}
\tag{11}
$$

where Π^{ki} is the viscous tensor ; $\vec{v}(t, \vec{x})$, $p(t, \vec{x})$ and $\rho(t, \vec{x})$ are the velocity, pressure and the fluid mass density respectively. Considering $i = l = k = x$ for one dimensional Cartesian case, we have

$$
\frac{\partial v_x}{\partial t} + v_x \frac{\partial v_x}{\partial x} = -\frac{1}{\rho}\frac{\partial p}{\partial x} + \frac{1}{\rho}\left(\zeta + \frac{4}{3}v \right)\frac{\partial^2 v_x}{\partial x^2}
\tag{12}
$$

$$
\frac{\partial \rho_B}{\partial t} + v_x \frac{\partial \rho_B}{\partial x} + \rho_B \frac{\partial v_x}{\partial x} = 0
\tag{13}
$$

The mass density and the baryon density are related to each other through $\rho = M\rho_B$, where M is the nucleon mass. The first law of thermodynamic at zero temperature results in

$$
d\varepsilon = \mu_B d\rho_B
\tag{14}
$$

So the chemical potential μ_B is

$$\mu_B = \frac{d\varepsilon}{d\rho_B} \tag{15}$$

Substitution of (14) and (15) into the Gibbs equation at zero temperature leading

$$d\varepsilon + dp = \rho_B d\mu_B + \mu_B d\rho_B \tag{16}$$

This yields to

$$dp = \rho_B d\mu_B \tag{17}$$

and finally we have

$$dp = \rho_B d\left(\frac{\partial \varepsilon}{\partial \rho_B}\right) \tag{18}$$

$$\frac{\partial p}{\partial x} = \rho_B \frac{\partial}{\partial x}\left(\frac{\partial \varepsilon}{\partial \rho_B}\right) \tag{19}$$

Replacing Equations (18) and (19) into (12) results in

$$\rho_B \frac{\partial v_x}{\partial t} + v_x \frac{\partial v_x}{\partial x} = -\frac{1}{M}\rho_B \frac{\partial}{\partial x}\left(\frac{\partial \varepsilon}{\partial \rho_B}\right) + \frac{1}{M}\left(\zeta + \frac{4}{3}v\right)\frac{\partial^2 v_x}{\partial x^2} \tag{20}$$

And using (8) we have

$$
\begin{aligned}
\rho_B \left(\frac{\partial v_x}{\partial t} + v_x \frac{\partial v_x}{\partial x}\right) &= -\frac{1}{M}\left(-7.5 + \frac{g_\omega^2}{m_\omega^2} + 16\frac{m_\sigma^2}{g_\sigma^2} + 15.2\frac{\kappa}{g_\sigma^3} + 10.84\frac{\lambda}{g_\sigma^4}\right) \\
&\quad \rho_B \frac{\partial \rho_B}{\partial x} + \left(127.56\frac{\kappa}{g_\sigma^3} + 182.1\frac{\lambda}{g_\sigma^4}\right) \\
&\quad \rho_B^2 \frac{\partial \rho_B}{\partial x} + \left(764.76\frac{\lambda}{g_\sigma^4}\right)\rho_B^3 \frac{\partial \rho_B}{\partial x} \\
&\quad -5.42\rho_B^{\frac{5}{3}}\frac{\partial \rho_B}{\partial x} + 2.9\rho_B^{\frac{2}{3}}\frac{\partial \rho_B}{\partial x} \\
&\quad +0.33\rho_B^{-\frac{1}{3}}\frac{\partial \rho_B}{\partial x} + \frac{1}{M}\left(\zeta + \frac{4}{3}v\right)\frac{\partial^2 v_x}{\partial x^2}
\end{aligned} \tag{21}
$$

This is the Navier-Stokes equation for the hadron phase [15].

III. Head-on collision in hadronic gas

There are two different types of interaction between solitons in one-dimensional collisions. In an overtaking collision, they move in the same direction with different velocities. Solitons receive a phase shift after the interaction, however their shapes remain almost unchanged. This type of collision can be studied using the inverse scattering method. The other one is head-on collision. It occurs when two solitary waves propagate in the opposite directions. In this situation, in addition to the phase shifts, trajectories of colliding solitons are changed after the collision as well.

Phase shift and the trajectories of interacting solitary waves after collision have been studied by many authors using several methods [24, 26, 27].

The extended version of Poincare-Lighthill-Kuo (PLK) approach based on the standard perturbation method is a well-known and powerful technique which can be used in head on collision interactions. This technique generally is called Reductive Perturbation Method

(RPM) [26, 27, 33]. In this method the nonlinearities, dissipative and dispersive effects are preserved in the wave equations. Conventionally a head on collision problem can be studied by introducing the stretched coordinates

$$
\begin{cases}
\xi = \sigma\left(x - c_1 t\right) + \sigma^2 P_0\left(\eta, \tau\right) + \sigma^3 P_1\left(\eta, \xi, \tau\right) + \cdots \\
\eta = \sigma\left(x + c_2 t\right) + \sigma^2 Q_0\left(\xi, \tau\right) + \sigma^3 Q_1\left(\eta, \xi, \tau\right) + \cdots \\
\qquad\qquad \tau = \sigma^3 t
\end{cases}
\tag{22}
$$

where ξ and η denote the trajectories of two localized waves travelling to the right and left directions respectively and σ is a small expansion parameter. The variables c_1 and c_2 are unknown phase velocities which will be calculated. Initially the dimensionless variables for the baryon density, the fluid velocity and the pressure are defined as:

$$
\rho = \frac{\rho_B}{\rho_0} \;,\; v_x = \frac{v_x}{c_s} \;,\; p = \frac{p}{p_0}
\tag{23}
$$

Where ρ_0, c_s and p_0 respectively are the background baryon density, the speed of sound and the background pressure in the medium respectively where perturbation propagates. Equations (13) and (21) can be rewritten using (23) as follows

$$
\frac{\partial \rho}{\partial t} + c_s v_x \frac{\partial \rho}{\partial x} + c_s \rho \frac{\partial v_x}{\partial x} = 0
\tag{24}
$$

$$
\begin{aligned}
\rho\left(\frac{\partial v_x}{\partial t} + c_s v_x \frac{\partial v_x}{\partial x}\right) &= \frac{\rho_0}{Mc_s}\left(7.5 - \frac{g_\omega^2}{m_\omega^2} - 16\frac{m_\sigma^2}{g_\sigma^2} - 15.2\frac{\kappa}{g_\sigma^3} - 10.84\frac{\lambda}{g_\sigma^4}\right) \\
&\quad \rho\frac{\partial \rho}{\partial x} - \frac{\rho_0^2}{Mc_s}\left(127.56\frac{\kappa}{g_\sigma^3} + 182.1\frac{\lambda}{g_\sigma^4}\right) \\
&\quad \rho^2\frac{\partial \rho}{\partial x} - \frac{\rho_0^3}{Mc_s}\left(764.76\frac{\lambda}{g_\sigma^4}\right)\rho^3\frac{\partial \rho}{\partial x} \\
&\quad +5.42\frac{\rho_0^{\frac{5}{3}}}{Mc_s}\rho^{\frac{5}{3}}\frac{\partial \rho}{\partial x} + 2.9\frac{\rho_0^{\frac{2}{3}}}{Mc_s}\rho^{\frac{2}{3}}\frac{\partial \rho}{\partial x} \\
&\quad +0.33\frac{\rho_0^{-\frac{1}{3}}}{Mc_s}\rho^{-\frac{1}{3}}\frac{\partial \rho}{\partial x} + \frac{1}{M\rho_0}\left(\zeta + \frac{4}{3}\nu\right)\frac{\partial^2 v_x}{\partial x^2}
\end{aligned}
\tag{25}
$$

If the dimensionless baryon density and the fluid velocity are expanded around their equilibrium values, we have:

$$
\rho = 1 + \sigma^2 \rho_1 + \sigma^3 \rho_2 + \sigma^4 \rho_3 + \cdots
\tag{26}
$$

$$
v = \sigma^2 v_1 + \sigma^3 v_2 + \sigma^4 v_4 + \cdots
\tag{27}
$$

Substituting equations (26) and (27) into equations (24) and (25), neglecting the terms proportional to $\sigma^{\geq 3}$ for first non-zero order of equations (24) and (25) will lead to

$$
c_s\frac{\partial v_1}{\partial \xi} + c_s\frac{\partial v_1}{\partial \eta} - c_1\frac{\partial \rho_1}{\partial \xi} + c_2\frac{\partial \rho_1}{\partial \eta} = 0
\tag{28}
$$

$$
\begin{aligned}
-c_1\frac{\partial v_1}{\partial \xi} + c_2\frac{\partial v_1}{\partial \eta} &- \left[\frac{\rho_0}{Mc_s}\left(7.5 - \frac{g_\omega^2}{m_\omega^2} - 16\frac{m_\sigma^2}{g_\sigma^2} - 15.2\frac{\kappa}{g_\sigma^3} - 10.84\frac{\lambda}{g_\sigma^4}\right)\right. \\
&\quad \left. -\frac{\rho_0^2}{Mc_s}\left(127.56\frac{\kappa}{g_\sigma^3} + 182.1\frac{\lambda}{g_\sigma^4}\right) - \frac{\rho_0^3}{Mc_s}\left(764.76\frac{\lambda}{g_\sigma^4}\right)\right.
\end{aligned}
$$

$$+5.42\frac{\rho_0^{\frac{5}{3}}}{Mc_s} - 2.9\frac{\rho_0^{\frac{2}{3}}}{Mc_s}\rho^{\frac{2}{3}} - 0.33\frac{\rho_0^{-\frac{1}{3}}}{Mc_s}]$$

$$\left(\frac{\partial\rho_1}{\partial\xi} + \frac{\partial\rho_1}{\partial\eta}\right) = 0 \tag{29}$$

Dependencies of ρ_1 and v_1 to the ξ, η and τ can be considered as $\rho_1 = \rho_1^1(\xi,\tau) + \rho_1^2(\eta,\tau)$ and $v_1 = v_1^1(\xi,\tau) + v_1^2(\eta,\tau)$. If these expressions are inserted into equations (28) and (29) then we will have the following:

$$c_s\frac{\partial v_1^1}{\partial\xi} + c_s\frac{\partial v_1^2}{\partial\eta} - c_1\frac{\partial\rho_1^1}{\partial\xi} + c_2\frac{\partial\rho_1^2}{\partial\eta} = 0 \tag{30}$$

$$-c_1\frac{\partial v_1^1}{\partial\xi} + c_2\frac{\partial v_1^2}{\partial\eta} - \left[\frac{\rho_0}{Mc_s}\left(7.5 - \frac{g_\omega^2}{m_\omega^2} - 16\frac{m_\sigma^2}{g_\sigma^2} - 15.2\frac{\kappa}{g_\sigma^3} - 10.84\frac{\lambda}{g_\sigma^4}\right)\right.$$

$$-\frac{\rho_0^2}{Mc_s}\left(127.56\frac{\kappa}{g_\sigma^3} + 182.1\frac{\lambda}{g_\sigma^4}\right) - \frac{\rho_0^3}{Mc_s}\left(764.76\frac{\lambda}{g_\sigma^4}\right)$$

$$+5.42\frac{\rho_0^{\frac{5}{3}}}{Mc_s} - 2.9\frac{\rho_0^{\frac{2}{3}}}{Mc_s}\rho^{\frac{2}{3}} - 0.33\frac{\rho_0^{-\frac{1}{3}}}{Mc_s}]$$

$$\left(\frac{\partial\rho_1^1}{\partial\xi} + \frac{\partial\rho_1^1}{\partial\eta}\right) = 0 \tag{31}$$

In this way the fluid velocity becomes

$$v_1 = \frac{1}{c_s}\left(c_1\rho_1^1(\xi,\tau) - c_2\rho_1^2(\eta,\tau)\right) \tag{32}$$

and the phase velocities are obtained as

$$c_1^2 = c_2^2 = -\frac{\rho_0}{M}\left(7.5 - \frac{g_\omega^2}{m_\omega^2} - 16\frac{m_\sigma^2}{g_\sigma^2} - 15.2\frac{\kappa}{g_\sigma^3} - 10.84\frac{\lambda}{g_\sigma^4}\right)$$

$$+\frac{\rho_0^2}{M}\left(127.56\frac{\kappa}{g_\sigma^3} + 182.1\frac{\lambda}{g_\sigma^4}\right) + \frac{\rho_0^3}{M}\left(764.76\frac{\lambda}{g_\sigma^4}\right)$$

$$-5.42\frac{\rho_0^{\frac{5}{3}}}{M} + 2.9\frac{\rho_0^{\frac{2}{3}}}{M}\rho^{\frac{2}{3}} + 0.33\frac{\rho_0^{-\frac{1}{3}}}{M} \tag{33}$$

The second order equations respect to σ in (24) and (25) lead to the same results by replacing index "1" by "2" and vice versa. Inserting equations (32) and (33) into equations (24) and (25) and collecting third order terms with respect to σ results in:

$$\frac{\partial\rho_1^1}{\partial\tau} + \frac{\partial\rho_1^2}{\partial\tau} - c_1\frac{\partial\rho_3}{\partial\xi} + c_2\frac{\partial\rho_3}{\partial\eta} - 2c_2Q_{0\xi}\frac{\partial\rho_1^2}{\partial\eta} + 2c_1P_{0\eta}\frac{\partial\rho_1^1}{\partial\xi} +$$

$$c_s\frac{\partial v_3}{\partial\xi} + c_s\frac{\partial v_3}{\partial\eta} + 2c_1\rho_1^1\frac{\partial\rho_1^1}{\partial\xi} - 2c_1\rho_1^2\frac{\partial\rho_1^2}{\partial\eta} = 0 \tag{34}$$

and

$$-c_1\frac{\partial v_3}{\partial\xi} + c_2\frac{\partial v_3}{\partial\eta} + \frac{2c_1^2}{c_s}P_{0\eta}\frac{\partial\rho_1^1}{\partial\xi} + \frac{2c_2^2}{c_s}Q_{0\xi}\frac{\partial\rho_1^2}{\partial\eta} + \frac{c_1}{c_s}\frac{\partial\rho_1^1}{\partial\tau} - \frac{c_2}{c_s}\frac{\partial\rho_1^2}{\partial\tau}$$

$$-\frac{1}{M\rho_0}\left(\tilde{\zeta}+\frac{4}{3}\tilde{\upsilon}\right)\left[\frac{c_1}{c_s}\frac{\partial^2\rho_1^1}{\partial\xi^2}-\frac{c_2}{c_s}\frac{\partial^2\rho_1^2}{\partial\eta^2}\right]+\frac{c_1^2}{c_s}\frac{\partial\rho_3}{\partial\xi}+\frac{c_2^2}{c_s}\frac{\partial\rho_3}{\partial\eta}$$

$$+\left[\frac{\rho_0^2}{Mc_s}\left(127.56\frac{\kappa}{g_\sigma^3}+182.1\frac{\lambda}{g_\sigma^4}\right)+\frac{\rho_0^3}{Mc_s}\left(1529.52\frac{\lambda}{g_\sigma^4}\right)\right.$$

$$\left.-3.61\frac{\rho_0^{\frac{5}{3}}}{Mc_s}+0.97\frac{\rho_0^{\frac{2}{3}}}{Mc_s}-0.44\frac{\rho_0^{-\frac{1}{3}}}{Mc_s}+\frac{c_1^2}{c_s}\right]\left(\rho_1^1\frac{\partial\rho_1^1}{\partial\xi}+\rho_1^2\frac{\partial\rho_1^2}{\partial\eta}\right)$$

$$+\left[\frac{\rho_0^2}{Mc_s}\left(127.56\frac{\kappa}{g_\sigma^3}+182.1\frac{\lambda}{g_\sigma^4}\right)+\frac{\rho_0^3}{Mc_s}\left(1529.52\frac{\lambda}{g_\sigma^4}\right)\right.$$

$$\left.-3.61\frac{\rho_0^{\frac{5}{3}}}{Mc_s}+0.97\frac{\rho_0^{\frac{2}{3}}}{Mc_s}-0.44\frac{\rho_0^{-\frac{1}{3}}}{Mc_s}-\frac{c_1^2}{c_s}\right]$$

$$\left(\rho_1^1\frac{\partial\rho_1^2}{\partial\eta}+\rho_1^2\frac{\partial\rho_1^1}{\partial\xi}\right)=0 \tag{35}$$

where $\tilde{\zeta}$ and $\tilde{\upsilon}$ are small perturbation in viscosities. Differentiating equations (34) and (35) with respect to ξ and η and performing some calculations the following equations obtain:

$$\frac{\partial\rho_1^1}{\partial\tau}+\frac{c_1}{2}(3+\Re)\rho_1^1\frac{\partial\rho_1^1}{\partial\xi}-\frac{1}{2M\rho_0}\left(\tilde{\zeta}+\frac{4}{3}\tilde{\upsilon}\right)\frac{\partial^2\rho_1^1}{\partial\xi^2}=0 \tag{36}$$

$$\frac{\partial\rho_1^2}{\partial\tau}-\frac{c_1}{2}(3+\Re)\rho_1^2\frac{\partial\rho_1^2}{\partial\eta}-\frac{1}{2M\rho_0}\left(\tilde{\zeta}+\frac{4}{3}\tilde{\upsilon}\right)\frac{\partial^2\rho_1^2}{\partial\eta^2}$$

$$-2c_1(\Re-1)\rho_1^2\frac{\partial\rho_1^1}{\partial\xi}=0 \tag{37}$$

$$P_{0\eta}=\frac{1}{4}(1-\Re)\rho_1^2 \tag{38}$$

$$Q_{0\xi}=\frac{1}{4}(1-\Re)\rho_1^1 \tag{39}$$

where $\Re$ is introduced as following expression

$$\Re=\frac{1}{c_1^2}\left[\frac{\rho_0^2}{Mc_s}\left(127.56\frac{\kappa}{g_\sigma^3}+182.1\frac{\lambda}{g_\sigma^4}\right)+\frac{\rho_0^3}{Mc_s}\left(1529.52\frac{\lambda}{g_\sigma^4}\right)\right.$$

$$\left.-3.61\frac{\rho_0^{\frac{5}{3}}}{Mc_s}+0.97\frac{\rho_0^{\frac{2}{3}}}{Mc_s}-0.44\frac{\rho_0^{-\frac{1}{3}}}{Mc_s}\right] \tag{40}$$

Equations (36) and (37) are the Burgers equations in (ξ,τ) and (η,τ) space, $P_{0\eta}$ and $Q_{0\xi}$ are the phase shifts of the localized waves after their head on collision respectively. For two shock waves that moved toward each other in the (x,t) space we have

$$\frac{\partial\hat{\rho}_1^1}{\partial t}+c_1\frac{\partial\hat{\rho}_1^1}{\partial x}+\frac{c_1}{2}(3+\Re)\hat{\rho}_1^1\frac{\partial\hat{\rho}_1^1}{\partial x}-\frac{1}{2M\rho_0}\left(\zeta+\frac{4}{3}\upsilon\right)\frac{\partial^2\hat{\rho}_1^1}{\partial x^2}+$$

$$+\frac{1}{4}(1-\Re)\hat{\rho}_1^2\left[\frac{\partial\hat{\rho}_1^1}{\partial t}-c_1\frac{\partial\hat{\rho}_1^1}{\partial x}+\frac{1}{2M\rho_0}\left(\zeta+\frac{4}{3}\upsilon\right)\frac{\partial^2\hat{\rho}_1^1}{\partial x^2}\right]=0 \tag{41}$$

This is the Burgers equation for $\hat{\rho}_1^1\equiv\sigma^2\rho_1^1$, which is a small localized perturbation in the baryon density moving towards the right. The following equation describes a moving perturbation propagates in inviscid medium with $\zeta = \upsilon = 0$

$$\frac{\partial \widehat{\rho}_1^1}{\partial t} + c_1 \frac{\partial \widehat{\rho}_1^1}{\partial x} + \frac{c_1}{2} (3 + \Re) \widehat{\rho}_1^1 \frac{\partial \widehat{\rho}_1^1}{\partial x} + \frac{1}{4} (1 - \Re) \widehat{\rho}_1^2 \left[\frac{\partial \widehat{\rho}_1^1}{\partial t} - c_1 \frac{\partial \widehat{\rho}_1^1}{\partial x} \right] = 0 \qquad (42)$$

This equation is called the breaking wave equation. Similarly, the Burgers equation for $\widehat{\rho}_1^2 \equiv \sigma^2 \rho_1^2$ becomes

$$\frac{\partial \widehat{\rho}_1^2}{\partial t} - c_1 \frac{\partial \widehat{\rho}_1^2}{\partial x} - \frac{c_1}{2} (3 + \Re) \widehat{\rho}_1^2 \frac{\partial \widehat{\rho}_1^2}{\partial x} - \frac{1}{2M\rho_0} \left(\zeta + \frac{4}{3}\upsilon \right) \frac{\partial^2 \widehat{\rho}_1^2}{\partial x^2}$$
$$+ \frac{1}{4} (1 - \Re) \widehat{\rho}_1^1 \left[\frac{\partial \widehat{\rho}_1^2}{\partial t} + c_1 \frac{\partial \widehat{\rho}_1^2}{\partial x} + \frac{1}{2M\rho_0} \left(\zeta + \frac{4}{3}\upsilon \right) \frac{\partial^2 \widehat{\rho}_1^2}{\partial x^2} \right]$$
$$+ \left[c_1 (1 - \Re) + \frac{c_1}{2} (1 - \Re)^2 \left[\widehat{\rho}_1^1 - \widehat{\rho}_1^2 \right] \right] \widehat{\rho}_1^2 \frac{\partial \widehat{\rho}_1^1}{\partial x} = 0 \qquad (43)$$

which is a small perturbation in the baryon density that moves to the left. The breaking wave equation for $\widehat{\rho}_1^2$ moving in inviscid media can be found by inserting $\zeta = \upsilon = 0$ in the equation (43) as

$$\frac{\partial \widehat{\rho}_1^2}{\partial t} - c_1 \frac{\partial \widehat{\rho}_1^2}{\partial x} - \frac{c_1}{2} (3 + \Re) \widehat{\rho}_1^2 \frac{\partial \widehat{\rho}_1^2}{\partial x}$$
$$+ \frac{1}{4} (1 - \Re) \widehat{\rho}_1^1 \left[\frac{\partial \widehat{\rho}_1^2}{\partial t} + c_1 \frac{\partial \widehat{\rho}_1^2}{\partial x} \right]$$
$$+ \left[c_1 (1 - \Re) + \frac{c_1}{2} (1 - \Re)^2 \left[\widehat{\rho}_1^1 - \widehat{\rho}_1^2 \right] \right] \widehat{\rho}_1^2 \frac{\partial \widehat{\rho}_1^1}{\partial x} = 0 \qquad (44)$$

Now we have found the main equations which describe head-on collision of localized waves in dense hadronic media. The above equations don't have known analytical solutions and therefore we have to solve them numerically.

IV. Numerical Discussion

We can rewrite the Burgers equations (41) and (43) in the general form

$$\frac{\partial \widehat{\rho}}{\partial t} + c \frac{\partial \widehat{\rho}}{\partial x} + \alpha \widehat{\rho} \frac{\partial \widehat{\rho}}{\partial x} = \mu \frac{\partial^2 \widehat{\rho}}{\partial x^2} \qquad (45)$$

where $\alpha = \pm \frac{c_1}{2} (3 + \Re)$ and μ are the respective nonlinear and dissipative coefficients for hadron phase. The dissipative coefficient μ is related to the viscosity. Since the Burgers equations dont have any exact solution, we use localized solutions of the Korteweg-de Vries (KdV) equation as an initial condition for numerical calculations. Hence, localized soliton-like structure $\rho(x, t = 0) = A\,sech(\frac{x}{L})$ are proposed as initial condition, where the initial amplitude is A and L denotes its width. Time evolution of this soliton-like solution simulates the evolution of a localized perturbation in viscous hadronic gas. Phase velocities are important parameters in head on collision which are described by equation (33). This equation shows that they have intricate relation with the medium parameters. Figure 1 presents the phase velocity c_1 of colliding waves as a function ρ_0 . As might be expected, phase velocities increase as the medium density increases. This figure also shows that in a small range of background density, phase velocity is almost a linear function of ρ_0 . Phase shifts are the other important parameters in head on collision. Calculated values of these parameters in

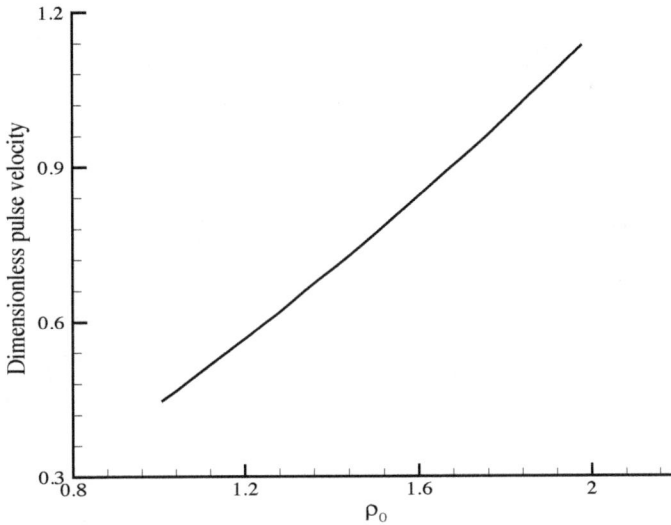

Figure 1: The phase velocity of propagated waves as a function of back ground density. Values of the other parameters have been given in the text

our problem have been presented by equations (38) and (39). These equations clearly show that they are negative because in hadron gas $\Re > 1$. Figure 2 demonstrates time evolution of the colliding waves during their interaction in a inviscid medium. Simulations clearly show that the phase velocities of the localized waves are almost the same. Figure 2 shows that the localized waves are changed into shock profiles while traveling in the medium. This figure also presents that the needed time to change a soliton into a shock profile becomes smaller when the initial waves have larger amplitude. Effects of viscosity on the evolution of waves during their collision have been presented in the figure 3. This figure clearly illustrates that the viscosity is able to control the creation of shock waves in the medium. Definitely the term with second order of derivation in the equations (41) and (43) reduces the non-linearity effects in a way that the shock profiles are created very late. In addition, viscosity causes a noticeable damping on the wave amplitude which is an important result. The traveling distance of localized perturbations in a medium helps us to find valuable information about the amount of viscosity of the medium. Comparing the traveling distance of such waves in media with different initial conditions give us a qualitative information about the shock profile. The figure 3 also shows that the amplitude and width of localized waves approximately remain constant in the viscid media.

V. Conclusions and Remarks

Propagation of solitary waves in cold and dense hadronic matter is studied in this work. It is expected to find such media in the core of neutron and compact stars. Hydrodynamic description and equation of state of such viscose fluid leaded us to find a nonlinear differential equation for propagation of localized perturbation which is called the Burgers equation. Viscosity is able to control the nonlinear effects so that it can postpone the shock profiles. It is shown that the amplitude and width of propagated solitary waves in viscose hadronic

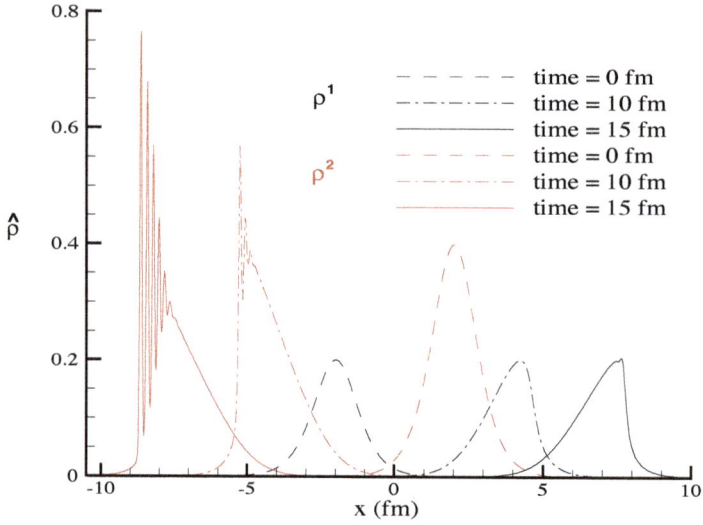

Figure 2: Wave profiles before and head on collision in non-viscous medium.

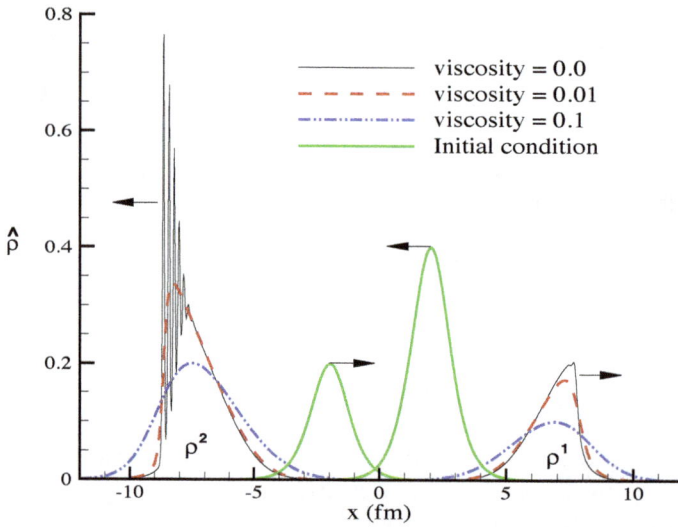

Figure 3: Wave profiles before and head on collision in viscous media with different viscosities.

medium remain approximately constant during and after the collision. The phase shifts of propagating waves after the collision are always negative. These phase shifts are functions of initial background density and characterize by the medium particles, but independent of the viscosity. It is interesting to investigate such media with finite temperature. For this purpose it is better to construct a more suitable EoS for this medium which can be studied in the future works. Since the EoS of the QGP phase in finite temperature is different, the same investigations for this medium should be considered.

References

[1] I. Mishustina, et al., Phase transition in compact stars due to a violent shock, Phys. Rev. C **91** (2015) 055806.

[2] P. K. Panda, H. S. Nataraj, Rotating compact star with superconducting quark matter, Phys. Rev. C **73** (2006) 025807.

[3] S. Weissenborn, et al., Hyperons and massive neutron stars: the role of hyperon potentials, Nucl. Phys. A **881** (2012) 62-77.

[4] J. Schaffner-Bielich, et al., Astrophysical implications of the QCD phase transition, PoS. Confinement. **8** (2009) 138.

[5] S. Schramm, et al., Structure and cooling of neutron and hybrid stars, arXiv:1202.5113 [astro-ph.SR].

[6] W. H. Y. Wang, et al., The third family of compact stars with the color-flavor locked quark core, Chin. Sci. Bull. **58** (2013) 3731-3734.

[7] R. E.Pudritz, et al., Shock interactions, turbulence and the origin of the stellar mass spectrum, Phil. Trans. R. Soc. A **371** (2013) 2003.

[8] J. Macher, J. Schaffner-Bielich, Phase transitions in Compact Stars, Eur. J. Phys. **26** (2005) 341-360.

[9] D. Logoteta, et al., Formation of hybrid stars from metastable hadronic stars,Phys. Rev. C**88** (2013) 5,055802.

[10] D. Logoteta, et al., Quark matter nucleation with a microscopic hadronic equation of state, Phys. Rev. C **85** (2012) 055807.

[11] H. R. Moshfegh, et al., Cold Hybrid star properties, AIP Conf. Proc.**1377** (2011) 405.

[12] G. Peilert, et al., Physics of high-energy heavy-ion collisions, Rep. Prog. Phys.**57** (1994) 533-602.

[13] S. Floerchinger, et al., A perturbative approach to the hydrodynamics of heavy ion collisions, Nucl. Phys. A**931** (2014) 965-9.

[14] I. Kozlov, et al., Signatures of collective behavior in small systems, Nucl. Phys. A**931** (2014) 1045-50.

[15] D.A. Fogaca, F.S. Navarra, L.G. Ferreira Filho, Viscosity, wave damping and shock wave formation in cold hadronic matter, Phys. Rev C **88** (2013) 025208.

[16] P. Huovinen, P.V. Ruuskanen, Hydrodynamic models for heavy ion collisions, Annu. Rev. Nucl. Part. Sci. **56** (2006) 163-206.

[17] A. Bazavov, The QCD equation of state, Nucl. Phys. A**931** (2014) 851-5.

[18] H. Niemi, Collective dynamics in relativistic nuclear collisions, Nucl. Phys. A **931** (2014) 227-37.

[19] D.A. Fogaca, F.S. Navarra, L.G. Ferreira Filho, Nonlinear waves in a Quark Gluon Plasma, Phys. Rev. C81 (2010) 055211.

[20] D.A. Fogaca, F.S. Navarra, L.G. Ferreira Filho, Kadomtsev-Petviashvili equation in relativistic fluid dynamics, Commun. Nonlinear. Sci. **18** (2013) 221-35.

[21] J. I. Kapusta, et al., Relativistic theory of hydrodynamic fluctuations with applications to heavy-ion collisions, Phys. Rev. C **85** (2012) 054906.

[22] P. Staig, E. Shuryak, The fate of the initial state fluctuations in heavy ion collisions , Phys. Rev. C **84** (2011) 034908.

[23] Z. Qiu, U. Heinz, Event-by-event shape and flow fluctuations of relativistic heavy-ion collision fireballs, Phys. Rev. C **84** (2011) 024911.

[24] H. Demiray, Head-on collision of solitary waves in fluid-filled elastic tubes, Appl. Math. Lett.**18** (2005) 941-50.

[25]T. Tsuboi, Phase shift in the collision of two solitons propagating in a nonlinear transmission line, Phys. Rev.A **40** (1989) 2753-5.

[26] E.F. El-Shamy, et al., Head-on collision of ion-acoustic solitary waves in multicomponent plasmas with positrons, Phys. Plasmas. **17** (2010) 082311.

[27] E.F. El-Shamy, W.A. Awad, on the characteristics of the head-on collision between two ion thermal waves in isothermal pair-ion plasmas containing charged dust grains, Chaos. Soliton. Fract. **45** (2012) 1520.

[28] W. Wen, G. Huang, Dynamics of dark solitons in superfluid Fermi gases in the BCS-BEC crossover, Phys. Rev.A **79** (2009) 023605.

[29] M.G. Paoli. D.P. Menezes, The importance of the mixed phase in hybrid stars built with the Nambu-Jona-Lasinio model, Eur. Phys. J. A **46** (2010) 413-20.

[30] D. Logoteta, et al., A chiral model approach to quark matter nucleation in neutron stars, Phys. Rev. D **85** (2012) 023003.

[31] I. Bombaci, et al., Metastability of hadronic compact stars, Phys. Rev. D **77** (2008) 083002.

[32] B.D. Serot, Building atomic nuclei with the Dirac equation, Int. J. Mod. Phys. A **19S1** (2004) 107-120.

[33] F. Verheest, et al., Head-on collisions of electrostatic solitons in nonthermal plasmas, Phys. Rev. E **86** (2012) 036402.

8

Cosmic Walls and Filaments Formation in Modified Chaplygin Gas Cosmology

S. Karbasi[1] · H. Razmi[2]

[1] Department of Physics, University of Qom, Qom, I. R. Iran;
    email: s.karbasi@stu.qom.ac.ir
[2] Department of Physics, University of Qom, Qom, I. R. Iran;
    email: razmi@qom.ac.ir

Abstract. We want to study the perturbation growth of an initial seed of an ellipsoidal shape in Top-Hat collapse model of the structure formation in the Modified Chaplygin gas cosmology. Considering reasonable values of the constants and the parameters of the model under study, we can show that a very small deviation from spherical symmetry (ellipsoidal geometry) in the initial seed leads to a final highly non-spherical structure which can be considered as a candidate for justifying the already known cosmological structures as cosmic walls and filaments.

Keywords: Structure formation; Chaplygin gas; Non-spherical collapse; Cosmic walls; Filaments

1 Introduction

As the Universe expands, overdense regions break away from the Hubble flow and the regions collapse and form virialized objects, such as cluster of galaxy, filaments and walls. The spherical-collapse model is a basic well-known model in the theory of cosmic structure formation [1]. Collapsed haloes form from the initial fluctuation field, leading to structure formation. universe expands, overdense regions expand sufficiently until they reach a maximum size; afterward, they collapse under the action of their own gravity. This has been already enough studied using both analytical models and numerical simulations, in the standard cosmological model [2-7]. In this model, the structures are assumed to have grown gravitationally from small, initially Gaussian density fluctuations in the early universe [8]. According to the real observations of large scale structure reasoning on the cosmic web formation of the universe, the complicated ellipsoidal collapse model has been under consideration [9-12]. As we know, cosmic walls, filaments, and galaxy clusters have non-spherical symmetry. There are a number of other necessary considerations and corrections in studying collapse models, among them are asymmetry of initial seeds, the influence of cosmic expansion and neighboring heavy structures. It is well known that the protohalo mass distribution misses its spherical symmetry due to the tidal forces affecting it during its evolution [13-17]. All the way, the final shape of the structures depends highly on the symmetry of their initial protohalos. Numerical simulations show that the virialized haloes can be considered as some triaxial objects [18-21]. Triaxial collapse models have been studied analytically with good results in the standard cosmology [22-23]. In this paper, we want to study ellipsoidal collapse model in Chaplygin gas (CG) cosmology. As we know, CG, as a strange fluid with an equation of state $P = -\frac{B}{\rho}$, $B > 0$, where ρ and P are the pressure and energy density respectively and B is a constant, plays the role of dark energy [24-25].

To be in good consistency with observational data, CG has been extended to the generalized Chaplygin gas (GCG) with the equation of state $P = -\frac{B}{\rho^\alpha}$ where $0 < \alpha < 1$ [26-29]. The modified Chaplygin gas (MCG) scenario, with the aim of considering both dark energy and dark matter, has been already considered with the equation of state $P = -A\rho - \frac{B}{\rho^\alpha}$ [30-32]. MCG model not only explain the evolution of the universe from early stage to the present time [32-36], but also can be used in justifying the inflation and radiation dominated state of the universe [35-38]. In what follows, considering the well known Top-Hat model in the large scale structure formation in GCG and MCG scenarios [39-41], we want to study ellipsoidal collapse model in MCG scenario.

2   Ellipsoidal collapse of Chaplygin gas

The ellipsoidal collapse model we study here is similar to the spherical one except that the shape of the initial seed is ellipsoidal. The background equations and all other necessary quantities and parameters (e.g. the density and the pressure) are as the same as what are in [39-41]:

$$\dot{\rho} = -3h(\rho + P) \qquad , \qquad \frac{\ddot{a}}{a} = -\frac{4\pi G}{3}\sum_j(\rho_j + 3P_j) \tag{1}$$

except that, in the perturbed region, we consider the following relations:

$$\dot{\rho}_c = -3h(\rho_c + P_c) \qquad , \qquad \frac{\ddot{r}}{r} = -2\pi G\sum_j(\rho_{c,j} + 3P_{c,j})b_i \tag{2}$$

where the index c refers to the initial formed seed, $h \equiv \frac{\dot{r}}{r} = H + \frac{\theta}{3a}, \theta \equiv \vec{\nabla} \cdot \vec{v}$ ($\vec{v}$ is the peculiar velocity) is the expansion rate for the initial seed, and 's are the coefficients usually introduced in the potential theory for homogeneous ellipsoids,

$$b_i = a_1 a_2 a_3 \int_0^\infty \frac{d\tau}{(x_i + \tau)\prod_{m=1}^{3}(x_m^2 + 1)^{\frac{3}{2}}} \tag{3}$$

where (a_1, a_2, a_3) are semiaxes of the ellipsoid and the coefficients satisfy $\sum_{i=1}^{3} b_i = 2$. If two of the axes are equal, the integrals are reduced to elementary functions. Two interesting cases are an oblate spheroid with $a_1 = a_2 > a_3$

$$b_1 = b_2 = \frac{\sqrt{1-e^2}}{e^2}\left[\frac{\sin^{-1}e}{e} - \sqrt{1-e^2}\right] \tag{4}$$

$$b_3 = \frac{2\sqrt{1-e^2}}{e^2}\left[\frac{1}{\sqrt{1-e^2}} - \frac{\sin^{-1}e}{e}\right] \tag{5}$$

and a prolate spheroid with $a_1 = a_2 < a_3$

$$b_1 = b_2 = \frac{1-e^2}{e^2}\left[\frac{1}{\sqrt{1-e^2}} - \frac{1}{2e}\ln\left(\frac{1+e}{1-e}\right)\right] \tag{6}$$

$$b_3 = \frac{2(1-e^2)}{e^2}\left[\frac{1}{2e}\ln\left(\frac{1+e}{1-e}\right) - 1\right] \tag{7}$$

where the eccentricity is defined as

$$e = \sqrt{1 - \frac{a_3^2}{a_1^2}} \tag{8}$$

Considering the density difference between the initial seed and the background as $\delta\rho$ and the density contrast as $\delta_j = \left(\frac{\delta\rho}{\rho}\right)_j$ where the index j refers to the Chaplygin gas or the baryons, the modified ellipsoidal evolution (the derivative with respect to the cosmic scale factor a) equation for δ_j is:

$$\delta_j' = -\frac{3}{a}(c_{eff,j}^2 - w_j)\delta_j - [1 + w_j + (1 + c_{eff,j}^2)\delta_j] \tag{9}$$

where $c_{eff} = \frac{\delta P}{\delta\rho}$ is the square of the effective sound speed and $w = \frac{P}{\rho}$ is the background state parameter. The corresponding modified ellipsoidal evolution (the derivative with respect to the cosmic scale factor a) equation for θ is:

$$\theta' = -\theta a - \theta^2 3 H a^2 - 94 H_0^2 H \left[\Omega_{\text{mcg}}[(1+\delta_{\text{mcg}})b_i - 23][\bar{c} + (1-\bar{c})a^{-3(1+A)(1+\alpha)}]^{\frac{1}{1+\alpha}}\right.$$

$$-\Omega_{\text{mcg}} 2A[\bar{c} + (1-\bar{c})a^{-3(1+A)(1+\alpha)}]^{\frac{1}{1+\alpha}} + 2\bar{c}(1+A)\Omega_{\text{mcg}}[\bar{c} + (1-\bar{c})a^{-3(1+A)(1+\alpha)}]^{\frac{\alpha}{1+\alpha}}$$

$$+\Omega_b a^3[(1+\delta_b)b_i - 23] + 3b_i[A\Omega_{\text{mcg}}[\bar{c} + (1-\bar{c})a^{-3(1+A)(1+\alpha)}]^{\frac{1}{1+\alpha}}(1+\delta_{\text{mcg}})$$

$$\left.-\bar{c}(1+A)\Omega_{\text{mcg}}[\bar{c} + (1-\bar{c})a^{-3(1+A)(1+\alpha)}]^{\frac{\alpha}{1+\alpha}}(1+\delta_{\text{mcg}})^\alpha]\right]. \tag{10}$$

where $\bar{c} = \frac{B}{(1+A)\rho_0^{1+\alpha}}$ (ρ_0 is the density at the present time), and with the same already known notation and parameters introduced in [39-41].

3   The method and the result

In this section, the equations (9) and (10) in the ellipsoidal collapse model under consideration are integrated out using the same parameters, equations, numerical values of the redshift and the density parameters and the Hubble constant, and the same computational programming previously used in spherical collapse model in MCG cosmology [41] with the same initial conditions except that here the parameter $\bar{c}$ is fixed at $\bar{c} = 0.99$ ideally and the constant A is assumed to have an arbitrary value $A = -0.02$ with $\delta_{MCG}(z = 100) = 3.5 \times 10^{-3}$ [40]. The constant α is assumed to have a value of $\alpha = 0.022$; this choice covers the accelerated expansion of the universe in MCG scenario. Assuming an initial small eccentricity of the value $e = 0.01$, the evolution of the rate of collapsed region, h, with respect to the redshift , has been shown in figure 1 (2) for a prolate (an oblate) spheroid structure.

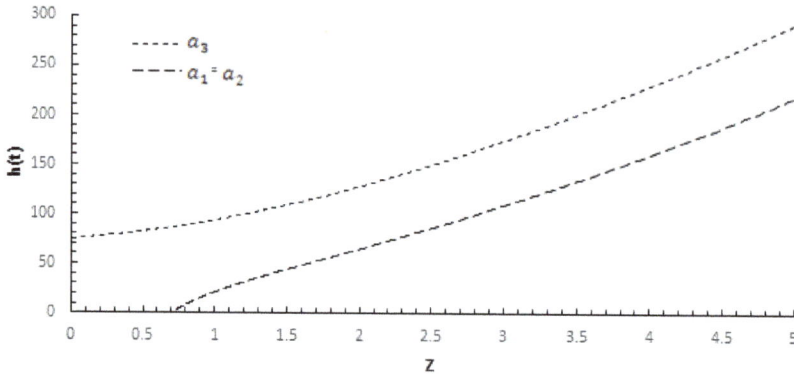

Figure 1: Evolution of collapsed region expansion rate, h, with respect to the redshift z for a prolate spheroid structure $a_1 = a_2 < a_3$.

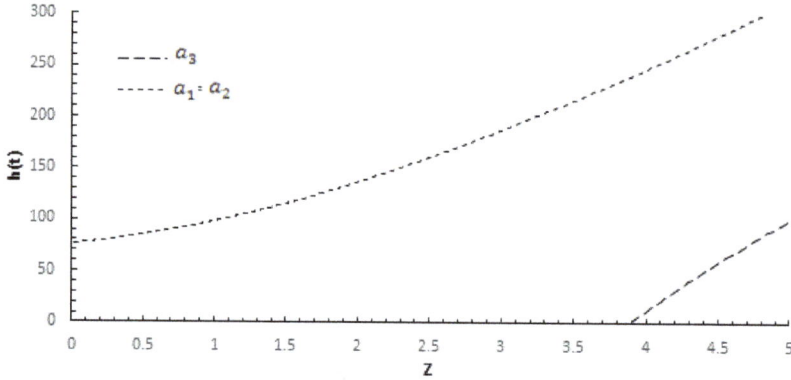

Figure 2: Evolution of collapsed region expansion rate, h, with respect to the redshift z for an oblate spheroid structure $a_1 = a_2 > a_3$.

According to the definition of the parameter h (the expansion rate of the seed), it is clear that when the value of this parameter approaches zero, the structure growth is stopped and collapses in itself. In this state, because of the contraction of the perturbed region, the structure under consideration is formed and its self gravitation makes rapid increase in its density; such a situation is known as the time of the structure formation. If there is an initial small difference between the main axes of the ellipsoid, there are possible situations where this small difference can grow well after enough and suitable time of evolution so that the three dimensional structure transforms to the two dimensional walls and even to the one dimensional filaments. In the figures 1 and 2, one can clearly see the formation of non-spherical structures like filaments and/or walls. As is seen, the shorter axis began to collapse at earlier times relative to the longer one which grows with a rate near to the Hubble rate. Although the elongation rate of the greater axis is near to the Hubble rate, the structure growth still continues; this is what one expects for the filaments (walls) formation which continually become elongated (wider) and denser. The square value of the adiabatic sound speed $c_s^2 = \frac{\partial P}{\partial \rho}$ has been shown in figure 3. As is seen, at lower than 1 values of the redshift, it is positive; this means that the adiabatic sound speed has a real value which guarantees the stability of the structure. In the early universe, at the higher values of z, the instability of the MCG fluid has a useful role in the structure formation [42].

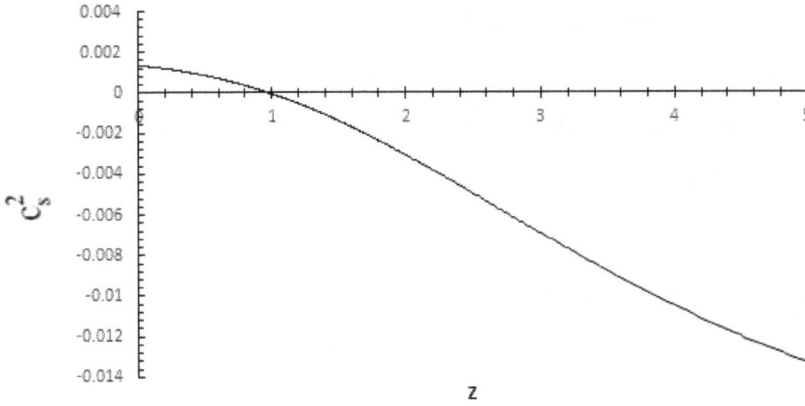

Figure 3: Evolution of the square value of the adiabatic sound speed c_s^2 with respect to the redshift

4   Conclusion

In this paper, we have studied perturbation growth in MCG-dominated universe in an ellipsoidal collapse model. As was seen, if the initial protohalo of the cosmic structure has a non-spherical (here ellipsoidal) geometry, then the final form of the structure can take elongated and/or wide shapes as filaments and/or walls. This has been shown in figures 1 and 2 schematically by analytical/numerical considerations with choosing particular values (which are consistent with the current theoretical and observational data) for the constant parameters appeared in the model. The stability of the system has been considered too. As is seen in figure 3, for $\alpha = 0.022$, the system is stable; because the adiabatic sound speed has real value at the present time.

References

[1] J. E. Gunn, J. R. Gott, Astrophys. J. 176, 1 (1972).

[2] C. Lacey, S. Cole, MNRAS, 262, 627 (1993).

[3] C. Lacey, S. Cole, MNRAS, 271, 676 (1994).

[4] G. Kauffmann, S. D. M. White, MNRAS, 261, 921 (1993).

[5] V. Springel, S. D. M. White, G. Tormen, G. Kauffmann, MNRAS, 328, 726 (2001).

[6] C. Giocoli, J. Moreno, R. K. Sheth, G. Tormen, MNRAS, 376, 977 (2007).

[7] F. Pace, J. C. Waizmann, M. Bartelmann, MNRAS, 406, 1865 (2010).

[8] A. G. Doroshkevich, Astrophysics, 6, 320 (1970).

[9] S. D. M. White, J. Silk, ApJ, 231, 1 (1979).

[10] M. Bartelmann, J. Ehlers, P. Schneider, Astron. Astrophys. 280, 351 (1993).

[11] D. J. Eisenstein, A. Loeb, ApJ, 439, 520 (1995).

[12] J. R. Bond, S. T. Myers, ApJS, 103, 1 (1996).

[13] R. K. Sheth, H. J. Mo, G. Tormen, MNRAS, 323, 1 (2001).

[14] R. K. Sheth, G. Tormen, MNRAS, 329, 61 (2002).

[15] V. Desjacques, MNRAS, 388, 638 (2008).

[16] T. Y. Lam, R. K. Sheth, MNRAS, 389, 1249 (2008).

[17] G. Rossi, R. K. Sheth, G. Tormen, MNRAS, 416, 248 (2011).

[18] G. Despali, G. Tormen, R. K. Sheth, MNRAS, 431, 1143 (2013).

[19] Y. P. Jing, Y. Suto, ApJ, 574, 538 (2002).

[20] B. Allgood, et all, MNRAS, 367, 1781 (2006).

[21] A. D. Ludlow, M. Borzyszkowski, C. Porciani, MNRAS, 445, 4 (2014).

[22] C. Angrick, M. Bartelmann, Aarton. Astrophys. 518, A38 (2010).

[23] G. Despali, G. Tormen, R. K. Sheth, MNRAS, 431, 1143 (2012).

[24] A. Y. Kamenshchik, U. Moschella, V. Pasquir, Phys. Lett. B, 511, 265 (2001).

[25] M. Makler, S. Q. Oliveira, I. Waga, Phys. Lett. B, 555, 1 (2003).

[26] M. C. Bento, O. Bertolami, A. A. Sen, Phys. Rev. D, 70, 083519 (2004).

[27] V. Gorini, A. Kamenshchik, U. Moschella, Phys. Rev. D, 67, 063509 (2003).

[28] U. Alam, V. Sahni, T. D. Saini, A. A. Starobinsky, MNRAS, 344, 1057 (2003).

[29] M. c. Bento, O. Bertolami, A. A. Sen, Phys. Rev. D, 66, 043507 (2002).

[30] U. Debnath, A. Banerjee, S. Chakraborty, Class. Quant. Grav. 21, 5609 (2004).

[31] J. D. Barrow, Nucl. Phys. B, 310, 743 (1998).

[32] H. Benaoum, hep-th/0205140.

[33] S. S. Costa, M. Ujevic, A. F. Santos, arXiv: gr-qc/0703140.

[34] S. Li, Y. G. Ma, Y. Chen, Int. J. Mod. Phys. D, 18, 1785 (2009).

[35] F. C. Santos, M. L. Bedran, V. Soares, Phys. lett. B, 646, 215 (2007).

[36] J. D. Barrow, Phys. Lett. B, 235, 40 (1990).

[37] R. Herrera, Gen. Rel. Grav. 41, 1259 (2009).

[38] H. Benaoum, hep-th/0205140.

[39] J. E. Gunn, J. R. Gott, Astrophys. J. 176, 1 (1972).

[40] R. A. A. Fernandes, J. P. M. de Carvalho, A. Yu. Kamenshchik, U. Moschella, A. da Silva, Phys. Rev. D, 85, 083501 (2012).

9

Spin and Isospin Asymmetry, Equation of State and Neutron Stars

Mohsen Bigdeli · Nariman Roohi · Mina Zamani
Department of Physics, University of Zanjan, P.O. Box 45195-313, Zanjan, Iran

Abstract. In the present work, we have obtained the equation of state for neutron star matter considering the influence of the ferromagnetic and antiferromagnetic spin state. We have also investigated the structure of neutron stars. According to our results, the spin asymmetry stiffens the equation of state and leads to high mass for the neutron star.

Keywords: Neutron stars, Spin asymmetry, Equation of state

1 Introduction

Neutron stars are hyper-dense and magnetized laboratories for investigating strange phenomena in the nuclear and particle physics. Pulsars and magnetars are two kinds of neutron stars with strong surface magnetic field. Actually the exact origin of this magnetic field is not yet known. In the interior of magnetars, the magnetic field strength may be even larger according to virial theorem [6] and such strong field may cause spin asymmetry. The occurrence of such strange phenomena can affect the equation of state (EOS) of neutron star matter. Theoretically, the equation of state has been applied to determine the maximum mass of a neutron star which should be in agreement with the precise observations. The accurate measurement of neutron star mass $M = 1.97 \pm 0.04 M_\odot$ in the system PSR J1614-2230 was one of the most important development in observational data [9]. This precise measurement is based on Shapiro delay in neutron star-white dwarf binary [12]. Another well-measured massive neutron star is PSR J0348+0432, with mass about $M = 2.01 \pm 0.04 M_\odot$ [1]. Next, there is an evidence that the black widow pulsar PSR B1957+20 might have even larger masses approximately $M_{PSR} = 2.4 M_\odot$ [17]; however, one have to consider the uncertainties in this mass estimation. Finally, the largest mass $2.1 M_\odot \leq M_{NS} \leq 2.7 M_\odot$ has been given for the gamma-ray black widow pulsar PSR J1311-3430 by simple heated light curve fits [16]. These massive neutron stars require the equation of state of the system to be rather stiff. Therefore, theoretical approaches should confirm these observational data.

Recently, several studies used different theoretical approaches showed the stiff EOS for the neutron star matter. Gandolfi et al. [10] have used quantum Monte Carlo techniques and calculated the equation of state of neutron star matter with realistic two- and three-nucleon interactions. Their calculation resulted $M_{max} < 2.2 M_\odot$ for neutron star mass. They have also used Auxiliary Field Diffusion Monte Carlo technique by incorporating semi-phenomenological Hamiltonian including a realistic two-body interaction and many-body forces [11]. They found the maximum mass of neutron star lies in the range 2.2-2.5 times of solar mass. Some other attempts by Partha Roy Chowdhury showed the rotating star mass is around $(1.93\text{-}1.95) M_\odot$ [7] . They have applied a pure nucleonic equation of state for neutron star matter. Shen et al. [14] have constructed a new equation of state for a wide range of temperatures, densities and proton fractions to be used in astrophysical simulations of neutron stars. They have predicted that the maximum mass of neutron star is about $2.77 M_\odot$ with a radius of about 13.3 km. Sun et al. [15] have investigated neutron star structure using EOS which has provided by density dependent relativistic Hartree-Fock theory. Their results showed that maximum mass of neutron stars lies in the range $(2.45 - 2.49) M_\odot$. More recently, we gained $M_{NS} = 1.991 M_\odot$ by applying the Lowest Order Constrained Variational (LOCV) method and using UV_{14}+TNI potential [4].

In this article, we investigate some physical properties of polarized neutron star matter using the LOCV method and the AV_{18} potential. This modern equation of state is derived from an accurate many-body calculation and is based on the cluster expansion of the energy functional. Moreover, we obtain the particles abundance, equation of state and the structure of neutron stars. Finally, we compare our results by experimental data.

2   Formalism

We assume the neutron star matter as a charge neutral infinite system that is a mixture of leptons and interacting nucleons. The energy density of this system can be obtained as follows,

$$\varepsilon = \varepsilon_N + \varepsilon_l, \tag{1}$$

where $\varepsilon_N(\varepsilon_l)$ is the energy density of nucleons (leptons). In the following, we determine these energy densities in more details.

2.1   Energy density of leptons

The energy density of leptons, which are considered as noninteracting Fermi gas, is given by,

$$\varepsilon_{lep} = \sum_{l=e,\,\mu} \sum_{k \leq k_l^F} (m_l^2 c^4 + \hbar^2 c^2 k^2)^{1/2} . \tag{2}$$

In this equation, $k_l^F = (6\pi^2 \rho_l / \nu)^{1/3}$ is Fermi momentum of leptons and ν is degeneracy. For fully spin polarized matter, degeneracy is $\nu = 1$.

2.2   Energy density of spin polarized nucleon matter

The nucleonic part of neutron star matter is composed of neutrons and protons with densities ρ_n and ρ_p, respectively. The total number density of the system is

$$
\begin{aligned}
\rho &= \rho_p + \rho_n, \\
&= (\rho_p^{(\uparrow)} + \rho_p^{(\downarrow)}) + (\rho_n^{(\uparrow)} + \rho_n^{(\downarrow)}).
\end{aligned} \tag{3}
$$

The labels $(\uparrow)$ and $(\downarrow)$ are used for spin-up and spin-down nucleons, respectively. The following parameters can be used to identify a given spin-polarized state for the asymmetric nuclear matter,

$$\delta_p = \frac{\rho_p^{(\uparrow)} - \rho_p^{(\downarrow)}}{\rho_p}, \quad \delta_n = \frac{\rho_n^{(\uparrow)} - \rho_n^{(\downarrow)}}{\rho_n} \tag{4}$$

δ_p and δ_n are proton and neutron spin asymmetry parameters, respectively. In the fully ferromagnetic (FM) polarized nuclear matter, spin of all neutrons and protons are parallel,

$\delta_n = \delta_p = 1.0$, and in the antiferromagnetic (AFM) spin state, we have $\delta_n = \pm 1.0, \delta_p = \mp 1.0$. The asymmetry parameter which describes the isospin asymmetry of the system is defined as,

$$\beta = \frac{\rho_n - \rho_p}{\rho} = 1 - 2x_p \tag{5}$$

where $x_p = \rho_p/\rho$ is the proton fraction. Pure neutron matter is totally an asymmetric nuclear matter with $x_p = 0$, while for the symmetric nuclear matter $x_p = 1/2$. The energy density of spin-polarized asymmetrical nuclear matter, ε_{nucl} can be determined as,

$$\varepsilon_N = \rho(E + m), \tag{6}$$

where $m = 938.92$ MeV is the nucleon mass and E is the total energy per nucleon which is calculated by using the LOCV method as follows.

We adopt a trial many-body wave function of the form

$$\psi = \mathcal{F}\phi, \tag{7}$$

where ϕ is the uncorrelated ground state wave function of A independent nucleons (simply the Slater determinant of the plane waves) and $\mathcal{F} = \mathcal{F}(1 \cdots A)$ is an appropriate A-body correlation operator which can be replaced by a Jastrow form i.e.,

$$\mathcal{F} = \mathcal{S} \prod_{i>j} f(ij), \tag{8}$$

in which $\mathcal{S}$ is a symmetrizing operator. Now, we consider the cluster expansion of the energy functional up to the two-body term [8],

$$E_{nuc}([f]) = \frac{1}{A} \frac{\langle \psi | H | \psi \rangle}{\langle \psi | \psi \rangle} = E_1 + E_2. \tag{9}$$

The one-body term E_1 for an asymmetrical nuclear matter is

$$E_1 = \sum_{\tau=n,p} \sum_{\sigma=\uparrow,\downarrow} \sum_{k \leq k_{F_\tau^\sigma}} \frac{\hbar^2 k^2}{2m_\tau}, \tag{10}$$

where $k_{F_\tau^\sigma} = (6\pi^2 \rho_\tau^\sigma)^{1/3}$ is the fermi momentum of each component of spin-polarized asymmetric nuclear matter. The two-body energy E_2 is

$$E_2 = \frac{1}{2A} \sum_{ij} \langle ij | \nu(12) | ij - ji \rangle, \tag{11}$$

where

$$\nu(12) = -\frac{\hbar^2}{2m} [f(12), [\nabla_{12}^2, f(12)]] + f(12)V(12)f(12). \tag{12}$$

Here, $f(12)$ and $V(12)$ are the two-body correlation and potential. In our calculations, we use the AV_{18} two-body potentials [20]. Now, we minimize the two-body energy, Eq. (11), with respect to the variations in the correlation functions $f^{(k)}$, but subject to the normalization constraint [13, 2],

$$\frac{1}{A} \sum_{ij} \langle ij | h_{S_z,T_z}^2 - f^2(12) | ij \rangle_a = 0, \tag{13}$$

where in the case of spin polarized asymmetrical nuclear matter, the Pauli function $h_{S_z,T_z}(r)$ is as follows,

$$
h_{S_z,T_z}(r) \;=\; \begin{cases} \left[1 - 9\left(\dfrac{J_J^2(k_{F_\tau}^{(\sigma)}r)}{k_{F_\tau}^{(\sigma)}r}\right)^2\right]^{-1/2} & S_z = \pm 1, T_z = \pm 1 \\[4mm] 1 & otherwise \end{cases} \tag{14}
$$

From the minimization of the two-body cluster energy, we get a set of coupled and uncoupled Euler-Lagrange differential equations [5]. We can calculate the correlation functions by numerically solving these differential equations and then, using these correlation functions, the two body energy is obtained. Finally, we can compute the energy of the system.

2.3   URCA processes

Now, we investigate direct URCA processes in the spin polarized neutron star matter. In fully polarized ferromagnetic spin state, the nature of chemical equilibrium is mainly dominated by the following weak interaction processes,

$$
\begin{aligned}
n(\uparrow) &\rightarrow p(\uparrow) + l(\uparrow) + \bar{\nu}_l(\downarrow) \\
p(\uparrow) + l(\uparrow) &\rightarrow n(\uparrow) + \nu_l(\downarrow)
\end{aligned} \tag{15}
$$

Here, ν_l stands for the leptons neutrinos which leave the system without delay. In this case, the β-equilibrium conditions and charge neutrality of neutron star matter impose the following coupled constraints on our calculations,

$$
\begin{aligned}
\mu_e(\uparrow) = \mu_\mu(\uparrow) &= \mu_n(\uparrow) - \mu_p(\uparrow) \\
&= 4(1-2x_p)S_2(\rho,\delta_n=\delta_p=1) + 8(1-2x_p)^3 S_4(\rho,\delta_n=\delta_p=1)
\end{aligned} \tag{16}
$$

$$
\rho_p(\uparrow) = \rho_e(\uparrow) + \rho_\mu(\uparrow) \tag{17}
$$

where S_2 and S_4 are given by [3],

$$
\begin{aligned}
S_2(\rho,\delta_n,\delta_p) &= \frac{1}{2}\left(\frac{\partial^2 E(\rho,\delta_n,\delta_p)}{\partial\beta^2}\right)_{\beta=0} \\
S_4(\rho,\delta_n,\delta_p) &= \frac{1}{24}\left(\frac{\partial^4 E(\rho,\delta_n,\delta_p)}{\partial\beta^4}\right)_{\beta=0}.
\end{aligned} \tag{18}
$$

Similarly, The β-equilibrium and the charge neutrality conditions for fully anti-ferromagnetic spin polarized are,

$$
\begin{aligned}
\mu_e(\downarrow) = \mu_\mu(\downarrow) &= \mu_n(\downarrow) - \mu_p(\uparrow) \\
&= 4(1-2x_p)S_2(\rho,\delta_n=-\delta_p=1) + 8(1-2x_p)^3 S_4(\rho,\delta_n=-\delta_p=1)
\end{aligned} \tag{19}
$$

$$
\rho_p(\uparrow) = \rho_e(\downarrow) + \rho_\mu(\downarrow). \tag{20}
$$

We find the abundance of the particles by solving the coupled equations of charge neutrality and β-equilibrium conditions. Finally, we calculate the total energy and the equation of state of the neutron star matter.

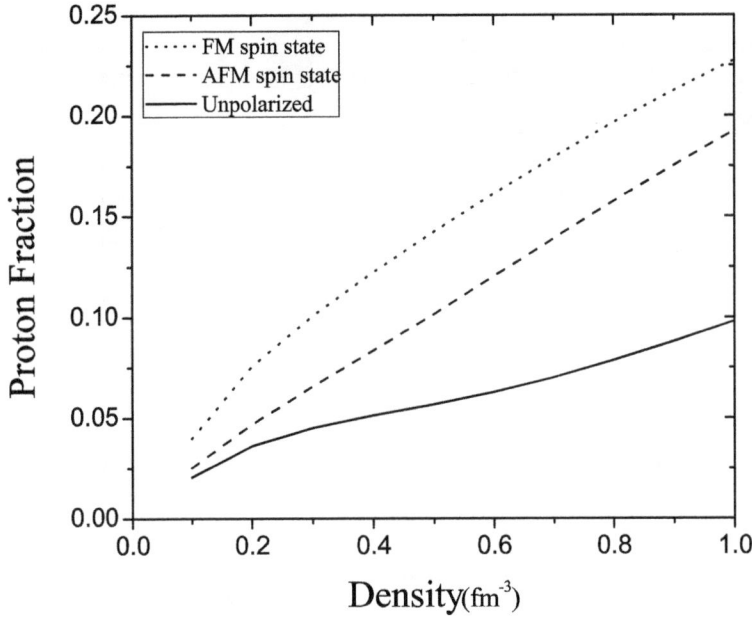

Figure 1: The proton fraction in the neutron star matter for different spin states.

3   Results and discussion

Figure 1 shows the proton fraction, x_p, versus the baryon number density, ρ, for unpolarized, ferromagnetic and antiferromagnetic spin state. It can be seen from this figure that the abundance of protons is an increasing function of both spin polarization and baryon density. Therefore, we can conclude that nuclear portion of spin polarized neutron star matter tend to be symmetric matter. It is also seen that for a given density, the highest value of proton fraction is gained for the ferromagnetic spin state.

In Figs. 2 and 3, we have presented the energy density, ε, and pressure of neutron star matter as a function of baryon number density, ρ, for unpolarized, ferromagnetic and antiferromagnetic spin state, respectively. Here, we have not considered the contribution of magnetic field. In these figures, we have also plotted the energy density and pressure of the fully polarized neutron matter (PNM), i.e. $\beta = 1, \delta_n = 1$. As we can see, the energy density and pressure increase by increasing both of spin and isospin asymmetry parameters. we have concluded that the spontaneous phase transition to ferromagnetic and antiferromanetic spin state does not occur. If such a transition existed, a crossing of the energies of different polarizations would have been observed at some density, indicating that the ground state of the system would be ferromagnetic or antiferromagnetic from that density on. As can be seen in these figures,there is no sign of such a crossing. Our results can be compared with those of Vidana's [18, 19]. Also, it is clear from these figures that the EOS of spin polarized neutron star matter is stiffer than unpolarized matter.

Now, we can investigate the structure of neutron star by using the equation of state and integrating the TOV equation. A summary of our results for the maximum mass, radius, central energy density and central baryon density of neutron star predicted from different

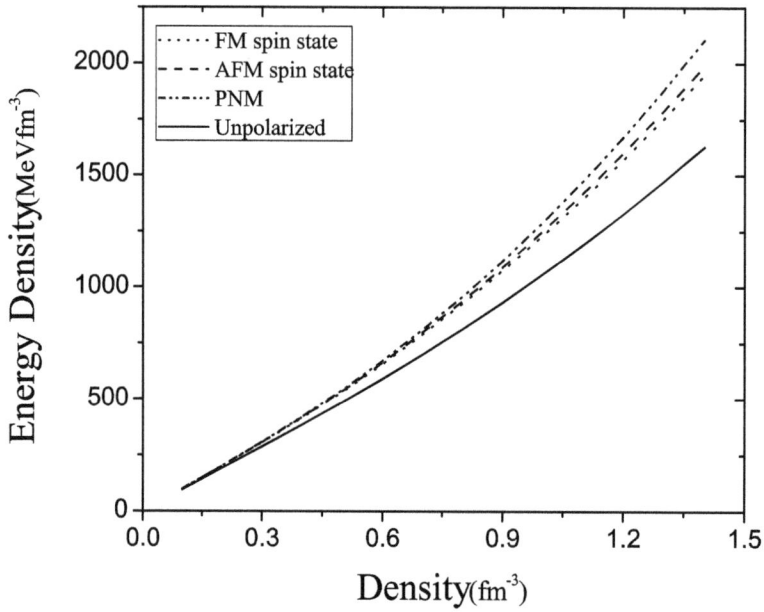

Figure 2: The energy density of neutron star matter versus baryon number density for for different spin states and fully polarized neutron matter.

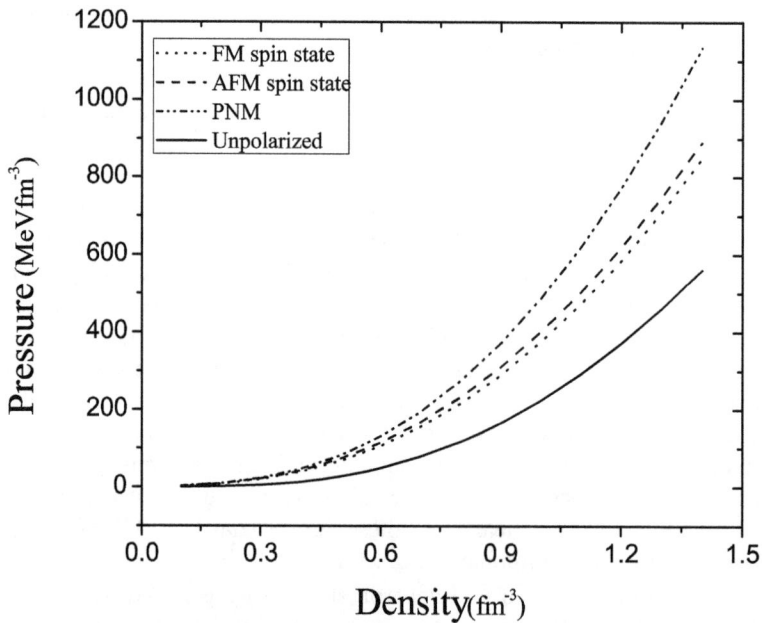

Figure 3: As a Fig. 2 but for pressure.

Table 1: Maximum mass, radius, central energy density and central baryon density of neutron star. The gravitational mass is given in solar mass ($M_\odot$).

EOS	M_{max}	R (km)	ϵ_c (10^{14} g/cm^3)	ρ_c (fm^{-3})
NSM [12]	1.63	8.04	-	-
FM-NSM	1.83	10.24	30.28	1.27
AFM-NSM	1.88	10.54	28.67	1.2
PNM	1.99	10.8	27.14	1.13

equations of state is given in table 1. We can conclude from this table that the more asymmetric is the neutron star matter, the higher maximum mass.

4   Summary and Conclusions

The purpose of this paper is investigating the influence of spin polarization on the equation of state of neutron star matter and, consequently, the structure of neutron star. We have used the lowest order constrained variational (LOCV) method by employing the AV$_{18}$ potentials for nucleon-nucleon interaction. We conclude that the equation of state become stiffer by considering spin polarization, and it yields to high maximum mass for neutron stars.

Acknowledgment

We wish to thank Research Councils of the University of Zanjan.

References

[1] Antoniadis, J. & et al. 2013, Science, 340, 448

[2] Bigdeli, M. 2010, Phys. Rev. C, 82, 054312

[3] Bigdeli, M. 2013, Int. J. Mod. Phys. E, 22, 1350054

[4] Bigdeli, M., & Elyasi, S. 2015, Eur. Phys. J. A, 51, 38

[5] Bordbar, G. H., & Modarres, M. 1998, Phys. Rev. C, 57, 714

[6] Chandrasekhar, S., & Fermi, E. 1953, ApJ, 118, 116

[7] Chowdhury, P. R. 2010, PoS NICXI, 175

[8] Clark, J. W., & Chao, N. C. 1969, Lett. Nuovo Cimento, 2, 185

[9] Demorest, P. B., Pennucci, T., Ransom, S. M., Roberts, M. S. E., & Hessels, J. W. T. 2010, Nature, 467, 1081

[10] Gandolfi, S., Carlson, J., & Sanjay Reddy 2012, Phys. Rev. C, 85, 032801

[11] Gandolfi, S., Yu. Illarionov, A., Fantoni, S., Miller, J. C., Pederiva, F., & Schmidt, K. E. 2010, MNRAS, 404, 35

[12] Lee, CH. 2012, EPJ Web of Conferences, 20, 04002

[13] Owen, J. C., Bishop, R. F., & Irvine, J. M. 1977, Nucl. Phys. A, 277, 45

[14] Shen, G., Horowitz, C. J., & Teige, S. 2011, Phys. Rev. C, 83, 035802

[15] Sun, B. Y., Long, W. H., Meng, J., & Lombardo, U. 2008, Phys. Rev. C, 78, 065805

[16] Romani, R. W., Filippenko, A. V., Silverman, J. M., Bradley Cenko, S., Greiner, J., Rau, A., Elliott, J., & Pletsch, H. J. 2012, ApJL, 760, 36

[17] van Kerkwijk, M. H., Breton, R. P., & Kulkarni, S. R. 2011, ApJ, 728, 95

[18] Vidana, I., & Bombaci, I. 2002, Phys. Rev. C, 66, 045801

[19] Vidana, I., Polls, A. & Ramos, A. 2002, Phys. Rev. C, 65, 035804

[20] Wiringa, R. B., Stoks, V. G. J., & Schiavilla, R. 1995, Phys. Rev. C, 51, 38

10

Formation of Large Structures in the Acceleration Universe with a Hybrid Expansion Law

Neda Amjadi · Vahid Abbasvand · D.M Jassur

Astrophysics Department, Physics Faculty, University of Tabriz, Tabriz, Iran

Abstract. In the current paper, we have studied the effect of dark energy on formation where dark energy exists in the background. For this purpose, we used both WMAP9 and Planck data to study how the radius changes with redshift in these models. We used different data sets to fix the cosmological parameters to obtain a solution for a spherical region under collapse. The mechanism of structure formation for dark and baryonic matter is different. When processed by gravitational instability, density perturbations have given rise to collapsed dark matter structures, called halos. These dark matter halos offer the backdrop for the subsequent formation of all collapsed baryonic structures, including stars, galaxies, and galaxy clusters. In Planck Data for ΛCDM, with the presence of dark energy in the background, the formation of baryonic matter is delayed. Therefore, it is a factor for the largening of the baryonic matter radius. Accompanying dark energy is entailing an increment of dark matter virial radius. For WACDM Data, dark energy alongside time-dependent parameter of state and baryon acoustic oscillations are the reasons for the delay of dark matter formation and the radius reduction. Due to the lack of data without baryonic acoustic waves in the background, we are left unable to delineate its impact on the structures. In $WCDM(BAO + H_0)$ and $WCDM(H_0)$, the lack of BAO shows a critical role in the delaying of baryonic matter structure formation. Respectively, it causes growing virial radius of dark matter. BAO, without taking dark energy into accounts, is the reason for the increasing and decresing of radius of dark and baryonic matter. It also delays baryonic matter formation. In $\Lambda CDM(BAO + H_0)$ and $\Lambda CDM(H_0)$, We have studied ΛCDM data for standard model under two circumstances: (a) $\Lambda CDM(BAO + H_0)$, (b) $\Lambda CDM(H_0)$ data. Dark energy in this data delays formation and intensifies virial radius of baryonic matter. Our studies show WCDM and ΛCDM have the same effect on formation if we do not consider dark energy in BG. Planck data, in comparing with WMAP, has important role in describing standard model.

Keywords: Dark Energy – Scalar Fields – Baryonic Acoustic Waves – Standard Model

1 Introduction

In 1998, two teams studying distant Type Ia supernovae presented independent evidence that our universe is currently accelerating [1, 2]. The physical origin of cosmic acceleration remains a deep mystery. The accelerated expansion of the Universe,discovered in 1998, has raised fascinating questions for cosmology and physics as a whole. Two different approaches are proposed for this problem: (i) Dark Energy models that modify the stress-energy content of the Universe, adding an additional component with equation of state w=-1. That is, we modify the right-hand side of the Einstein equations. (ii) The Modified Gravity category corresponds to modifying the left-hand side. For example General Relativity (GR) by modifying the Einstein-Hilbert action [3, 4, 5].

The candidates of dark energy include cosmological constant and a variety of scalar field models. The models based upon modified theories of gravity are faced with the challenges posed by the local physics. Large scale modification of gravity essentially implicates extra degrees of freedom which might influence local physics where the Einstein theory of gravity is approving with observations. By giving a priori a cosmic history, specifying either the equation of state (EoS) or the scale factor $"a"$, we can always construct a scalar field potential which would imitate the desired result. Similar reconstruction can be executed in scalar tensor theories [6, 7].

The dynamics of realistic Universe are illustrated by an EoS parameter which behaves differently at different epochs. For instance, in general relativistic description of the dynamics of the spatially flat RW space-time, the fluids with constant EoS parameter $w > -1$ give rise to a power-law expansion ($a \propto t^{\frac{2}{3(1+w)}}$) of the Universe and for an exponential expansion $a \propto e^{kt}$, where $k > 0$ is a constant; it is required that $w = -1$. The solution of the Einstein's field equation in the presence of a single fluid with a constant EoS parameter gives a relation for the EoS parameter of a fluid. The discovery of cosmic acceleration is debatably one of the most important developments in modern cosmology [8, 9].

Most dark energy modelling using scalar fields has followed the Quintessence pattern of a slowly rolling canonical scalar field. However, there has been increasing interest in loosening the assumption of a canonical kinetic term. In its most general form, this idea is known as k-essence [10].

Tachyon dark energy has been explored by many authors, e.g.[6, 11, 12]. Bagla et al (2003) focused on two specific choices of Tachyon potential, and carried out numerical analysis of the cosmological evolution in order to constrain them against supernova data and the growth rate of large-scale structure [6]. Copeland et al (2005) studied a wider range of potentials, concentrating mainly on analytical inspection of attractor behavior and the critical point structure without making comparison to specific observations [12]. By studying a wide range of potentials and testing them directly against current observational constraints, they aim to combine some of the positive features of each analysis [13].

Parsons and Barrow studied the behavior of the scale factor in the context of inflation in the early Universe [14]. They pointed out that Einstein's field equations in the presence of self-interacting scalar field are invariant under the constant rescaling of the scalar field, and then they generated the HEL (Hybrid Expansion Law) behavior from power-law expansion. They also showed that such an expansion of the Universe can be represented as a Friedmann Universe in the presence of imperfect fluid. Akarsu et al (2014) study HEL expansion in the context of the history of the Universe after the inflation took place, and mainly investigate whether this law could be used for explaining the evolution of the Universe starting from the radiation- or matter-dominated Universe to the currently accelerating Universe. They also carried out the effective fluid and the single scalar field reconstruction using Quintessence, Tachyon and Phantom fields, which can capture HEL in the framework of general relativity [15].

In cosmology, baryon acoustic oscillations (BAO) refers to regular, periodic fluctuations in the density of the visible baryonic matter (BM) of the Universe. BAO matter clustering provides a $"standard ruler"$ for length scale in cosmology in the same way that supernova experiments provide a $"standard candle"$ for astronomical observations. BAO measurements help cosmologists understand the nature of dark energy better by constraining cosmological existing BAO in the background [15].

The effect of the dynamics of dark energy on the growth rate of the large scale structures in the framework of LCDM and MOND is investigated. For variable dark energy model, increasing the bending parameter b causes the structure viralizes at lower redshift with larger radius. Therefore, the variable dark energy model put off the spherical collapse to the later times. The case of the low-density model has an intermediate behavior such that the virialization redshift in this model corresponds with b = 0.4 in variable dark energy model. Finally, we compared the virialization of structures under the variable dark energy model with the recent results of MONDian N-body simulations. We showed that the various models of simulation are consistent with the variable dark energy model with different bending parameter [16, 17].

The process of the structure formation in the presence of dark energy models must be studied linear theory in order to address the growth of structures in linear regime and the effect of dark energy on matte power spectrum and variance, which is then, must be used in Press- Schechter kind of study of the nonlinear structures [18]. There is ample evidence that galaxies reside in extended halos of dark matter(DM) which forms through gravitational instability. Density perturbations grow linearly until they reach a critical density, after which they turn around from the expansion of the Universe and collapse to form virialized dark matter halos. These halos continue to grow in mass and size, either by accreting material from their neighborhood or by merging with other halos. Some of these halos may survinve as bound entities after merging into a bigger halo, thus giving rise to a population of subhalos. The illustrated process shows the formation of a dark matter halo in a numerical simulation of structure formation in a CDM cosmology. It also shows how a small volume with small perturbations initially expands with the Universe. As time proceeds, small-scale perturbations grow and collapse to form small halos. At a later stage, these small halos merge together to form a single virialized DM halo with an ellipsoidal shape, which reveals some substructure in the form of DM subhalos[19].

Within this paper, we intend to investigate whether a simple scale factor obtained by multiplying power-law and exponential law, which we will call hybrid expansion law, could triumph in explaining the observed Universe. We have used scalar fields for investigating structures formation and the effect dark energy has on it.

Here follows the outline: In Sec. 2, we will inspect the potential and EoS in scalar fields such as Quintessence, Tachyon and Phantom. In Sec. 3, we will exhibit how the presence of dark energy affects structure formation by using Planck, WACDM, and WCDM data. We will finalize the paper summarizing the results in the conclusion section.

2   Scalar fields

Scalar field models have played a vital role in cosmological studies for nearly half a century. Those assumed scalar fields have appeared in different cosmological research aspects to resolve various cosmological problems [21], such as driving inflation, time variable cosmological constant explaination, and so on. The scalar fields have played one other essential role for the past fifteen years as a candidate for dark energy proceeding the discovery of the accelerating expansion of universe. There are so many phenomenological dark energy models of scalar fields, such as Quintessence, Phantom, quintom and the scalar fields with non-canonical kinetic energy term [22, 23].

To study the dynamical evolution of those scalar field models and their cosmological implications with a phase-plane analysis is a very useful and common method. However, most studies only focus on the Quintessence models (including Phantom, Quintessence, and quintom) with unique exponential potential and Tachyon models (including Phantom Tachyon) with inverse square potential. Correspondingly, the dynamical systems are two dimensional autonomous systems with those particular forms of potentials [24, 25].

2.1   Quintessence field

Most cosmological models implicitly assume that matter and dark energy interact only gravitationally. In the absence of an underlying symmetry that would suppress a matter-dark energy coupling (or interaction), there is no a priori reason for dismissing it. Cosmological models in which dark energy and matter do not evolve separately but interact with one another were first introduced to justify the small value of the cosmological constant [26]. Recently, various proposals at the fundamental level, including field Lagrangians, have been advanced to account for the coupling. Scalar field Lagrangians coupled with matter do not generate scaling solutions with a long enough DM dominated period as required by structure formation. The phenomenological model we are going to discuss was constructed to account for late acceleration in the framework of Einstein's relativity and to significantly alleviate the mentioned coincidence problem and escapes the limits imposed by it[27, 28].

Most of the dark energy studies are carried out within the Quintessence pattern of a slowly rolling canonical scalar field with a potential. For that reason, we will first consider the Quintessence realization of the HEL. In general relativity, the effective energy density and EoS parameter of the fluid follow as below [20]:

$$\rho_{eff}(t) = 3(\frac{\alpha}{t} + \frac{\beta}{t_0})^2 \tag{1}$$

$$W_{eff} = \frac{2}{3}\frac{\alpha}{t^2}(\frac{\alpha}{t} + \frac{\beta}{t_0})^{-2} - 1 \tag{2}$$

The potential and EoS parameter as a function of time (t) are then given by the following expression:

$$\dot{\phi}^2(t) = \frac{2\alpha}{t^2} \tag{3}$$

$$V(t) = 3(\frac{\alpha}{t} + \frac{\beta}{t_0})^2 - \frac{\alpha}{t^2} \tag{4}$$

$$W(t) = \frac{\frac{2\alpha}{t^2}}{3(\frac{\alpha}{t} + \frac{\beta}{t_0})^2} - 1 \tag{5}$$

2.2   Tachyon field

Quintessence pattern relies on the potential energy of scalar fields to drive the late time acceleration of the Universe. On the other hand, it is also possible to relate the late time acceleration of the Universe to the kinetic term of the scalar field by relaxing its canonical kinetic term. This idea is known as k-essence [29]. Tachyon fields can be taken as a particular case of k-essence models with Dirac-Born-Infeld (DBI) action and can also be motivated by the string theory [30]. That item together with $p = w\rho$, give the following relations:

$$\dot{\phi}^2 = \frac{\frac{2\alpha}{t^2}}{3(\frac{\alpha}{t} + \frac{\beta}{t_0})^2} \tag{6}$$

$$V(t) = 3(\frac{\alpha}{t} + \frac{\beta}{t_0})^2 \sqrt{1 - \frac{\frac{2\alpha}{t^2}}{3(\frac{\alpha}{t} + \frac{\beta}{t_0})^2}} \tag{7}$$

$$w = \frac{\frac{2\alpha}{t^2}}{3(\frac{\alpha}{t} + \frac{\beta}{t_0})^2} - 1 \tag{8}$$

2.3 Phantom field

Quintessence and Tachyon fields investigated in the previous two subsections can yield EoS parameters $w \geq -1$. However, present observations allow slight Phantom values for the EoS parameter, i.e., $w < -1$. Sources behaving as a Phantom field can arise in braneworlds, Brans-Dicke scalar-tensor gravity and may be motivated by S-brane constructions in the string theory [15, 20]. On the other hand, the Phantom energy can, in general, be simply described by a scalar field with a potential $V(\phi)$ like the Quintessence dark energy, yet with a negative kinetic term [31]. Accordingly, the energy density and pressure of the Phantom field can be given by

$$\rho = -\frac{1}{2}\dot{\phi}^2 + V(\phi) \tag{9}$$

$$P = -\frac{1}{2}\dot{\phi}^2 - V(\phi) \tag{10}$$

where ϕ is the Phantom field with potential $V(\phi)$. We rescale time as $t \longrightarrow t_s - t$, where t_s is a sufficiently positive reference time. Thus, the HEL ansatz [1] becomes

$$a = a_0(\frac{t_s - t}{t_s - t_0})^\alpha e^{[\beta(\frac{t_s - t}{t_s - t_0} - 1)]} \tag{11}$$

The effective EoS parameter and energy density ρ are respectively:

$$\rho_{eff}(t) = 3(\frac{\alpha}{t_s - t} + \frac{\beta}{t_s - t_0})^2 \tag{12}$$

$$w_{eff} = \frac{2}{3}\frac{\alpha}{(t_s - t)^2}(\frac{\alpha}{t_s - t} + \frac{\beta}{t_s - t_0})^{-2} - 1 \tag{13}$$

Thus, we can get following equations from Eqs $(14 - 18)$:

$$\dot{\phi}^2 = \frac{-2\alpha}{(t_s - t)^2} \tag{14}$$

$$V(t) = 3(\frac{\alpha}{t_s - t} + \frac{\beta}{t_s - t_0})^2 - \frac{\alpha}{(t_s - t)^2} \tag{15}$$

$$w = \frac{\frac{2\alpha}{(t_s-t)^2}}{3(\frac{\alpha}{t_s-t} - \frac{\beta}{t_s-t_0})^2} \tag{16}$$

where α and β are non-negative constants ($\alpha = 0.488, \beta = 0.444$).

3   Results & Discussion

In the current paper, we have investigated scalar field models for ansatz that are produced with power-law and exponential type of functions. We have also carried out the evolution of large scale structures in a single scalar field reconstruction using Quintessence, Tachyon and Phantom fields and compared them with the standard model (ΛCDM). Following that purpose, we have applied dark energy effect on formation. We have used two numerical data, namely WMAP9 and Planck to study how the radius changes with redshift in these models. It helps us obtain virilization of models, which agrees to structure formation in Cosmological observations. Results are regularized in three subsections. First, we will investigate Planck data first, then WACDM and finally WCDM.

3.1   Planck Data

Cosmological observations prior to Planck were consistent with the simplest models of inflation within the slow-roll paradigm. Planck data are remarkably consistent with the predictions of the base ΛCDM cosmology. In the following section, we will use Planck numerical data [33] to investigate dark energy effect on structure formation in background under two circumstances: (a) radiation and matter(dark and light separately), (b) radiation, matter and dark energy , in the BG.

Since the HEL model predicts the beginning of universe with radiation and the current accelerating phase of the universe at the same time. However, from the CMB test it does not accommodate the matter-dominated era properly unless we consider the parameter α. Thus, with the current form of HEL, radiation alone cannot construct structure [15].

According to Newton's gravity law, the radius of dark and baryonic matter can be given by

$$\frac{dr_b}{da} = \frac{1}{aH((-r_b^2 H^2 \delta_b + \frac{2GM}{r_b})^{-\frac{1}{2}})}$$
$$\frac{dr_d}{da} = \frac{1}{aH((-3r_d^2 H^2 \delta_d + \frac{2GM}{r_d})^{-\frac{1}{2}})} \tag{17}$$

Where $a_0 = 1$ and δ is matter density contrast. Now using Planck data [32, 33] and Harrison-Zeldovich spectrum data [16], we obtain numerical data that shows in Tab.1.

In the other case, Friedmann equation encompasses radiation, matter and energy. There, the numerical value of Planck data is placed in ΛCDM and scalar field models follow as Tab.1.

We plotted radius evolution of scalar fields and ΛCDM models in Fig.1 by considering the values of model parameters given in Table 3 from Planck data. According to virial theorem, when the kinetic energy of structure abates, the total energy becomes potential energy. In this phase, the structure spends its maximum radius. For comparison, all of the models are in a plot which shows all models have similar behavior for DM. In other words, Planck data are compatible with the predictions of the base ΛCDM cosmology. By using this data, standard model can be explained as structure formation. Hence, dark energy in the background makes a delay in formation of BM and intensifies its radius. Accompanying the existence of dark energy is entailing an intensification of DM virial radius. For scalar field models, dark energy is a factor to the accretion of DM radius. However, BM cannot be constructed with such data.

Table 1: The mass of baryonic and DM ($10^{12}M_\odot$) with radiation and dark energy in background in three models. Using Planck numerical data to investigate dark energy effect on structure formation in background under two circumstances (a) and (b).

	$Model$	Com	$R_{Max}(Kpc)$	$R_{vir}(Kpc)$	Z_{Max}	Z_{vir}
a	Plank	DM	156.268±2.315	81.809±1.404	5.7±0.020	3.4±0.012
		BM	467.657±4.428	250.161±2.705	1.3±0.007	0.45±0.001
b	ΛCDM	DM	156.263±2.315	83.247±1.407	5.7±0.020	3.4±0.012
		BM	467.750±4.429	251.441±2.709	1.25±0.007	0.35±0.001
b	Phantom	DM	156.263±2.315	83.247±1.407	5.7±0.001	3.4±0.012
		BM	421.7±4.421	421.7±4.421	2.05±0.009	2.05±0.009
b	Tachyon &	DM	156.263±2.315	83.247±1.407	5.7±0.020	3.4±0.012
	Quintessence	BM	422.494±4.429	422.494±4.429	2.05±0.009	2.05±0.009

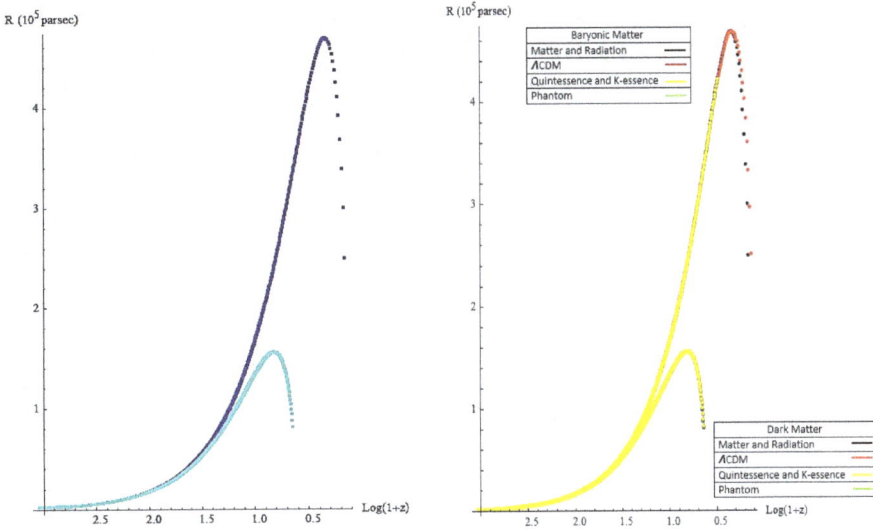

Figure 1: <u>Left</u>: Radial evolution of BM (dark blue) and DM (light blue) for mass $= 10^{12}M_\odot$ from Planck Data. In this model, radiation and matter exist in background. <u>Right</u>: Diagrams show radial evolution of BM and DM for ΛCDM model (red), Phantom model (green), Tachyon and Quintessence models (yellow). Black curve shows matter and radiation in background. Dots relate virialization of model. In BM, this compatible is interrupted for $\log(1+Z) < 0.5$.

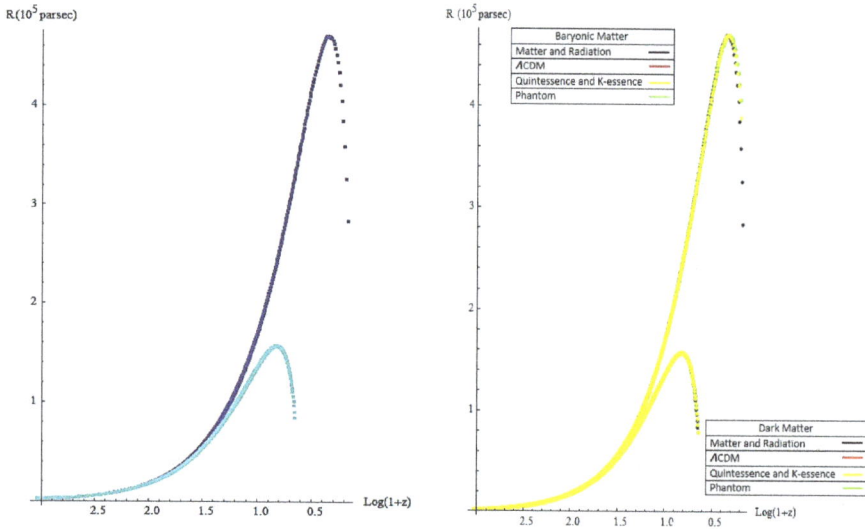

Figure 2: The same as in Figure 1 but by using WACDM Data. Black curve shows matter and radiation in background. Dots relate virialization of model.

3.2   WACDM Data

A major challenge for cosmology today is to elucidate the nature of the dark energy driving the accelerated expansion of the Universe. Among cosmological models, there is one that depicts flat universe with dark energy where equation of state is time-dependent. Scalar field models are time-dependent. Therefore, we analyzed redshifts in which parameter of state has a negative value. For Tachyon and Quintessence models, this value is $z < 2.273$. However, in the case of the Phantom scenario, the ansatz and Hubble parameter diverge as $t \longrightarrow t_s$, and thus expose the Universe to Big Rip. We supposed that in flat universe, radiation, matter and energy exist in the background. Tab.2 shows numerical data that is given from WACDM data.

Table 2: The same as in Table 1 but by using WACDM Data.

	$Model$	$Comp$	$R_{Max}(Kpc)$	$R_{vir}(Kpc)$	Z_{Max}	Z_{vir}
a	Scalar	DM	157±2.316	84.087±1.410	5.74±0.020	3.44±0.012
	field	BM	469.245±4.431	283.47±3.070	1.34±0.007	0.49±0.001
b	Phantom	DM	157±2.316	78.914±1.397	5.74±0.020	3.39±0.011
		BM	470.181±4.432	406.64±4.360	1.24±0.007	0.59±0.001
b	Tachyon &	DM	157±2.316	78.914±1.397	5.74±0.020	3.39±0.011
	Quintessence	BM	469.967±4.431	388.905±4.115	1.29±0.007	0.59±0.001

Since DM is constructed earlier than BM, for investigating dark energy effect on structure, we compare redshift virilism with a state wherein energy is not in the background. Our study shows that dark energy alongside time-dependent parameter of state and baryon acoustic oscillations. These are the factor for the reducing of DM radius. Due to lack of data on the absence of baryonic acoustic waves in the background, we are left unable to delineate its impact on the structures. The absence mentioned is on the account of the simultaneous existence of dark energy and BAO in the background.

3.3   WCDM Data

WCDM data has investigated flat universe with dark energy by observing the effect of baryon acoustic oscillations. The BAO angular scale serves as a standard ruler and allows us to map out the expansion history of the Universe after last scattering. The BAO scale, which is extracted from galaxy redshift surveys, provides a constraint on the late-time geometry and breaks degeneracies with other cosmological parameters.

3.3.1 $WCDM(BAO + H_0)\&WCDM(H_0)$

Within the following subsection, first, we studied WCDM data for scalar fields under two circumstances: (a) WCDM ($H_0 + BAO$), (b) WCDM (H_0) data then, inspected the presence of dark energy in the background. Tab.3 , Tab.4 and Fig.3 show numerical data that is given from WCDM data.

Table 3: The same as in Table 1 but by using WCDM($H_0 + BAO$) Data.

	$Model$	Com	$R_{Max}(Kpc)$	$R_{vir}(Kpc)$	Z_{Max}	Z_{vir}
a	Scalar	DM	156.783±2.317	81.955±1.405	5.71±0.020	3.41±0.012
	field	BM	469.13±4.430	255.459±2.715	1.31±0.007	0.46±0.001
b	Phantom	DM	156.778±2.317	83.553±1.408	5.71±0.020	3.41±0.012
		BM	469.056±4.430	466.667±4.427	1.26±0.006	1.01±0.005
b	Tachyon &	DM	156.778±2.317	83.553±1.408	5.71±0.020	3.41±0.012
	Quintessence	BM	468.786±4.429	464.351±4.421	1.31±4.421	1.01±0.005

Table 4: The same as in Table 1 but by using WCDM(H_0) Data.

	$Model$	Com	$R_{Max}(Kpc)$	$R_{vir}(Kpc)$	Z_{Max}	Z_{vir}
a	Scalar	DM	157.337±2.318	81.578±1.403	5.8±0.022	3.45±0.012
	field	BM	470.402±4.432	281.74±2.773	1.35±0.007	0.5±0.001
b	Phantom	DM	157.374±2.318	85.616±1.411	5.75±0.021	3.45±0.012
		BM	471.203±4.439	312.244±3.001	1.25±0.007	0.4±0.001
b	Tachyon &	DM	157.374±2.318	85.616±1.411	5.75±0.021	3.45±0.012
	Quintessence	BM	470.832±4.433	275.387±2.755	1.3±0.007	0.4±0.001

The presence of baryon acoustic oscillations plays a critical role in the postponing of dark and baryonic matter structure formation. Respectively, it causes increasing and decreasing virial radius of dark and baryonic matter. Dark energy, without taking BAO into accounts, is the reason for the declining of BM radius. If we consider both of them, we will be facing an increment of DM radius.

3.3.2 $\Lambda CDM(BAO + H_0)\&\Lambda CDM(H_0)$

Through this instance, we studied ΛCDM data for standard model under two circumstances: (a) $\Lambda CDM(BAO + H_0)$, (b) $\Lambda CDM(H_0)$ data. Then, we suppose that in flat universe, radiation, matter and energy exist in the background. Tab.5 and Fig.4 show this success.

Dark energy in this data grows virial radius of both of them. The structure formation in a standard model is dependent on BAO. In other words, BAO is a necessity factor for constructing structure in the standard model.

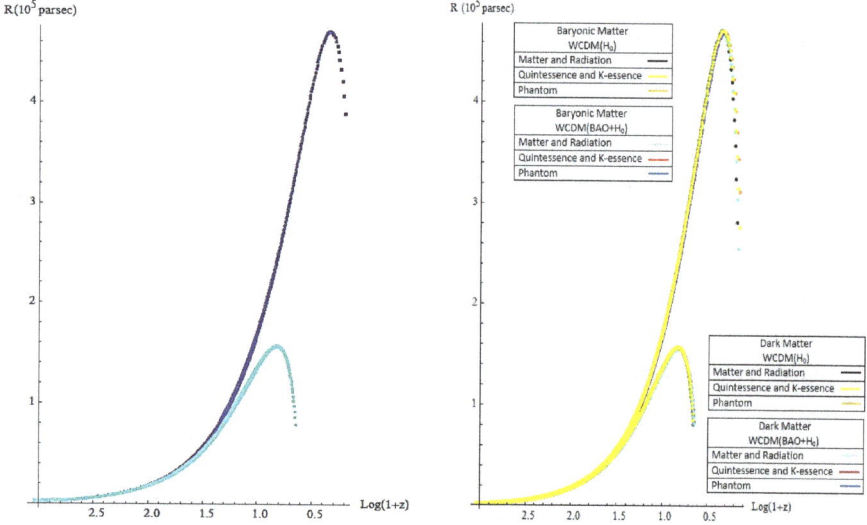

Figure 3: The same as in Figure 1 but by using <u>Left:</u> $WCDM(BAO + H_0)\&WCDM(H_0)$ Data. <u>Right:</u> $WCDM(BAO + H_0)\&WCDM(H_0)$ Data for Phantom, Tachyon and Quintessence models that radiation, matter and dark energy exist in background. Dots relate virialization of model.

Table 5: The same as in Table 1 but by using ΛCDM Data.

	$Data$	Com	$R_{Max}(Kpc)$	$R_{vir}(Kpc)$	Z_{Max}	Z_{vir}
a	ΛCDM	DM	157.111±2.316	80.160±1.398	5.77±0.022	3.42±0.012
	$(BAO + H_0)$	BM	470.175±4.430	260.364±2.601	1.32±0.007	0.47±0.001
a	ΛCDM	DM	157.55±2.319	78.948±1.389	5.79±0.022	3.44±0.012
	(H_0)	BM	471.371±4.433	268.541±2.696	1.34±0.007	0.49±0.001
b	ΛCDM	DM	157.149±2.316	84.078±1.409	5.72±0.022	3.42±0.012
	$(BAO + H_0)$	BM	470.144±4.432	271.715±2.720	1.27±0.007	0.37±0.001
b	ΛCDM	DM	157.545±2.317	645.903±2.544	5.79±0.021	-1.91±0.001
	(H_0)	BM	471.203±4.431	914.531±10.117	1.29±0.007	-1.91±0.001

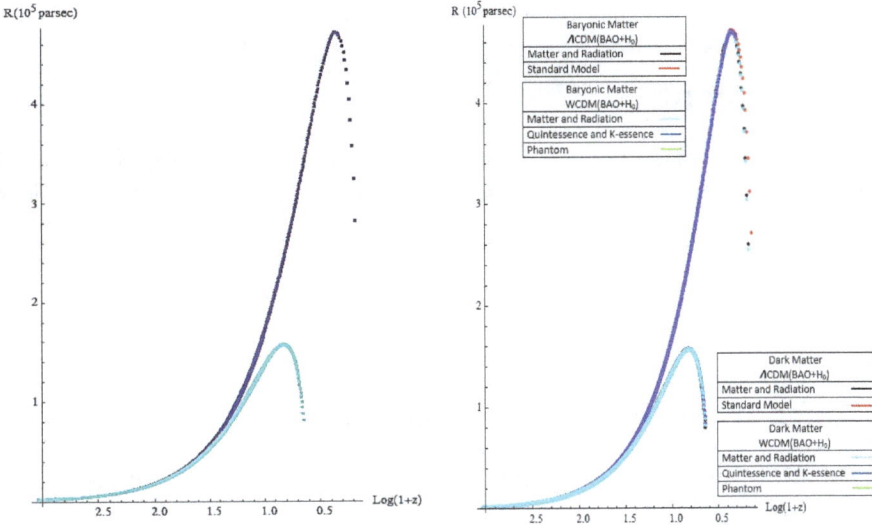

Figure 4: The same as in Figure 1. <u>Left</u>:$\Lambda CDM(BAO + H_0)$&$\Lambda CDM(H_0)$ Data. <u>Right</u>: ΛCDM Data.

4   Conclusion

In this paper, We have examined the hybrid form of scale factor, namely, a product of power law and an exponential function, which provides a simple mechanism of transition from decelerating to accelerating phase. Two numerical data are utilized to study how the radius of structures can remark in these models, as the virialization of structures depends strongly on the background. We used different data sets to fix the cosmological parameters to get a solution for a spherical region under collapse. We dealt with the problem of the structure formation in the framework of Cosmological Models under the influence of dark energy and BAO.

The mechanism of structure formation for DM and BM is different. Due to hierarchical structure formation, because of gravitational instability, density perturbations have given rise to collapsed DM structures, is called halos. These DM halos provide the backdrop for the subsequent formation of all collapsed baryonic structures, including stars, galaxies, and galaxy clusters.

We showed that the dark energy dominated background delays the virialization of structures and formation of BM which proceeds a larger structures. In other word, it causes an increase of virial radius of DM without deferment. Using Planck data in scalar field models shows that dark energy is an element to the intensification of dark matter radius. However, BM cannot be constructed with such data. Our study on WACDM data shows that the presence of dark energy alongside time-dependent parameter of state and baryon acoustic oscillations are shrinking reasons of radius for DM and absence of structure for BM. Due to lack of data on the absence of baryonic acoustic waves in the background, we are left unable to delineate its impact on the structures.

Finally on WCDM data, the presence of baryon acoustic oscillations plays an influential role in postponing. In turn, it causes decrease and increase to virial radius of dark and baryonic matter and the absence of BAO delays formation and abates virial radius of BM. The structure formation in a standard model is dependent on BAO. WCDM and ΛCDM have the same effect on formation if we do not consider dark energy in BG. Planck data, in comparing with WMAP, has important role in describing standard model.

Acknowledgment

The WMAP mission is made possible by the support of the NASA Science Mission Directorate. This research has made use of NASA's Astrophysics Data System Bibliographic Services.

References

[1] Riess et al. 1998, The Astronomical, 116, 1009

[2] Perlmutter et al. 1999, The Astronomical, 517, 565

[3] Frieman J.A., Turner M.S. and Huterer D. 2008, Annual Review of Astronomy and Astrophysics, 46, 385

[4] Sami M. 2009, Current Science, 97, 887

[5] Nojiri Sh. and Odintsov S.D. 2011, Physics Reports, 505, 59

[6] Bagla J.S., Jassal H.K. and Padmanabhan T. 2003, Physical Review D, 67, 11

[7] Kamenshchik A.Yu., Tronconi A., Venturi G. and Vernov S.Yu. 2011, Physics Letters B, 702, 191

[8] Zeldovich Ya.B. 1962, SOVIET PHYSICS JETP, 14, 1609

[9] Barrow J.D. 1978, Nature, 272, 211

[10] Armendariz-Picon C., Mukhanov V. and Steinhardt P.J. 2000, Physical Review Letters, 85, 4438

[11] Garousi M.R., Sami M. and Tsujikawa Sh. 2004, Physical Review D, 70, 15

[12] Copeland J., Garousi M.R., Sami M., Tsujikawa Sh. 2005, Physical Review D, 71,

[13] Wang Yun. and Mukherjee Pia. 2006, Astrophysical Journal, 650, 1

[14] Parsons P. and Barrow J.D. 1995, Classical and Quantum Gravity, 12, 1715

[15] Akarsu O. et al. 2014, Cosmology and Astroparticle Physics, 2014, 18

[16] Malekjani M., Rahvar S. and Jassur D.M. 2009, New Astronomy, 14, 398

[17] Malekjani M., Rahvar S. and Haghi H. 2009, ApJ, 694, 1220

[18] Amendalo L. and Tsujikawa Sh. 2010, Cambridge University press, chapter 11-12

[19] Houjun Mo, Bosch F., and White S. The book" Galaxy Formation and Evolution" Cambridge University press 2010

[20] Eisenstein D.J. 2005, New Astronomy Reviews, 49, 360

[21] Ratra B. and Peebles P.J.E. 1988, Physical Review D, 37, 3406

[22] Copeland E.J., Sami M. And Tsujikawa S. 2006, International Journal of Modern Physics D, 15, 1753

[23] Li. Miao et al. 2011, Communications in Theoretical Physics, 56, 525

[24] Coley A.A. 2003, 291 Kluwer Academic Publishers

[25] Wei F. and Hui-Qing Lu. 2010, European Physical Journal C, 68, 5677

[26] Wetterich C. 1988, Nuclear Physics B, 302, 668

[27] Chimento L.P., Jakubi A.S., avn D. and Zimdahl PW. 2003, Physical Review D, 67, 29

[28] Amendola L., Quartin M., Tsujikawa S. and Waga I. 2006, Physical Review D, 74, 14

[29] Armendariz-Picon C., Mukhanov V. and Steinhardt P.J. 2000, Physical Review Letters, 85, 4438

[30] Gibbons G.W. 2002, Physics Letters B, 537, 1

[31] Caldwell R.R. 2002, Physics Letters B, 545, 23

[32] Planck Collaboration and Ade P.A.R., Aghanim N., Armitage-Caplan C., Arnaud M., Ashdown M., Atrio-Barandela F., Aumont J., Baccigalupi C. and Banday A.J. et al. 2014, Astronomy & Astrophysics, 571

[33] WMAP Cosmological Parameters Model, LAMBDA-Data Products , http://lambda.gsfc.nasa.gov/, NASA.

11

DASTWAR: A Tool for Completeness Estimation in Magnitude-size Plane *

Ali Koohpaee[1] · Mehdi Khakian Ghomi[2]

[1] Department of Energy Engineering and Physics,
Amirkabir University of Technology, Tehran, Iran
email: akoohpaee@gmail.com

[2] Department of Energy Engineering and Physics,
Amirkabir University of Technology, Tehran, Iran

Abstract. Today, great observatories around the world, devote a substantial amount of observing time to sky surveys. The resulted images are inputs of source finder modules. These modules search for the target objects and provide us with source catalogues. We sought to quantify the ability of detection tools in recovering faint galaxies regularly encountered in deep surveys. Our approach was based on completeness estimation in magnitude - size plane. The adopted method was incorporating artificial galaxies. We improvised a software that estimates completeness in a given interval of magnitude and size. The software generates artificial galaxies and iteratively inserts them to the source finder modules input image. Evaluating the ratio of the number of detected to the number of inserted artificial galaxies provides us with means to estimate completeness. Completeness estimation is helpful in selecting unbiased samples.

Keywords: Galaxies: structure, galaxies: size, magnitude

1   Introduction

The past two decades have seen the growing number of imaging surveys of the extragalactic sky. Deep field optical/NIR imaging surveys such as COSMOS [19], HUDF [1], CANDELS [16] and GOODS [14] have become the frontier of astronomical studies in various topics. Moreover, within the next few years, imaging surveys with unprecedented depth and area (e.g. LSST [18] and Euclid [21]) will take place.

Any imaging survey is restricted and biased in its sampling of the galaxy population by a number of selection effects (e.g. [10, 11, 17, 4]). The visibility of a particular galaxy depends both on its intrinsic properties (e.g. brightness, light profile, apparent scale size) and the nature of the survey imaging data (e.g. exposure time and sky brightness) [4].

When CCD is used in acquisition of imaging data, outcome images will be of digital type. Acquired images, after passing the process of data reduction, will be given to source finder modules. These modules will identify the sources targeted by the survey and provide us with their photometric and structural properties.

In the case of galaxies, surface brightness is seen to be a key factor in their detectability by source finder modules. Generally, galaxies with lower surface brightness are harder to detect [14, 2, 4]. However, detection is a complex process and surface brightness is not the only factor in determining the detectability of a particular galaxy. For instance, blending with other sources, image artefacts as well as structural properties such as morphological type and position angle can be mentioned as factors playing a role in their detectability [3]. There is no strict low surface brightness threshold, above which, the detectability of galaxies is assured. We expect that, as the surface brightness of a particular galaxy decreases, the probability of its detection also reduces. Completeness parameter has been defined to quantify the probability of detection [29, 20, 31, 7]. This parameter is defined as the ratio of the number of detected objects to the total number of objects present in the image.

The prospect of forthcoming imaging surveys with unprecedented depth and area testifies to the significance of automated and efficient modules to evaluate their completeness. In this paper, we describe a software package which is improvised to estimate completeness of galaxy detection as a function of apparent magnitude and half-light radius. This paper is organized as follows. In section 2, the methodology of the software is described. Section 3 is devoted to the study of the usage and efficiency of software. Ultimately, summary will be presented in section 4.

2 Methods

We based the completeness estimation procedure on incorporation of artificial galaxies. The core of DASTWAR[1] is an IRAF script written in IRAF command language. The software generates artificial galaxies and iteratively inserts them to the source finder modules input image (cf. [29, 20, 14, 4, 6]). Next, it utilizes the source finder module to detect the inserted artificial galaxies. By comparing the extracted catalog to the catalog of artificial galaxies inserted to the input image, completeness would be estimated. Completeness is defined as the ratio of the number of extracted artificial galaxies to the number of artificial galaxies present in the image. DASTWAR performs completeness estimation as a function of apparent magnitude and half-light radius of artificial galaxies (e.g. [14, 2, 4]). Inserting artificial galaxies to the observed image preserves any observational artifacts and sky noise when quantifying the probability of detection [6].

The software makes use of IRAF artdada package for generating artificial galaxies [2]. This package has been widely used to simulate galaxies in deep images (e.g. [14, 4]). Simulated galaxies are of either early-type or late-type morphology, respectively obeying de Vauculeurs [9] and Exponential [12] surface brightness laws. The package enables the generation of artificial galaxies in a given bin of apparent magnitude and half-light radius. The software makes use of SExtractor [3] as source finder module. SExtractor is the standard detecting tool in extracting galaxies based on deep optical/NIR images. SExtractor isolates sources in the image given as input, and carries out photometric and structural measurements. Also, a catalogue of detected sources along with their photometric and structural parameters is returned at the end.

The workflow of the procedure is depicted in Fig. 1. The procedure starts with obtaining the values of the input parameters and the input image, which are provided by the user (see Table 1). Next, the software initiates generating the simulated images. A simulated image is a modified version of the input image. This modified version is constructed by inserting the artificial galaxies to the input image. The software utilizes IRAF artdata package to produce artificial galaxies according to the prescriptions of the user indicated by values of input parameters. Among the parameters which could be set by the user are the number of artificial galaxies to insert, their apparent magnitude and half-light radius tolerance, their morphological distribution and their inclination tolerance (see Table 1). It should be noted that, the software would not modify the input image before inserting artificial galaxies.

---

[1]Dastwar (pronounced Dastoor in present day persian) is the persian word for adviser.
[2]ftp://iraf.noao.edu/iraf/docs/glos210b.ps.Z

Figure 1: Workflow of the software is illustrated (see text).

Hence, the user has to decide whether to mask real sources or to leave them intact in the input image.

A two-dimensional point spread function, provided by the user, will be convolved with artificial galaxies before inserting them to the input image. Artificial galaxies will be uniformly distributed throughout a subregion of the input image defined by the user. Also their apparent magnitude and half-light radius will be uniformly distributed in the apparent magnitude and half-light radius tolerances indicated by the user. For subsequent referencing, properties of the inserted artificial galaxies, including their positions, apparent magnitudes and half-light radii will be saved in a catalog.

The process of generating the simulated images is iterative. The outcome of each iteration is a single simulated image (see Fig. 2). The total number of iterations will be set by giving value to the appropriate parameter. Willing to end up with an adequate completeness estimation necessitates balance between the number of artificial galaxies in each simulated image and the total number of iterations. When sufficient number of simulated images are generated, software moves forward to the next step.

As a result of complex observing strategies which are at work in deep imaging surveys (e.g. dithering [22]), yielded images are not generally associated with flat edges. Also, these images normally result from stacking a number of slide images on top of one another. In consequence, the edges of the obtained images are usually indented and a set of pixels in the image array are seen to have zero value. In such an image, the entire area is not covered by data. If user notifies the software of the partial data-coverage in the input image, a mask image will be created. It is an image with the same width and height as the input image, in which the partial area covered by data is indicated. This mask image will be multiplied with each of the simulated images. In this way, the analogy of the data-covered area in the input image and the simulated images is assured. If user has not warned about the partial data-coverage in the input image, software leaps to the next step.

Now, all is at hand to start source detection. For this task, DASTWAR makes use of SExtractor [3]. SExtractor will be executed on each of the simulated images based on the input parameters set by the user. Hence, for each simulated image, we will be provided by a catalog of detected sources along with their photometric and structural properties.

Amongst the measured quantities for each detected source is the pixel position of the center. These positions will be used to crossmatch the SExtractor provided catalog with the catalog of artificial galaxies inserted to each simulated image. The radius of crossmatch will be set by the user. An artificial galaxy is designated as recovered if centroid of a unique detected source falls within its circle of crossmatch. For each simulated image, recovered galaxies will be listed in a new catalogue. Each line of this catalogue represents an artificial galaxy which was successfully detected by SExtractor.

At this point, for each of the simulated images, two catalogs are at hand. The first one is the catalog of inserted artificial galaxies and the second one is the catalog of recovered artificial galaxies. By accumulating the catalogs of inserted artificial galaxies into one catalog, we end up in the master catalog of artificial galaxies. This master catalog enlists all of the artificial galaxies inserted to the set of simulated images. In the same manner master catalog of recovered artificial galaxies is constructed. The latter catalog embraces the list of all artificial galaxies which are already detected by the source finder module. Comparing these two master catalogs enable us to quantify the degree of completeness. DASTWAR estimates completeness as a function of artificial galaxies' apparent magnitude and half-light radius.

Table 1: Input parameters of DASTWAR are listed in this table. IRAF artdada package parameters are marked with star.

Parameter	Data Type	Description
nrun	int	Number of iterations
ngal	int	Number of galaxies generated in each iteration
xcormin	real	Minimum X coordinate of artificial galaxies
ycormin	real	Minimum Y coordinate of artificial galaxies
xcormax	real	Maximum X coordinate of artificial galaxies
ycormax	real	Maximum Y coordinate of artificial galaxies
magmin	real	Upper magnitude limit for artificial galaxies
magmax	real	Lower magnitude limit for artificial galaxies
minrad	real	Minimum half-light ratio of artificial galaxies
maxrad	real	Maximum half-light ratio to artificial galaxies
*efrac	real	Fraction of early-type galaxies
*axisrat	real	Minimum axis ratio for early-type galaxies
*srefrac	real	Late-type/early-type radius at a given magnitude
*abs	real	Absorption in edge-on late-type galaxies
inpimage	char	Input image name
wimage	char	Weight image name
psf	char	PSF image name
*poinoi	bool	Add Poisson noise?
*rad	real	Seeing radius/scale (pixels)
*psfar	real	Star/PSF axial ratio
*psfpa	real	Star/PSF position angle
*magzp	real	Magnitude zero point
*ccdgain	real	Gain
*ccdreadnoise	real	CCD Read noise
seconfig	char	Name of Sextractor configuration file
sennw	char	Name of Sextractor Neural network/weights file
separam	char	Name of Sextractor Parameter file
seconv	char	Name of Sextractor convolution kernel
dx_gal	int	X-range for coverage tests (artificial galaxies)
dy_gal	int	Y-range for coverage tests (artificial galaxies)
dx_sex	int	X-range for coverage tests (detected sources)
dy_sex	int	Y-range for coverage tests (detected sources)
magbin	int	Number of magnitude bins for completeness estimation
sizebin	int	Number of size bins for completeness estimation
covcheck	bool	Apply coverage checks?
crop	bool	Apply cropping?
clean	int	Delete additional files made?
pixsize	real	Image pixel scale (arcsec/pix)
maxdist_arcs	real	Cross-match radius (arcsec)
compoutput	char	Completeness matrix file name
*background	real	Default background
*nxc	int	Number of PSF centers per pixel in X
*nyc	int	Number of PSF centers per pixel in Y
*nxsub	int	Number of pixel subsamples in X
*nysub	int	Number of pixel subsamples in Y
*nxgsub	int	Number of galaxy pixel subsamples in X
*nygsub	int	Number of galaxy pixel subsamples in Y
*dyrange	real	Profile intensity dynamic range
*psfrange	real	PSF convolution dynamic range

Figure 2: Five examples of simulated images are shown in this figure. In the uppermost left mosaic, we have shown the input image and the remaining mosaics illustrate examples of simulated images. The input image is a 512×512 pixels cutout of v2.0 images of HST/ACS in southern GOODS field which is acquired in F850LP band.

Table 2: Five lines of DASTWAR's output file are given as example.

Label 1	Label 2	Magnitude	Half-light radius	Completeness	Detected	Inserted
31	10	20.95	0.9189	0.842105	32	38
31	11	21.05	0.9189	0.932203	55	59
31	12	21.15	0.9189	0.886364	39	44
31	13	21.25	0.9189	0.903846	47	52
31	14	21.35	0.9189	0.857143	48	56

As was noted earlier, galaxies will be uniformly distributed throughout a subregion of the input image that is defined by user. Occasionally, an inserted artificial galaxy would reside in a position too close to the edge of the image or edge of the data-covered area. In such instances, the light profile of the artificial galaxy may become cropped; a phenomenon which usually results in its erroneous detection. When user warns DASTWAR of the possibility of existence of such sources, software attempts to identify them. This is done by defining a rectangular mask for each artificial galaxy, width and length of which is to be determined by the user. Center of this mask will be coincided to the center of each of the inserted artificial galaxies. Inspecting values of the pixels residing inside the mask would characterize the distance between the object and the edges. When an object is identified as being too close to the edges, it will be marked with edge-grazing flag.

The software proceeds to compute completeness as a function of apparent magnitude and half-light radius. Completeness is defined as the ratio of the number of detected artificial galaxies to the number of inserted galaxies not marked with edge-grazing flag. Completeness is estimated as a function of artificial galaxies apparent magnitude and half-light radius. Accordingly, inserted and recovered artificial galaxies are enumerated in bins of apparent magnitude and half-light radius. The plane of apparent magnitude and half-light radius is divided to two dimensional bins. Number of these bins will be determined by user and completeness will be assessed specifically in each bin.

The output of the software is a text file, each line of which provides the result of completeness estimation for each of the two dimensional bins. The first two columns contain two labels which uniquely designate every two dimensional bin. In the third and fourth columns magnitude and half-light radius of the center of two dimensional bin are given respectively. Completeness value for the two dimensional bin is written in the fifth column. Finally, in sixth and seventh column, the number of detected and inserted artificial galaxies for each bin are given.

3 Example

In this section, we intend to demonstrate the usage and efficiency of the software. The inspection is based on Hubble Space Telescopes data acquired during GOODS [4] survey [14].

---

[4]Great Observatories Origins Deep Survey

The input image given to the software for completeness estimation has been selected from southern GOODS field and covers nearly 25 square arcminutes. We have used images taken in F850LP band which is the band normally utilized for detection [8]. We based our study on version v2.0 of HST/ACS data [13].

The number of iterations was set to 165 and in each iteration 1500 artificial galaxies generated. In total, 247500 artificial galaxies were used in the procedure. Artificial galaxies uniformly populate the apparent magnitude range $20 \leq m \leq 30$ (AB magnitude system measured in F850LP band) and half-light radius range $0.01 \leq r_{50} \leq 1.5$ arcseconds. The F850LP-band magnitude zero point was set to 24.862 which is obtained from this URL [6] [30]. The apparent magnitude range was divided to 100 bins while the half-light radius range was intersected to 50 bins. The average number of inserted galaxies in each two-dimensional bin is 50.

The fraction of early-type galaxies to late-type galaxies was set to one. Also, the ratio between half-flux scale radii of late-type and early-type galaxies at a given magnitude was equal to one. For early-type galaxies the axial ratio (b/a) was randomly selected in the range $0.3 \leq b/a \leq 1.0$. For late-type galaxies, inclinations range uniformly between 0 and 90 degrees. We did not apply internal absorption correction.

As was noted in §2, before being inserted to input image, surface brightness profile of the artificial galaxy is to be convolved with appropriate point spread function. For the present study, point spread function is inferred from detailed examination of surface brightness profiles of spectroscopically confirmed stars. Using the updated version (v2) of the GOODS-MUSIC catalogue [15, 28], we selected 138 objects with spectroscopic redshift quality flag < 2 and zspec = 0 as stars (cf. [5]). Moreover, we included 63 stars in GOODS-South field identified by [23] based on low resolution spectra acquired in PEARS survey [24]. Of 172 unique stars thus spotted throughout the southern GOODS field, 45 stars reside within the region covered in our input image. Surface brightness profile of 42 stars of the selected sample were used to construct point spread function. For construction of point spread function, we utilized `IRAF DAOPHOT` package.

When convolved with point spread function, artificial galaxies inserted to the input image without additional Poisson noise. As a result, for bright objects, the noise is slightly underestimated while for faint objects this shortcut does not affect the results as the background completely dominates (cf. [14]). We used the `SExtractor` configuration files which were optimized for detection based on v2.0 of HST/ACS images of southern GOODS field in F850LP band and are publicly available through this URL [7]

The crossmatch radius used to isolate the recovered artificial galaxies was 0.15 arcsec. Recalling the value of 0.03 arcsec/pix for pixel scale, 0.15 arcsec is equivalent to 5 pixels in the input image. The crossmatch radius was made conservatively small to ensure that the chance of erroneous matches to existing objects is negligible (cf. [4]). The width and length of the rectangular mask defined to identify edge-grazed galaxies were equally set to 3 pixels.

The resulting distribution of completeness in the plane of apparent magnitude and half-light radius is shown in Fig. 3. Contours of constant completeness are illustrated. It is seen that an increase in apparent magnitude in constant half light radius is associated with a decrease in completeness. The same behavior is seen when half-light radius is increased in fixed apparent magnitude. Such a trend introduces surface brightness as a key factor in detectability of galaxies.

Moreover, it should be noted that constant completeness contours tend to fainter apparent magnitudes as half-light radius becomes smaller. For instance, at $r_{50} = 1.5$ arcsec,

---

[6] http://archive.stsci.edu/pub/hlsp/goods/v2/h_goods_v2.0_rdm.html
[7] http://archive.stsci.edu/pub/hlsp/goods/catlog_r2/h_r2.0z_readme.html

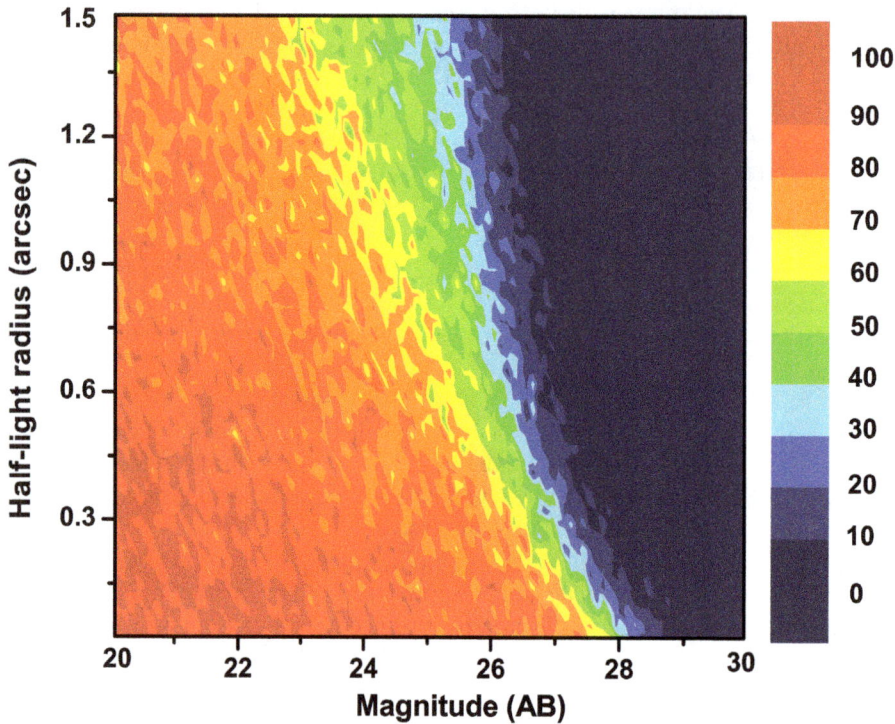

Figure 3: Contour plot representing completeness values in a plane of F850LP magnitude and half-light radius (in units of arcsec.) is shown.

the 70% completeness contour is located at $m \sim 23$ while at $r_{50} = 0.15$ arcsec the same contour is seen at $m \sim 27$. This fact also testifies to the crucial role of surface brightness in detectability of galaxies. It is also seen that the space between the adjacent contours are not uniform. In high surface brightness areas, the contours are more apart compared to low surface brightness regions. This fact reflects that, in spite of its importance, surface brightness is not the only factor that influences the detectability of galaxies.

In a similar study which dates back to 2004 [14], Giavalisco et al. assessed completeness limits based on version v0.5 of HST/ACS images acquired in southern GOODS field in F850LP band. They adopted an analogous method for estimation of completeness as a function of apparent magnitude and half-light radius (see Fig. 4 in [14]). In their study, IRAF artdata package was utilized for generating artificial galaxies and SExtractor was used as source finder module. Their artificial galaxies uniformly populated the magnitude range $20 m 28$ (AB magnitude system measured in F850LP band) and the range of galaxy half-light radius was $0.01 \leq r_{50} \leq 1.5$ arcseconds. Morphologically, half of their generated galaxies were of early-type morphology and the remaining half were of late-type morphology. Early-type galaxies had a uniform axial ratio distribution in the range $0.3 \leq b/a \leq 0.9$. Internal absorption was ignored and no additional Poisson noise was at work.

The methodological homology noticed between the two studies enables the comparison of results and interpretation of the differences in terms of differences in the inputs. Comparison between the two completeness distributions reveals the shifting of completeness contours toward fainter apparent magnitudes in our distribution. Hence, our study implies a higher completeness at a given apparent magnitude in constant half-light radius.

Such a difference can be attributed to the difference in depth of the input images used for the two studies. Our study was based on version v2.0 [13], a significant improvement upon the previous v1.0 release, which is itself an improved version of v0.5 release of GOODS reduced HST/ACS images. Version v1.0 data release provided data acquired as part of the original GOODS HST/ACS program [14]. Version 2.0 augments this with additional data acquired on the two GOODS fields during the search for high redshift Type Ia supernovae carried out during Cycles 12 and 13 (Program ID 9727, P.I. Saul Perlmutter, and 9728, 10339,10340, P.I. Adam Riess [27, 26, 25]. As a result of the additional data, the v2.0 mosaics offer roughly twice the exposure time in the F850LP band compared to version v1.0 images.

4   Summary

Within the next few years, imaging surveys with unprecedented depth and area will revolutionize our vision of the extragalactic sky. Nevertheless, any imaging survey is restricted and biased in its sampling of the galaxy population. Completeness parameter, which quantifies the probability of detection, has proved to be a useful and conventional parameter in assessing the bias in sampling of galaxies. Given the prospect of forthcoming imaging surveys, automated and efficient modules to evaluate their completeness are demanded.

Throughout this paper, we described a software package, named DASTWAR, which was improvised to estimate completeness of galaxy detection as a function of apparent magnitude and half-light radius parameters. The software generates artificial galaxies and iteratively inserts them into the input image and then utilizes source finder module to detect them. Comparing the extracted catalog with the catalog of artificial galaxies inserted to input image, yields completeness in the magnitude-size plane (see section 2).

In order to demonstrate the efficiency of the software, we utilized it for completeness estimation on the basis of version v2.0 HST/ACS data in southern GOODS field. In total, 247500 artificial galaxies were generated and used in the procedure. Distribution of completeness values in the magnitude-size plane shows that an increase in apparent magnitude in constant half-light radius is associated with a decrease in completeness and the same trend is noticed when half-light radius is increased in fixed apparent magnitude. We interpreted the mentioned trend as an evidence of pivotal role of surface brightness in determining detectability of galaxies.

We also compared the resulted completeness distribution with the corresponding distribution given by [14]. Comparison between the two completeness distributions revealed the shifting of completeness contours toward fainter apparent magnitudes in our distribution. Such a difference was expected, given the substantially higher depth of v2.0 compared to v0.5 HST/ACS data in southern GOODS field. We envisage that the improvised software would be effective in estimating completeness and helpful in quantifying the biases in sampling of the galaxy population.

Acknowledgment

The cornerstone of the software was laid down when A. K. had been collaborating with School of Astronomy of Institute for Studies in Basic Sciences (IPM), Tehran, Iran. He would like to express his gratitude to Reza Mansouri, Bahram Mobasher, Habib Khosroshahi, Saeed Tavasoli and administrators of the school for their financial and scientific support.

References

[1] Beckwith, S. V. W., Stiavelli, M., Koekemoer, A. M., et al. 2006, AJ, 132, 1729

[2] Bouwens, R. J., Illingworth, G. D., Blakeslee, J. P., Broadhurst, T. J., & Franx, M. 2004, ApJL, 611, L1

[3] Bertin, E., & Arnouts, S. 1996, A&AS, 117, 393

[4] Cameron, E., & Driver, S. P. 2007, MNRAS, 377, 523

[5] Cameron, E., Carollo, C. M., Oesch, P. A., et al. 2011, ApJ, 743, 146

[6] Chiu, I., Desai, S., & Liu, J. 2016, Astronomy and Computing, 16, 79

[7] Chiu, I., Dietrich, J. P., Mohr, J., et al. 2016, MNRAS, 457, 3050

[8] Dahlen, T., Mobasher, B., Dickinson, M., et al. 2010, ApJ, 724, 425

[9] de Vaucouleurs, G. 1948, Annales d'Astrophysique, 11, 247

[10] Disney, M. J. 1976, Nature, 263, 573

[11] Disney, M.; Phillipps, S., 1983, MNRAS, 205, 1253

[12] Freeman, K. C. 1970, ApJ, 160, 811

[13] Giavalisco, M., and the GOODS Team, 2008, in preparation

[14] Giavalisco, M., Ferguson, H. C., Koekemoer, A. M., et al. 2004, ApJL, 600, L93

[15] Grazian, A., Fontana, A., de Santis, C., et al. 2006, A&A, 449, 951

[16] Grogin, N. A., Kocevski, D. D., Faber, S. M., et al. 2011, ApJS, 197, 35

[17] Impey, C., & Bothun, G. 1997, ARA&A, 35, 267

[18] Ivezic, Z. et al., 2008, arXiv:0805.2366

[19] Koekemoer, A. M., Aussel, H., Calzetti, D., et al. 2007, ApJS, 172, 196

[20] La Barbera, F., de Carvalho, R. R., Kohl-Moreira, J. L., et al. 2008, PASP, 120, 681

[21] Laureijs, R., Amiaux, J., Arduini, S., et al. 2011, arXiv:1110.3193

[22] Mutchler, M., Published online at http://www.stsci.edu/institute/conference/cal10/ proceedings, p.69

[23] Pirzkal, N., Burgasser, A. J., Malhotra, S., et al. 2009, ApJ, 695, 1591

[24] Pirzkal, N., Xu, C., Malhotra, S., et al. 2004, ApJS, 154, 501

[25] Riess, A. G., Strolger, L.-G., Casertano, S., et al. 2007, ApJ, 659, 98

[26] Riess, A. G., Li, W., Stetson, P. B., et al. 2005, ApJ, 627, 579

[27] Riess, A. G., Strolger, L.-G., Tonry, J., et al. 2004, ApJ, 607, 665

[28] Santini, P., Fontana, A., Grazian, A., et al. 2009, A&A, 504, 751

[29] Stetson, P. B. 1987, PASP, 99, 191

[30] Windhorst, R. A., Cohen, S. H., Hathi, N. P., et al. 2011, ApJS, 193, 27

[31] Zenteno, A., Song, J., Desai, S., et al. 2011, ApJ, 734, 3

12

Spherically Symmetric Solutions in a New Braneworld Massive Gravity Theory

Amir Asaiyan[1] · Kourosh Nozari[2]

[1] Department of Physics, Faculty of Basic Sciences, University of Mazandaran, P.O.Box 47416-95447, Babolsar, Iran ;
email: a.asaiyan@umz.ac.ir (Corresponding Author)

[2] email: knozari@umz.ac.ir

Abstract. In this paper, a combination of the braneworld scenario and covariant de Rham-Gabadadze-Tolley (dRGT) massive Gravity theory is proposed. In this setup, the five-dimensional bulk graviton is considered to be massive. The five dimensional nonlinear ghost-free massive gravity theory affects the 3-brane dynamics and the gravitational potential on the brane. Following the solutions with spherical symmetry on the brane, the full field equations together with the generalized Israel-Darmois junction conditions on the brane and their weak field limits are presented in details. Generally, the theory has four Stückelberg fields along with the components of physical metric. Although analytical solutions of these equations are impossible in general, by considering some simplifying assumptions, two classes of four-dimensional spherically symmetric solutions on the brane with different background Stückelberg fields are obtained.

Keywords: Braneworld Gravity, Massive Gravity, Black Holes

1 Introduction

The accelerating expansion of the Universe has forced us to challenge with our understanding of the fundamental physics [1, 2, 3]. In the last two decades, there has been considerable interest in theories of gravitation that modify the Einstein's gravity at very large distance scales. These theories could explain the present day acceleration, without including a cosmological constant or an exotic matter content. Adding one or even many extra spatial dimensions to the 4D Einstein's theory of gravity may lead to the interesting phenomenological results. The Braneworld model is an extra dimensional theory, in which our universe is a 3-brane embedded in a five-dimensional spacetime called the bulk [4, 5]. All matter fields reside on the brane, but gravitons can travel into the extra dimension. The Dvali-Gabadadze-Porrati (DGP) model [6] is an interesting braneworld model in which the bulk is empty, the extra dimension is infinitely large. Also a 4D Einstein-Hilbert term in the braneworld action exists. The model has attractive results from cosmological viewpoint because gravity on the brane is weakened and becomes five-dimensional at large scales, $r \gg r_c$ (where r_c is the DGP crossover distance), while on small scales, gravity is effectively bounded to the brane and 4D dynamics is regained. It contains a self-accelerating branch of the solutions which can explain late time cosmic speed up [7, 8, 9]. From the 4D perspective, gravity on the brane is mediated by an infinite number of Kaluza-Klein (KK) modes that have not discontinuities. The 4D Einstein-Hilbert term on the brane will suppress the wave functions of heavier KK modes, so that they do not participate in the gravitational interactions on the brane at observable distances [10]. The 4D gravity on the brane is mediated by a massless zero mode, whereas the couplings of the heavy KK modes to ordinary matter are suppressed.

Due to the cosmological constant problem, we should look for a technically natural way of describing cosmic acceleration. The massive gravity theories are other kinds of modified gravity theories, in which a small graviton mass may lead to an IR modification of gravity with an accelerated expansion without a small cosmological constant. The recent experiments GW150914 and GW151226 [11, 12] by LIGO, were able to detect the gravitational waves and put an upper limit on the graviton mass, i.e. $m < 1.2 \times 10^{-22}$ eV [13]. At the linearized level, the Fierz-Pauli (FP) graviton mass term is the only Lorentz-invariant mass term which after quantization does not generate ghosts in flat space [14]. However, choosing a Fierz-Pauli mass term for the graviton will lead to the well known vDVZ discontinuity [15, 16]. The coupling of the longitudinal polarization of the massive graviton to trace of the energy-momentum tensor in the limit of zero graviton mass is responsible for this discontinuity, such that the tensor structure of the gravitational interaction deviates from that of Einstein gravity. To restore the continuity of Fierz-Pauli massive gravity theory at graviton mass $m = 0$, two different approaches have been proposed. The first one, which was first pointed out by Vainshtein [17, 18], is to consider nonlinear effects. The other way is to consider a curved maximally symmetric spacetime (dS or AdS) with $\frac{m}{H} \to 0$ [19, 20]. In 1972, Vainshtein noted that there is a radius r_V, known as Vainshtein radius, around a massive source, inside of it the linear approximation breaks down and at massless limit r_V goes to infinity [17]. Therefore, the nonlinear terms are important in the limit $m \to 0$. However, Boulware and Deser argued that the non-linear terms cause a scalar field with wrong sign kinetic term, known as Boulware-Deser (BD) ghost [21]. At the classical level, this scalar may not be a problem due to non-linear effects [17, 18], but at the quantum level the theory becomes strongly coupled [22] at energy scale $\Lambda_5 \equiv (m^4 M_P)^{1/5}$. By adding higher order operators, this scale can be raised to order $\Lambda_3 \equiv (m^2 M_P)^{1/3}$.

In 2010, de Rham and Gabadadze studied generic extensions of the Fierz-Pauli Lagrangian by higher-order interactions of the massive spin-2 fluctuation $h_{\mu\nu}$ [23]. Their analysis went to quintic order in the longitudinal component of $h_{\mu\nu}$ and demonstrated that its interactions could in fact be made ghost-free in a decoupling limit. The decoupling limit analysis relies heavily on the aforementioned Goldstone boson analogously suggested by Arkani-Hamed, Georgi and Schwartz [22]. de Rham, Gabadadze and Tolley (dRGT) [24] completed their investigations by presenting a nonlinear theory of massive gravity whose decoupling limit is ghost-free for all nonlinear self-interactions of the longitudinal component [24, 25, 26, 27]. The dRGT theory is the unique ghost-free theory for massive graviton and new kinetic interactions are not consistent [28, 29]. See [30, 31, 32] for recent reviews on all aspects of massive gravity and bimetric theories. In the context of the dRGT non-linear covariant massive gravity model [23, 24], some self-accelerating solutions have been discovered [33, 34, 35, 36, 37, 38]. Dynamics of the scalar mode of a massive graviton in four-dimensions has been studied in detail in [36], showing that a non-trivial configuration for this field leads to self-acceleration. Scalar fluctuations around these self-accelerating configurations are proved to be free of ghosts.

It is worthwhile to note that one way in which a massive graviton naturally arises is higher dimensional scenarios. A theory of gravity with compactified extra dimensions can be viewed as a four dimensional theory of multiple gravitons, i.e. KK modes. An alternative to the KK paradigm was the ADD model [39, 40] in which one (or more) extra dimension could emerge from a theory of a finite number of massive gauge fields or gravitons living in four dimensions. Their idea, named "Dimensional Deconstruction", can be viewed as taking a five dimensional gauge or gravity theory and discretizing the extra dimension(s).

It has been shown that Dimensional Deconstruction is equivalent to a truncation of the KK tower at the nonlinear level [28]. It has been shown that the DGP model is closely related to massive gravity. In this model, the 4D graviton propagator on the brane in the Gaussian normal coordinates is similar to the propagator for 4D massive gravity with graviton mass $m^2 = (\frac{1}{r_c})\sqrt{-\Box}$, where $r_c \equiv (M_p^2/2M_5^3)$ is the DGP crossover length scale and $\Box$ is the four-dimensional d'Alembertian. In other words, the graviton acquires a soft mass, or resonance effectively, in the DGP model. The induced gravity term in the brane action acts as a kinetic term for a 4D graviton while the bulk Einstein-Hilbert term acts as a gauge invariant mass term. Therefore, the vDVZ discontinuity problem is also present in the DGP model. Here, the massless limit converts to the limit $r_c \to \infty$. As argued by Vainshtein, at distances smaller than the radius r_V, the linearization breaks down and by considering non-linear effects, we can restore the predictions of GR on the brane [17, 18, 41, 42]. However, the DGP model has some consistency problems. The normal branch of the DGP theory is free of ghosts and instabilities, but the self-accelerating branch is completely unstable [43, 44, 45]. The DGP model has strong interactions at energy scale $\Lambda \sim (M_p/r_c^2)^{1/3}$. From the 4D point of view, there is an extra scalar degree of freedom π that contributes to the extrinsic curvature of the brane as $K_{\mu\nu} \propto \partial_\mu\partial_\nu\pi$ [43, 44]. Indeed, this scalar is a brane bending mode that interacts strongly at momenta of order Λ. In the decoupling limit of the DGP model, in which Λ is kept fixed, only the π sector exists and all other degrees of freedom decouple. This limit reduces to the cubic Galileon for the helicity-0 mode π [46].

The works done by Gabadadze and de Rham before proposing the interesting dRGT theory have shown that the introduction of the spurious extra dimension provides a geometrical interpretation of massive gravity, for which non-linearities can be tracked down explicitly [47, 48]. By studying massive gravity from extra dimensional point of view, we can better understand certain aspects of the dRGT theory [23, 24] and its bigravity [49] and multi-gravity [50] extensions. In 2009, Gabadadze considered an extension of GR by an auxiliary non-dynamical extra dimension and showed that the obtained gravitational equations could have a self-accelerated solution, which is due to a new mass parameter m. The auxiliary dimension gives an extrinsic curvature to the 4D space-time and the extrinsic curvature is responsible for creating the mass term. The special structure $[K]^2 - [K^2]$ arose from the Gauss equation for the bulk Ricci scalar ensures the Fierz-Pauli structure which is ghost-free at the linearized level [47]. de Rham and Gabadadze [48, 51] verified that the theory in the decoupling limit is free of the Boulware-Deser ghost to cubic order. In ref [28], it was shown that the ghost-free models of massive gravity and their multi-graviton extensions can follow from considering higher dimensional extension of GR in the Einstein-Cartan form on a discrete extra dimension. Indeed, discretizing the extra dimension in the vielbein language can automatically generate the square root structure characteristic of the dRGT model, i.e. $\mathcal{K}_\mu^\nu$, [28]. Indeed, the expression for the discretized extrinsic curvature coincides with $\mathcal{K}_\mu^\nu$.

By considering the above arguments, now giving a mass to the graviton in Higher-dimensional theories and exploring the overall effects of massive gravity and extra dimension could be interesting from theoretical and phenomenological viewpoints. The final results may have some relations with the multi-metric theories and then lead to physically interesting predictions. In 2004, Chacko et al., considered a braneworld setup in warped anti-de Sitter spacetime (Randall-Sundrum (RS) two-brane model [52]) with a mass term for the graviton on the infrared brane [53]. The predictions of this theory coincide with the results of GR at distances smaller than the infrared scale but at longer distances a theory of massive gravity exists. However, in the low energy limit of the theory, there is a ghost, which corresponds to the radion field. In Ref. [54], both of the bulk and the brane mass terms were introduced in the action of the RS two-brane model to quadratic order to modify the profile of the graviton zero-mode in the extra dimensions. It was found that for a particular choice of parameters, there is an IR-peaked zero-mode, i.e. the graviton can be localized on the IR brane. In 2014, a braneworld scenario has been investigated in which the infinite-volume bulk graviton was massive [55]. The bulk graviton can be as heavy as the bulk Planck scale which is much larger than the inverse Hubble size. The 4D induced gravity term on the brane shields the brane matter from both strong bulk gravity and large bulk graviton mass. Higher-dimensional gravity at large distances are not obtained on the brane in this setup and at distances above the bulk Planck length scale, the 4D graviton on the brane acquires a small mass. The author of [55] considered a mass potential that arose via the gravitational Higgs mechanism, such that a general quadratic potential in terms of perturbation tensor h_{AB} was introduced in the bulk action. In this extension of the DGP model, even for the case of ghost-free Fierz-Pauli bulk mass term, the 4D tensor structure on a 3-brane could be obtained [55]. Here, the key point is that the trace $h \equiv \eta^{AB} h_{AB}$ is perturbatively a ghost. However, it was shown that the non-perturbative Hamiltonian is bounded from below and there is no ghost in full nonlinear theory [56, 57, 58].

With these detailed preliminaries which are necessary for a reader to understand forthcoming arguments in this paper, we consider a combination of the DGP braneworld and dRGT massive gravity models, by introducing a five dimensional nonlinear ghost-free potential in the bulk action. In this setup, our universe is a 3-brane embedded in a 5D bulk where the extra spatial dimension is large. A 5D ghost-free massive gravity theory propagates nine degrees of freedom (DOF) and the extra four DOFs added to the five DOFs of 5D massless graviton, which is effectively equivalent to a 4D softly massive graviton, are the extra polarizations of the 5D massive graviton. We considered the induced gravity term on the brane action, because this term in the DGP setup acts as a kinetic term for the 4D graviton. The 5D extension of dRGT theory is free of ghosts and we want to explore the effects of this nonlinear theory on the brane dynamics and the effective 4D gravitational interactions on the brane. For this purpose, the full 5D field equations and their weak field limits have been studied. Our focus is on the solutions with spherical symmetry on the brane. The full nonlinear equations of motion in the presence of the unknown stückelberg fields are generally very complicated to solve for analytical solutions, unless we consider some simplifying assumptions. So, to have some intuition and to be more clarified, we have adopted step by step some reasonable and simplifying assumptions to find a class of four-dimensional spherically symmetric solutions on the brane. We considered two simplified linear theories in both unitary and non-unitary gauges and found in both cases a flat solution on the brane with different background Stückelberg fields. In non-unitary gauge we restricted ourself to special choices of the free parameters of the theory. We note that general massive braneworld solutions, resulting from the full nonlinear theory, should reduce to the massless braneworld solution in the limit of zero bulk graviton mass as has been studied in [59]. We are attempting to follow new approaches, such as solving the nonlinear field equations numerically or finding the effective 4D field equations on the brane [60, 61], to examine the Vainshtein mechanism in our model.

2   Braneworld Massive Gravity

In braneworld scenarios, we assume that our (1+3)-dimensional spacetime is a domain wall embedded in a five-dimensional spacetime called the bulk [4, 5]. All matter fields live on the brane but only gravitons can travel into the bulk. In the DGP braneworld model, the bulk is empty, the extra dimension is infinitely large and a 4D Einstein-Hilbert term exists on the brane action [6]. In our braneworld massive gravity model, we introduce a mass potential to the bulk action, which is a 5D extension of the dRGT's 4D nonlinear ghost-free massive gravity theory [23, 24]. We consider a 3-brane Σ embedded in the five-dimensional massive bulk $\mathcal{M}$. The total action is

$$S = \frac{M_5^3}{2} \int_{\mathcal{M}} d^5X \sqrt{-g} \left({}^{(5)}R + m_g^2 \mathcal{U}(g, \mathcal{K}) \right) + S_{brane}, \tag{1}$$

where S_{brane} is the 3-brane action defined as

$$S_{brane} = \frac{M_p^2}{2} \int_{\Sigma} d^4x \sqrt{-q} \, {}^{(4)}R + \int_{\Sigma} d^4x \sqrt{-q} \, \mathcal{L}_4^{matt} + \int_{\Sigma} d^4x \sqrt{-q} \, \frac{K}{\kappa_5^2}. \tag{2}$$

g_{AB} is the 5D bulk metric with corresponding Ricci tensor given by ${}^{(5)}R_{AB}$. X^A, $A = 0, 1, 2, 3, 5$ are the coordinates in the bulk. The brane has induced metric $q_{\mu\nu}$ with corresponding Ricci tensor ${}^{(4)}R_{\mu\nu}$. $\mathcal{L}_4^{matt}$ is the matter Lagrangian localized on the brane. We note also that the bulk Planck mass M_5 and the 4-dimensional Planck scale M_p are defined as $\kappa_5^2 = 8\pi G_{(5)} = M_5^{-3}$ and $\kappa_4^2 = 8\pi G_{(4)} = M_p^{-2}$. $\mathcal{U}$ is a dimensionless "potential" for the metric g_{AB} that makes bulk graviton massive, where the dimension-full parameter m_g sets the graviton mass scale. This potential depends on three dimensionless arbitrary parameters α_3, α_4 and α_5 and is composed of four parts,

$$\mathcal{U}(g, \mathcal{K}) = \sum_{n=2}^{5} \alpha_n \mathcal{U}_n(\mathcal{K}) = \mathcal{U}_2 + \alpha_3 \mathcal{U}_3 + \alpha_4 \mathcal{U}_4 + \alpha_5 \mathcal{U}_5, \tag{3}$$

where $\alpha_2 = 1$. The tensor $\mathcal{K}_A^B$ is

$$\mathcal{K}_A^B = \delta_A^B - \sqrt{g^{BC}(g_{CA} - H_{CA})} = \delta_A^B - \sqrt{g^{BC} f_{ab} \partial_C \phi^a \partial_A \phi^b}. \tag{4}$$

The potential (3) is unique and no further polynomial terms can be added to the action without introducing the BD ghost [23, 24, 25, 26, 27]. The sum is finite and stops at $n = 5$, since the total derivative combinations vanish for $n > D = 5$ [24, 31]. It was shown that this is the most general potential for a ghost-free theory of massive gravity [62]. f_{ab} is the fiducial (or reference) metric, which we assume to be the Minkowski metric, $\eta_{\mu\nu}$, and ϕ^a are the Stückelberg scalar fields introduced to give a manifestly diffeomorphism invariant description [22]. Under a diffeomorphism $\delta X^A = \xi^A(X)$, the Stückelberg fields ϕ^0, ϕ^i ($i = 1, 2, 3, 5$) transform as simple scalars. The tensor h_{AB} represents the fluctuations of bulk metric about Minkowski reference metric, $h_{AB} = g_{AB} - \eta_{AB}$, and H_{AB} corresponds to the covariantization of metric perturbations, defined as $H_{AB} = g_{AB} - \partial_A \phi^a \partial_B \phi^b \eta_{ab}$. The square root is formally understood as $\sqrt{W}_C^A \sqrt{W}_B^C = W_B^A$. The four polynomial terms $\mathcal{U}_2$, $\mathcal{U}_3$, $\mathcal{U}_4$, and $\mathcal{U}_5$ depend on the metric g and Stückelberg fields ϕ^a as

$$\mathcal{U}_2 = [\mathcal{K}]^2 - [\mathcal{K}^2], \tag{5}$$

$$\mathcal{U}_3 = \frac{1}{3}[\mathcal{K}]^3 - [\mathcal{K}^2][\mathcal{K}] + \frac{2}{3}[\mathcal{K}^3], \tag{6}$$

$$\mathcal{U}_4 = \frac{1}{12}[\mathcal{K}]^4 - \frac{1}{2}[\mathcal{K}^2][\mathcal{K}]^2 + \frac{2}{3}[\mathcal{K}^3][\mathcal{K}] + \frac{1}{4}[\mathcal{K}^2]^2 - \frac{1}{2}[\mathcal{K}^4], \tag{7}$$

$$\mathcal{U}_5 = \frac{1}{60}[\mathcal{K}]^5 - \frac{1}{3}[\mathcal{K}^3][\mathcal{K}^2] + \frac{1}{3}[\mathcal{K}^3][\mathcal{K}]^2 - \frac{1}{6}[\mathcal{K}^2][\mathcal{K}]^3 - \frac{1}{2}[\mathcal{K}][\mathcal{K}^4] + \frac{1}{4}[\mathcal{K}][\mathcal{K}^2]^2 + \frac{2}{5}[\mathcal{K}^5], \quad (8)$$

where the square brackets are defined as

$$[\mathcal{K}] \equiv \mathrm{tr}\mathcal{K}_A^B, \quad [\mathcal{K}]^2 \equiv \left(\mathrm{tr}\mathcal{K}_A^B\right)^2, \quad [\mathcal{K}^2] \equiv \mathrm{tr}\mathcal{K}_C^B\mathcal{K}_A^C. \quad (9)$$

We chose a coordinate y for the extra dimension so that our 3-brane is localized at $y = 0$. Variation of the action (1) with respect to the bulk metric leads to the modified 5D field equations in the bulk as [56, 57, 58]

$$^{(5)}G_{AB} + m_g^2 \overline{X}_{AB} = \kappa_5^2 \, {}^{(loc)}T_{AB}\delta(y), \quad (10)$$

where $\overline{X}_{AB}$ is the effective energy-momentum tensor due to the graviton mass and expressed as

$$\overline{X}_{AB} = X_{AB} + \sigma Y_{AB}, \quad (11)$$

with

$$X_{AB} = -\frac{1}{2}(\alpha \mathcal{U}_2 + \beta \mathcal{U}_3)g_{AB} + \tilde{X}_{AB}, \quad (12)$$

$$\tilde{X}_{AB} = \mathcal{K}_{AB} - [\mathcal{K}]g_{AB} - \alpha\left\{\mathcal{K}_{AB}^2 - [\mathcal{K}]\mathcal{K}_{AB}\right\} + \beta\left\{\mathcal{K}_{AB}^3 - [\mathcal{K}]\mathcal{K}_{AB}^2 + \frac{\mathcal{U}_2}{2}\mathcal{K}_{AB}\right\}, \quad (13)$$

$$Y_{AB} = -\frac{\mathcal{U}_4}{2}g_{AB} + \tilde{Y}_{AB}, \quad (14)$$

$$\tilde{Y}_{AB} = \frac{\mathcal{U}_3}{2}\mathcal{K}_{AB} - \frac{\mathcal{U}_2}{2}\mathcal{K}_{AB}^2 + [\mathcal{K}]\mathcal{K}_{AB}^3 - \mathcal{K}_{AB}^4. \quad (15)$$

The new parameters α, β, and σ are defined as $\alpha = 1 + \alpha_3$, $\beta = \alpha_3 + \alpha_4$, $\sigma = \alpha_4 + \alpha_5$, and the indices are raised and lowered by the "physical" metric g_{AB}, so that $\mathcal{K}_{AB} = g_{AC}\mathcal{K}_B^C$, $\mathcal{K}_{AB}^2 = g_{AD}\mathcal{K}_C^D\mathcal{K}_B^C$, etc.

The effective localized energy-momentum tensor on the brane including the contribution from the induced 4D Einstein-Hilbert term on the brane is

$$^{(loc)}T_{AB} = g_A^\mu g_B^\nu(-\frac{1}{\kappa_4^2})\sqrt{\frac{-q}{-g}}\left(^{(4)}G_{\mu\nu} - \kappa_4^2 \, {}^{(4)}T_{\mu\nu}\right). \quad (16)$$

where $^{(5)}G_{AB}$ and $^{(4)}G_{AB}$ denote the Einstein tensors constructed from the bulk and the brane metrics respectively. The tensor $q_{AB} = g_{AB} - n_A n_B$ is the induced metric on the brane Σ with n_A the normal vector on this hypersurface. The field equations in the bulk ($y \neq 0$) take the following form

$$^{(5)}G_{AB} = {}^{(5)}R_{AB} - \frac{1}{2}\,{}^{(5)}R\,g_{AB} = -m_g^2 \tilde{X}_{AB}. \quad (17)$$

Moreover, if the components of $\tilde{X}_{AB}$ be continuous across $y = 0$, the following modified (due to the presence of induced gravity on the brane) Israel-Darmois junction conditions, as a boundary condition for the field equations in the bulk, would be obtained

$$[K_\mu^\nu] - \delta_\mu^\nu[K] = -\kappa_5^2 \, {}^{(loc)}T_\mu^\nu = \left(\frac{\kappa_5^2}{\kappa_4^2}\right)\,{}^{(4)}G_\mu^\nu - \kappa_5^2 \, {}^{(4)}T_\mu^\nu, \quad (18)$$

where $K_{\mu\nu} = \frac{1}{2}\partial_y(g_{\mu\nu})$ is the extrinsic curvature of the brane and brackets denote jump across the brane $(y = 0)$. We assume a $\mathbf{Z}_2$-symmetry on reflection around the brane, thus the Israel-Darmois junction conditions become

$$\overline{K}_\mu^{\ \nu} - \overline{K}\delta_\mu^{\ \nu} = r_c\,^{(4)}G_\mu^{\ \nu} - \frac{\kappa_5^2}{2}\,^{(4)}T_\mu^{\ \nu}, \tag{19}$$

where $r_c = \frac{\kappa_5^2}{2\kappa_4^2} = \frac{M_P^2}{2M_5^3}$ is the well-known DGP crossover distance, and by definition $\overline{K}_\mu^{\ \nu} = K_\mu^{\ \nu}(y = 0^+) = -K_\mu^{\ \nu}(y = 0^-)$.

After presentation of general field equations in the proposed setup, now we seek for some spherically symmetric solutions on the brane.

3   Spherically Symmetric Solutions

Here, we consider the static spherically symmetric configurations on the brane and our concentration is on the issue of braneworld black holes, i.e. finding the bulk and the brane metric when a spherically symmetric energy-momentum distribution is localized on the brane. In our previous work [59], black hole solutions in warped DGP braneworld model with a cosmological constant term in the bulk were obtained (see [63, 64, 65, 66] for further black hole solutions in braneworld scenarios). We found a 5D black string solution for the bulk metric, which reduces to 4D Schwarzschild-AdS solution on the brane. The 4D AdS curvature radius is proportional to r_c, therefore the Schwarzschild solution is recovered on the brane in the limit $r_c \to \infty$ [59]. As we already noted, the DGP model is closely related to massive gravity and the 4D graviton propagator on the brane is similar to the propagator for 4D massive graviton. In the dRGT theory with a Minkowski reference metric, a class of non-bidiagonal Schwarzschild-dS solutions was found in [33, 34]. In this theory, for a special choice of free parameters of the action, the Schwarzschild-dS type of black hole solutions was obtained in ref [67, 35], where the mass term behaves similar to the cosmological constant term in GR. For this choice of parameters, the Bianchi identity is automatically satisfied for a certain diagonal and time-independent metrics in spherical polar coordinates, whereas the kinetic terms for both the vector and scalar fluctuations vanish in the decoupling limit. Although it was shown that the linearized solutions of GR can be reproduced below the Vainshtein radius in a certain region of parameter space, the metric here is accompanied by nontrivial backgrounds for the Stückelberg fields. The vector and scalar modes A^μ and π of massive gravitons are the nonunitary parts of the background Stückelberg fields [35], i.e. $x^\mu - \phi^\mu = (m\,A^\mu + \partial^\mu\pi)/\Lambda^3$. For reviewing the black hole solutions in massive gravity see refs. [68, 69, 70, 71].

All of these papers have focused only on the four-dimensional dRGT theory [23, 24], in which only the usual graviton terms, $\mathcal{U}_i$ $(i = 2 - 4)$, are considered. For spherically symmetric solutions in extra dimensional setups, some types of black hole solutions for dRGT massive gravity with their thermodynamical properties have been investigated in d-dimensional spacetimes $(d \geq 3)$ in refs. [72, 73, 74, 75, 76, 62]. The behavior of massive graviton terms for some cosmological solutions such as the FLRW, Bianchi type I, and also Schwarzschild-Tangherlini-(A)dS metrics in a specific five-dimensional nonlinear massive gravity and bigravity models have been clarified in Refs. [62, 77]. In ref. [78], it was argued that giving a space-dependent mass to the 5D graviton, which depends on the extra-dimensional coordinate, can localize Einstein gravity on a 3-brane embedded in a 5D

Minkowski space. They focused on the quadratic Fierz-Pauli Lagrangian for 5D metric perturbations and explored the linearized equations of motion for 4D scalar, vector and tensor modes. They showed that there is no ghost on the brane and conserved matter on the brane does not couple to the scalar massless mode. The nonlinear extension of the theory has not been studied yet.

We want to find a 4D spherically symmetric solution for our nonlinear massive braneworld setup and separately determine the effects of bulk graviton mass term and also the large extra dimension on the gravitational interactions on the brane. We expect that the predictions of GR and DGP model be reproduced in appropriate limits, i.e. $m \to 0$ limit for recovering the DGP results and $r_c \to \infty$ limit in addition to the previous one for recovering GR on the brane. The issues of the vDVZ discontinuity and the Vainshtein mechanism to resolve it should be carefully studied. The effects of bulk nonlinear terms and the brane bending modes play important roles in these limits. To obtain black hole solutions in a braneworld scenario, generally there are two different approaches. In the first approach, as we explained in last section, dynamics and geometry of the whole bulk spacetime are primarily considered; then the dynamics on the brane is extracted by using the Israel-Darmois matching conditions. The second approach is to obtain the effective four-dimensional field equations on the brane firstly and then try to extend these solutions into the bulk [60, 61]. Here, we will follow the first approach. Therefore, to choose an appropriate 5D line element which is spherically symmetric on the brane, we review the 4D black hole solutions of the original dRGT theory. In this case, the ansatz for the static spherically symmetric solutions is the same as in GR. The only subtlety consists in getting the correct configuration for the four scalar fields. Regarding the vacuum solution of the theory, ($\phi^a = x^\mu \, \delta_\mu^a$ and $g_{\mu\nu} = \eta_{\mu\nu}$), the spherically symmetric line element and the four scalar fields for 4D massive gravity models can be written as follows

$$ds^2 = -\alpha(r)dt^2 + 2\delta(r)dtdr + \beta(r)dr^2 + \chi(r)\Big(d\theta^2 + \sin^2(\theta)d\varphi^2\Big), \tag{20}$$

$$\phi^0 = t + h(r), \quad \phi^i = \phi(r)\frac{x^i}{r}. \tag{21}$$

In the unitary gauge, the scalar fields are $\phi^a = x^a = (t, r\sin\theta\cos\phi, r\sin\theta\sin\phi, r\cos\theta)$. Therefore, in this gauge $h(r) = 0$ and $\phi(r) = r$. The field configuration is invariant under two residual coordinate transformations. The first one is an arbitrary change of the radial coordinate $r \to \tilde{r} = \tilde{r}(r)$, which allows to set either $\chi(r) = r^2$ or $\phi(r) = r$. The second one is the redefinition of the time variable $t \to \tilde{t} = t + \tau(r)$, which allows to cancel either $\delta(r)$ or $h(r)$. In our five dimensional braneworld theory, we can choose a coordinate system in which the brane is located at $y = 0$ and the 5D metric with spherical symmetry on the brane are as follows

$$ds_5^2 = -e^{\nu(r,y)}dt^2 + e^{\lambda(r,y)}dr^2 + r^2 e^{\mu(r,y)}d\Omega^2 + dy^2, \tag{22}$$

where the 5D Stückelberg fields are

$$\phi^0 = t, \quad \phi^i = \phi(r)\frac{x^i}{r}, \quad \phi^5 = y. \tag{23}$$

As compared to ordinary Braneworld theories, this configuration contains an additional radial function $\phi(r)$, which should be determined. The matter content of the 3-brane universe is considered to be a localized spherically symmetric untilted perfect fluid (e.g. a star) with

$$^{(4)}T_{\mu\nu} = (\rho + p)u_\mu u_\nu + pq_{\mu\nu}, \tag{24}$$

where u^μ stands for the 4-velocity of the fluid and $\rho = p = 0$ for $r > R$. Nevertheless, since we want to obtain static black hole solutions outside the star (that is, for $r > R$), in these regions the brane is empty. With the ansatz (22) and (23), the components of $\mathcal{K}_A^B$ would take the following form

$$\mathcal{K}_A^B = \text{diag}\left(1 - (e^{-\frac{\nu}{2}}), 1 - (\phi' e^{-\frac{\lambda}{2}}), 1 - (\frac{\phi}{r} e^{-\frac{\lambda}{2}}), 1 - (\frac{\phi}{r} e^{-\frac{\lambda}{2}}), 0\right). \tag{25}$$

By using these components, we can obtain the total derivative combinations $\mathcal{U}_2$, $\mathcal{U}_3$, $\mathcal{U}_4$ and $\mathcal{U}_5$. We have found that the term $\mathcal{U}_5$ vanishes for this configuration. Consequently, the components of $\overline{X}_{AB}$ can be obtained analytically although their expressions are so lengthy. The Einstein tensor components are nonlinear and second order in terms of ν, λ, μ and their partial derivatives. To find some analytical solutions, firstly we consider the *weak-field regime* (i.e. far enough from the source localized on the brane). In this respect, we will find solutions in the regimes where $|\nu|$, $|\lambda|$ and $|\mu|$ are small quantities compared to unity; that is, $|\nu|, |\lambda|, |\mu| \ll 1$. By adopting this assumption, we linearize our field equations with respect to these functions. Now, by putting the metric (22) into the bulk field equations (17) and keeping only the leading-order terms, we obtain the (tt), (rr), $(\theta\theta)$, (yy), and (ry) components of the bulk field equations respectively as follows:

$$2(\mu - \lambda) + 2r^2 \mu_{rr} + 6r\mu_r - 2r\lambda_r + r^2(\lambda_{yy} + 2\mu_{yy})$$

$$+2m_g^2 r^2 \left\{ \left[3 + 3\alpha + \beta - (1 + 2\alpha + \beta)(\phi' + 2\frac{\phi}{r}) + (\alpha + \beta)(\frac{\phi^2}{r^2} + 2\frac{\phi\phi'}{r}) - \beta\frac{\phi'\phi^2}{r^2} \right](1 + \nu) \right.$$

$$\left. + \left[(1 + 2\alpha + \beta)\frac{\phi'}{2} - (\alpha + \beta)\frac{\phi\phi'}{r} + \beta\frac{\phi'\phi^2}{2r^2} \right]\lambda + \left[(1 + 2\alpha + \beta)\frac{\phi}{r} - (\alpha + \beta)(\frac{\phi\phi'}{r} + \frac{\phi^2}{r^2}) + \beta\frac{\phi'\phi^2}{r^2} \right]\mu \right\} = 0, \tag{26}$$

$$2(\lambda - \mu) - 2r\mu_r - 2r\nu_r - r^2(\nu_{yy} + 2\mu_{yy}) + 2m_g^2 r^2 \left\{ \left[-(\alpha + 2) + 2(\alpha + 1)\frac{\phi}{r} - \alpha\frac{\phi^2}{r^2} \right](1 + \lambda) \right.$$

$$\left. + \left[-\frac{1}{2}(1 + 2\alpha + \beta) + (\alpha + \beta)\frac{\phi}{r} - \frac{1}{2}\beta\frac{\phi^2}{r^2} \right]\nu + \left[-(\alpha + 1)\frac{\phi}{r} + \alpha\frac{\phi^2}{r^2} \right]\mu \right\} = 0, \tag{27}$$

$$-r^2(\nu_{rr} + \mu_{rr}) - r\nu_r - 2r\mu_r + r\lambda_r - r(\nu_{yy} + \lambda_{yy} + \mu_{yy}) + 2m_g^2 r^2 \left\{ -(\alpha + 2) + (\alpha + 1)(\frac{\phi}{r} + \phi') - \alpha\frac{\phi\phi'}{r} \right.$$

$$+ \left[-\frac{1}{2}(1 + 2\alpha + \beta) + \frac{1}{2}(\alpha + \beta)(\frac{\phi}{r} + \phi') - \frac{1}{2}\beta\frac{\phi\phi'}{r} \right]\nu + \left[-\frac{1}{2}(1 + \alpha)\phi' + \frac{1}{2}\alpha\frac{\phi\phi'}{r} \right]\lambda$$

$$+ \left[-2 - \alpha + 3\beta - 3\sigma + (1 + \alpha - 5\beta + 11\sigma)\frac{\phi}{2r} + (1 + \alpha - \beta + \sigma)\phi' + (-\alpha + \beta - 3\sigma)\frac{\phi\phi'}{2r} - \frac{5}{2}\sigma\frac{\phi^2}{r^2} + \frac{1}{2}\sigma\frac{\phi'\phi^2}{r^2} \right]\mu \right\} = 0, \tag{28}$$

$$2(\lambda - \mu) - 2r^2 \mu_{rr} - r^2\nu_{rr} + 2r(\lambda_r - 3\mu_r - \nu_r) + 2m_g^2 r^2 \left\{ -(3 + 3\alpha + \beta) + (1 + 2\alpha + \beta)(\phi' + 2\frac{\phi}{r}) - (\alpha + \beta)(\frac{\phi^2}{r^2} + 2\frac{\phi'\phi}{r}) \right.$$

$$+\beta\frac{\phi'\phi^2}{r^2}+\left[-\frac{1}{2}(1+3\alpha+3\beta+\sigma)+\frac{1}{2}(\alpha+2\beta+\sigma)(\phi'+2\frac{\phi}{r})-(\sigma+\beta)(\frac{\phi\phi'}{r}+\frac{\phi^2}{2r^2})+\frac{1}{2}\sigma\frac{\phi'\phi^2}{r^2}\right]\nu$$

$$+\left[-\frac{1}{2}(1+2\alpha+\beta)\phi'+(\alpha+\beta)\frac{\phi\phi'}{r}-\frac{1}{2}\beta\frac{\phi'\phi^2}{r^2}\right]\lambda+\left[-(1+2\alpha+\beta)\frac{\phi}{r}+(\alpha+\beta)(\frac{\phi'\phi}{r}+\frac{\phi^2}{r^2})-\beta\frac{\phi'\phi^2}{r^2}\right]\mu\Bigg\}$$

$$=0\,,\tag{29}$$

$$(\lambda-\mu)=r\mu_r+\frac{1}{2}r\nu_r+f(r)\,,\tag{30}$$

where $f(r)$ is an arbitrary function of r. The subscripts y and r in these relations represent partial differentiation with respect to y and r respectively. Prime in ϕ' denotes derivative with respect to r. In addition to the generalized field equations (17), the Bianchi identities lead to the constraint:

$$m_g^2\ \left.\nabla^A\overline{X}_{AB}\right)=0\,,\tag{31}$$

where ∇^A denotes the covariant derivative with respect to physical metric g_{AB}. In the cases $m_g\neq0$, the linearized form of these constraints for $B=1$ and $B=4$ are respectively as follows

$$2\alpha(\frac{\phi^2}{r^2}-\frac{\phi\phi'}{r})+2(1+\alpha)(\phi'-\frac{\phi}{r})+\left[-\frac{1}{2}(1+2\alpha+\beta)+(\alpha+\beta)\frac{\phi}{r}-\frac{1}{2}\beta\frac{\phi^2}{r^2}\right]r\nu_r+\left[-(\alpha+1)\frac{\phi}{r}+\alpha\frac{\phi^2}{r^2}\right]r\mu_r$$

$$+\left[(\alpha+\beta)(\phi'-\frac{\phi}{r})+\beta(\frac{\phi^2}{r^2}-\frac{\phi\phi'}{r})\right]\nu+\left[(\alpha+1)(\frac{\phi}{r}-\phi')+2\alpha(\frac{\phi\phi'}{r}-\frac{\phi^2}{r^2})\right]\mu=0\,,\quad(32)$$

$$\frac{\partial}{\partial y}(\overline{X}_{44})=\left[-\frac{1}{2}(1+3\alpha+3\beta+\sigma)+\frac{1}{2}(\alpha+2\beta+\sigma)(\phi'+2\frac{\phi}{r})-(\sigma+\beta)(\frac{\phi\phi'}{r}+\frac{\phi^2}{2r^2})+\frac{1}{2}\sigma\frac{\phi'\phi^2}{r^2}\right]\nu_y$$

$$+\left[-\frac{1}{2}(1+2\alpha+\beta)\phi'+(\alpha+\beta)\frac{\phi\phi'}{r}-\frac{1}{2}\beta\frac{\phi'\phi^2}{r^2}\right]\lambda_y+\left[-(1+2\alpha+\beta)\frac{\phi}{r}+(\alpha+\beta)(\frac{\phi'\phi}{r}+\frac{\phi^2}{r^2})-\beta\frac{\phi'\phi^2}{r^2}\right]\mu_y$$

$$=0\,,\tag{33}$$

where other components of the constraint (31) are satisfied automatically. Contrary to the easy DGP model, which we studied in our previous paper [59], the presence of graviton mass terms in the 5D field equations (26)-(30) makes it more difficult to find an exact solution. The linearised form of the Israel-Darmois matching conditions (19) will lead to the following boundary conditions (on the brane) for the filed equations in the bulk

$$-\frac{1}{2}\Big(2\mu_y+\lambda_y\Big)|_{y=0^+}=r_c\left[-\frac{1}{r^2}\Big(\mu-\lambda+3r\mu_r+r^2\mu_{rr}-r\lambda_r\Big)\right]\,,\tag{34}$$

$$-\frac{1}{2}\Big(2\mu_y+\nu_y\Big)|_{y=0^+}=r_c\left[-\frac{1}{r^2}\Big(\mu-\lambda+r\mu_r+r\nu_r\Big)\right]\,,\tag{35}$$

$$-\frac{1}{2}\Big(\nu_y+\lambda_y+\mu_y\Big)|_{y=0^+}=r_c\left[-\frac{1}{2r}\Big(r\nu_{rr}+r\mu_{rr}+2\mu_r+\nu_r-\lambda_r\Big)\right]\,.\tag{36}$$

Note that these equations are hold on the brane outside our spherical object, where ρ and p are zero. Solving the linearized bulk field equations (26)-(30) with constraints (32) and

(33) (resulting from the Bianchi identities), is a very difficult task in non-unitary gauges. Therefore, here we consider some additional simplifying assumptions for the theory. The first assumption is that we find solutions in the unitary gauge, i.e. $\phi(r) = r$. In this gauge, the linearized form of all the higher order combinations, $\mathcal{U}_2$, $\mathcal{U}_3$ and $\mathcal{U}_4$ vanish such that

$$\overline{X}_{AB} = \mathcal{K}_{AB} - [\mathcal{K}]g_{AB} = \mathrm{diag}\ \left(\frac{1}{2}\lambda + \mu\right), -\left(\frac{1}{2}\nu + \mu\right), -\frac{1}{2}r^2(\nu + \lambda + \mu),\right.$$

$$\left. -\frac{1}{2}r^2\sin^2(\theta)(\nu + \lambda + \mu), -\frac{1}{2}(\nu + \lambda + 2\mu)\right). \tag{37}$$

Therefore, in the unitary gauge, the free parameters of the theory are absent in the field equations and effectively the Fierz-Pauli mass term is rebuilt. In this situation, the equations that should be solved are simplified to the following system of partial differential equations

$$2(\mu - \lambda) + 2r^2\mu_{rr} + 6r\mu_r - 2r\lambda_r + r^2(\lambda_{yy} + 2\mu_{yy}) + m_g^2 r^2(\lambda + 2\mu)) = 0\,, \tag{38}$$

$$2(\lambda - \mu) - 2r\mu_r - 2r\nu_r - r^2(\nu_{yy} + 2\mu_{yy}) - m_g^2 r^2(\nu + 2\mu) = 0\,, \tag{39}$$

$$-r^2(\nu_{rr} + \mu_{rr}) - r\nu_r - 2r\mu_r + r\lambda_r - r(\nu_{yy} + \lambda_{yy} + \mu_{yy}) - m_g^2 r^2(\nu + \lambda + \mu) = 0\,, \tag{40}$$

$$2(\lambda - \mu) - 2r^2\mu_{rr} - r^2\nu_{rr} + 2r(\lambda_r - 3\mu_r - \nu_r) - m_g^2 r^2(\nu + \lambda + 2\mu) = 0\,, \tag{41}$$

$$(\lambda - \mu) = r\mu_r + \frac{1}{2}r\nu_r + f(r)\,, \tag{42}$$

where $f(r)$ is an arbitrary function. The constraint equations (32) and (33) in the unitary gauge are represented by the following equations

$$\nu_r + 2\mu_r = 0, \tag{43}$$

$$\nu_y + \lambda_y + 2\mu_y = 0. \tag{44}$$

The Israel-Darmois junction conditions on the brane are independent of the gauge and are the same as before, that is, Eqs. (34)-(36). The three free parameters of the theory α, β and σ do not exist in the unitary gauge. The general solution of the bulk field equations with the mentioned assumptions that satisfies the constraint equations are obtained as follows

$$\lambda = \mu = a\cos(m_g y) + b\sin(m_g y), \tag{45}$$

$$\nu = -3\mu = -3\left(a\cos(m_g y) + b\sin(m_g y)\right), \tag{46}$$

where a and b are integration constants. By putting these solutions into the Israel-Darmois junction conditions, we see that b should be zero. Therefore, the linearized theory in the unitary gauge leads to the following line element on the brane

$$ds_4{}^2 = -(1 - 3a)dt^2 + (1 + a)dr^2 + r^2(1 + a)d\Omega^2\,. \tag{47}$$

Actually, this solution after the coordinates redefinition $(t, r) \to (t', r')$, where $t' = (\sqrt{1 - 3a})\,t$ and $r' = (\sqrt{1 + a})\,r$, reduces to the 4D flat Minkowski metric. But, this coordinates transformation leads to the appearance of the temporal component of the Stückelberg fields as

$$\phi^0 = (1 - \eta)t', \quad \eta = 1 - \frac{1}{\sqrt{1 - 3a}}, \tag{48}$$

and the scalar mode of massive graviton resulting from this Stückelberg field is $\pi = \frac{1}{2}\eta\Lambda^3 t'^2$. The final result is a flat 3-brane solution which is accompanied by the obtained scalar mode.

For the second simplifying assumption, we decided to work in non-unitary gauges. In this case, the free parameters of the theory (α, β, σ) play important roles in characterizing the properties of the solution, such as the (A)dS curvature scale. Moreover, the unknown scalar field $\phi(r)$ is coupled nonlinearly with other unknown metric components which can make the field equations more difficult to solve. We should determine a consistent scalar field $\phi(r)$ together with other unknown functions from field equations. Here, we consider three additional simplifying assumptions. The first one is to assume that the functional $\mu(r, y)$ be just a function of the extra dimension y, which in the non-unitary gauge $(\phi(r) \neq r)$ it could be a reasonable assumption. In this situation, solving the field equations could be slightly more easier. Moreover, we can restrict ourself to specific choices of the free parameters. The second assumption is to consider the case $\alpha = \beta = \sigma = 0$, which is equivalent to the choices $\alpha_3 = -\alpha_4 = \alpha_5 = -1$. By this assumption, the effective energy-momentum tensor $\overline{X}_{AB}$ takes the Fierz-Pauli structure, i.e. $\overline{X}_{AB} = \mathcal{K}_{AB} - [\mathcal{K}]g_{AB}$. However, the expression of it's components are not the same as eq. (37), which resulted in the unitary gauge. For this special choices of the free parameters, the components of $\overline{X}_{AB}$ takes the following form

$$\overline{X}_{00} = 3 - \phi' - 2\frac{\phi}{r} + (3 - \phi' - 2\frac{\phi}{r})\nu + \frac{1}{2}\phi'\lambda + \frac{\phi}{r}\mu, \tag{49}$$

$$\overline{X}_{11} = -2 + 2\frac{\phi}{r} - \frac{1}{2}\nu - \frac{\phi}{r}\mu + (-2 + 2\frac{\phi}{r})\lambda, \tag{50}$$

$$\overline{X}_{22} = r^2\left(-2 + \frac{\phi}{r} + \phi' - \frac{1}{2}\nu - \frac{1}{2}\phi'\lambda + (-2 + \frac{\phi}{2r} + \phi')\mu\right), \tag{51}$$

$$\overline{X}_{33} = \sin^2(\theta)\overline{X}_{22}, \tag{52}$$

$$\overline{X}_{55} = -3 + 2\frac{\phi}{r} + \phi' - \frac{1}{2}\nu - \frac{1}{2}\phi'\lambda - \frac{\phi}{r}\mu, \tag{53}$$

where reduce to (37) for $\phi(r) = r$. The constraint equations (32) and (33) for these special choices of the parameters are

$$2(\phi' - \frac{\phi}{r}) - \frac{1}{2}r\nu_r - \phi\mu_r + (\frac{\phi}{r} - \phi')\mu = 0, \tag{54}$$

$$\nu_y + \phi'\lambda_y + 2\frac{\phi}{r}\mu_y = 0. \tag{55}$$

The scalar field $\phi(r)$ is yet stayed coupled with other fields which this makes finding the solutions of the field equations difficult. The third assumption we do is to linearize the field equations with respect to the scalar field by considering $\phi(r) \ll 1$ and ignoring the nonlinear terms in the above equations. Therefore, by imposing these three assumptions we reach to the following field equations that should be solved analytically

$$2(\mu - \lambda) - 2r\lambda_r + r^2(\lambda_{yy} + 2\mu_{yy}) + 2m_g^2 r^2(3 - \phi' - 2\frac{\phi}{r} + 3\nu) = 0, \tag{56}$$

$$2(\lambda - \mu) - 2r\nu_r - r^2(\nu_{yy} + 2\mu_{yy}) + 2m_g^2 r^2(-2 + 2\frac{\phi}{r} - \frac{1}{2}\nu - 2\lambda) = 0, \tag{57}$$

$$-r^2\nu_{rr} + r(\lambda_r - \nu_r) - r^2(\nu_{yy} + \lambda_{yy} + \mu_{yy}) + 2m_g^2 r^2(-2 + \frac{\phi}{r} + \phi' - \frac{1}{2}\nu - 2\mu) = 0, \tag{58}$$

$$2(\lambda - \mu) - r^2\nu_{rr} + 2r(\lambda_r - \nu_r) + 2m_g^2 r^2(-3 + \phi' + 2\frac{\phi}{r} - \frac{1}{2}\nu) = 0, \tag{59}$$

$$(\lambda - \mu) = \frac{1}{2}r\nu_r + f(r), \tag{60}$$

$$2(\phi' - \frac{\phi}{r}) - \frac{1}{2}r\nu_r = 0, \tag{61}$$

$$\nu_y = 0. \tag{62}$$

These equations are valid in the regions where ν, μ, λ and ϕ are very small. We obtained the following solutions for these linearized field equations

$$\nu(r, y) = a,$$

$$\lambda(r, y) = \mu(r, y) = \frac{3}{4}a,$$

$$\phi(r) = (1 + a)r, \tag{63}$$

where a is an integration constant. Note that these solutions are valid in the regions where the obtained $\phi(r)$ is very small, i.e $r \ll \left(\frac{1}{1+a}\right)$. However, the metric here is accompanied by a nontrivial spatial backgrounds for the Stückelberg fields, $\pi^i = x^i - \phi^i = -ax^i, (i = 1, 2, 3)$, where $x^i = (r\sin\theta\cos\phi, r\sin\theta\sin\phi, r\cos\theta)$. The corresponding 4D line element on the 3-brane is given by

$$ds_4^2 = -(1 + a)dt^2 + (1 + \frac{3}{4}a)dr^2 + r^2(1 + \frac{3}{4}a)d\Omega^2. \tag{64}$$

However, in this case the solution on the brane transforms also to the 4D flat Minkowski metric, after the coordinates redefinition $(t, r) \to (t', r')$ with $t' = \sqrt{1 + a}\,t$ and $r' = \sqrt{1 + \frac{3}{4}a}\,r$. Due to this coordinates transformation, the temporal and spatial components of the Stückelberg fields will take the following forms

$$\phi'^0 = \frac{1}{\sqrt{1 + a}}t', \tag{65}$$

$$\phi'^i = \frac{1 + a}{\sqrt{1 + \frac{3}{4}a}}x'^i. \tag{66}$$

Finaly, the scalar mode of massive graviton resulting from these Stückelberg fields is

$$\pi = \frac{\Lambda^3}{2}(\delta\,t'^2 + \gamma\,x'^2), \tag{67}$$

where the constants δ and γ are related to a via $\delta = 1 - \frac{1}{\sqrt{1+a}}$ and $\gamma = 1 - \frac{1+a}{\sqrt{1+\frac{3}{4}a}}$.

In this paper, we considered two simplified linear theories in both unitary and non-unitary gauges and found in both cases a flat solution on the brane with different background Stückelberg fields (after a coordinates redefinition). In non-unitary gauge, we restricted ourself to special choices of the free parameters of the theory. Finding a general analytical solution for the linear theory with arbitrary α, β and σ together with the unknown scalar field $\phi(r)$ and then screening the solution on the brane to be consistent with junction conditions is a very difficult and complicated procedure. However, in the regions where we should keep nonlinear terms in the field equations, solving them will be more intricate. In this situation, we can pursue alternative approaches, such as solving the equations numerically or finding the effective 4D field equations on the brane for the new braneworld massive gravity theory and then solving them analytically [60, 61]. We are working on these subjects and the outcomes after completion will be presented in another paper.

4   Summary

We know that a way in which a massive graviton can naturally arise is from higher dimensional scenarios, such as KK, ADD and DGP theories. It has been shown that there is a deep connection between the DGP braneworld gravity and massive gravity theories. The graviton in the DGP setup acquires effectively a soft mass and the induced gravity term in the brane action acts as a kinetic term for a 4D graviton, while the bulk Einstein-Hilbert term acts as a gauge invariant mass term. Studying massive gravity from extra dimensional point of view can be useful for better understanding of certain aspects of the dRGT massive gravity theory and its bigravity and multi-gravity extensions. This fact was the original motivation of this paper to construct an extension of massive gravity in the spirit of braneworld scenarios. We have constructed a combination of the braneworld scenario and covariant de Rham-Gabadadze-Tolley (dRGT) massive Gravity, where we suppose that the five-dimensional bulk graviton is massive. We considered a static 5D configuration with spherical symmetry on the brane, aimed at separately determining the effects of bulk graviton mass term and also the large extra dimension on the gravitational interactions on the brane. Then, by a detailed analytical treatment, the effects of the nonlinear ghost-free massive gravity on brane dynamics and effective gravitational potential on the brane are examined. In this manner, the full field equations and their weak field limits together with the generalized Israel-Darmois junction conditions on the brane are presented. This set of equations are so complicated to be solved analytically without some simplifying assumptions. For this reason, by adopting some simplifying assumptions, we were able to find two classes of four-dimensional spherically symmetric solutions on the brane in unitary and non-uniary gauges. Both of them were flat solutions on the brane with different background Stckelberg fields (after a coordinates redefinition). We note that general massive braneworld solutions should reduce to the massless braneworld solution in the limit of zero bulk graviton mass as has been studied in [59]. To restore the GR or the original DGP model on the brane, we should consider certain nonlinear terms in the bulk field equations and the brane junction conditions, which make the solving procedure more difficult (because of the bulk mass terms). We are attempting to follow alternative approaches, such as solving the field equations numerically or finding the effective 4D field equations on the brane for the new massive braneworld theory [60, 61], to examine the Vainshtein mechanism in our model. This issue in the absence of the Boulware-Deser ghost and also the instability issue are subject of our forthcoming work.

References

[1] Riess A. G., et al., 1998, Astron. J, vol. 116, no. 3, pp. 1009-1038.

[2] Perlmutter S., et al., 1999, Astron. J, vol. 517, no. 2, pp. 565-586.

[3] Spergel D. N., et al., 2007, Astrophys. J., Suppl. Ser., vol. 170, no. 2, pp. 377-466.

[4] Csaki C., 2004, "TASI Lectures on Extra Dimensions and Branes," *Shifman, M. (ed.) et al.: *From fields to strings*, vol. 2* pp. 967-1060, [arXiv:hep-ph/0404096].

[5] Maartens R. and Koyama K., 2010, Living. Rev. Relat., vol. 13, no. 5.

[6] Dvali G., Gabadadze G., and Porrati M., 2000, Phys. Lett. B, vol. 485, no. 1-3, pp. 208-214.

[7] Deffayet C., 2001, Phys. Lett. B, vol. 502, no. 1-4, pp. 199-208.

[8] Dvali G. R. and Gabadadze G., 2001, Phys. Rev. D, vol. 63, no. 6, p. 065007.

[9] Deffayet C., Dvali G. R., and Gabadadze G., 2002, Phys. Rev. D, vol. 65, no. 4, p. 044023.

[10] Dvali G. R., Gabadadze G., Kolanovic M., and Nitti F., 2001, Phys. Rev. D, vol. 64, no. 8, p. 084004.

[11] Abbott B. P., et al. [LIGO Scientific and Virgo Collaborations], 2016, Phys. Rev. Lett., vol. 116, no. 6, p. 061102.

[12] Abbott B. P., et al. [LIGO Scientific and Virgo Collaborations], 2016, Phys. Rev. Lett., vol. 116, no. 24, p. 241103.

[13] Abbott B. P., et al., [LIGO Scientific and Virgo Collaborations], 2016, Phys. Rev. Lett., vol. 116, no. 22, p. 221101.

[14] Fierz M. and Pauli W., 1939, Proc. R. Soc. Lond. Ser. A, vol. 173, no. 953, pp. 211-232.

[15] van Dam H. and Veltman M. J. G., 1970, Nucl. Phys. B, vol. 22, no, 2, pp. 397-411.

[16] Zakharov V. I., 1970, JETP. Lett., vol. 12, pp. 312-314.

[17] Vainshtein A. I., 1972, Phys. Lett. B, vol. 39, no. 3, pp. 393-394.

[18] Deffayet C., Dvali G., Gabadadze G., and Vainshtein A., 2002, Phys. Rev. D, vol. 65, no. 4, p. 044026.

[19] Porrati M., 2001, Phys. Lett. B, vol. 498, no. 1-2, pp. 92-96.

[20] Kogan I. I., Mouslopoulos S., and Papazoglou A. 2001, Phys. Lett. B, vol. 503, no. 1-2, pp. 173-180.

[21] Boulware D. G. and Deser S., 1972, Phys. Rev. D, vol. 6, no. 12, p. 3368.

[22] Arkani-Hamed N., Georgi H., and Schwartz M. D., 2003, Ann. Phys, vol. 305, no. 2, pp. 96-118.

[23] de Rham C. and Gabadadze G., 2010, Phys. Rev. D, vol. 82, no. 04, p. 044020.

[24] de Rham C., Gabadadze G., and Tolley A. J., 2011, Phys. Rev. D, vol. 106, no. 23, p. 231101.

[25] Hassan S. and Rosen R. A., 2012, Phys. Rev. Lett., vol. 108, no. 4, p. 041101.

[26] Hassan S., Rosen R. A., and Schmidt-May A., 2012, J. High. Energy. Phys., vol. 1202, p. 026.

[27] Hassan S. and Rosen R. A., 2012, J. High. Energy. Phys., vol. 1204, p. 123.

[28] de Rham C., Matas A., and Tolley A. J., 2014, Class. Quantum. Grav., vol. 31, no. 06, p. 065004.

[29] Gao X., 2014, Phys. Rev. D, vol. 90, no. 6, p. 064024.

[30] de Rham C., 2014, Living. Rev. Relat., vol. 17, no. 7.

[31] Hinterbichler K., 2012, Rev. Mod. Phys, vol. 84, no. 2, p. 671.

[32] Schmidt-May A. and von Strauss M., 2016, J. Phys. A: Math. Theor., vol. 49, no. 18, p. 183001.

[33] Koyama K., Niz G., and Tasinato G., 2011, Phys. Rev. Lett., vol. 107, no. 13, p. 131101.

[34] Koyama K., Niz G., and Tasinato G., 2011, Phys. Rev. D, vol. 84, no. 6, p. 064033.

[35] Nieuwenhuizen T., 2011, Phys. Rev. D, vol. 84, no. 2, p. 024038.

[36] de Rham C., Gabadadze G., Heisenberg L., and Pirtskhalava D., 2011, Phys. Rev. D, vol. 83, no. 10, p. 103516.

[37] D'Amico G., de Rham C. , Dubovsky S., Gabadadze G., Pirtskhalava D., and Tolley A. J., 2011, Phys. Rev. D, vol. 84, no. 12, p. 124046.

[38] Gumrukcuoglu A., Lin C., and Mukohyama S., 2011, J. Cosmol. Astropart., vol. 2011, p. 030.

[39] Arkani-Hamed N., Cohen A. G., and Georgi H., 2011, Phy. Rev. Lett., vol. 86, no. 21, p. 4757.

[40] Arkani-Hamed N., Cohen A. G., and Georgi H., 2001, Phy. Lett. B, vol. 513, no. 1-2, pp. 232-240.

[41] Gruzinov A., 2005, New. Astron., vol. 10, no. 4, pp. 311-314.

[42] Porrati M., 2002, Phys. Lett. B, vol. 534, no. 1-4, pp. 209-215.

[43] Luty M. A., Porrati M., and Rattazzi R., 2003, J. High. Energy. Phys., vol. 0309, p. 029.

[44] Nicolis A., and Rattazzi R., 2004, J. High. Energy. Phys., vol. 0406, p. 059.

[45] Koyama K., 2005, Phys. Rev. D, vol. 72, no. 12, p. 123511.

[46] Nicolis A., Rattazzi R., and Trincherini E., 2009, Phys. Rev. D, vol. 79, no. 6, p. 064036.

[47] Gabadadze G., 2009, Phys. Lett. B, vol. 681, no. 1, pp. 89-95.

[48] de Rham C. , 2010, Phys. Lett. B, vol. 688, no. 2-3, pp. 137-141.

[49] Hassan S. F. and Rosen R. A., 2012, J. High. Energy. Phys., vol. 1202, p. 126.

[50] Hinterbichler K. and Rosen R. A., 2012, J. High. Energy. Phys., vol. 1207, p. 047.

[51] de Rham C. and Gabadadze G., 2010, Phys. Lett. B, vol. 693, no. 3, pp. 334-338.

[52] Randall L. and Sundrum R., 1999, Phys. Rev. Lett., vol. 83, no. 17, p. 3370.

[53] Chacko Z., Graesser M. L., Grojean C., and Pilo L., 2004, Phys. Rev. D, vol. 70, no. 8, p. 084028.

[54] Gherghetta T., Peloso M., and Poppitz E., 2005, Phys. Rev. D, vol. 72, no. 10, p. 104003.

[55] Kakushadze Z., 2014, Acta. Phys. Pol. B, vol. 45, pp. 1671-1699.

[56] Kluson J., 2011, Class. Quantum. Grav., vol. 28, no. 15, p. 155014.

[57] Iglesias A. and Kakushadze Z., 2010, Phys. Rev. D, vol. 82, no. 12, p. 124001.

[58] Iglesias A. and Kakushadze Z., 2011, Phys. Rev. D, vol. 84, no. 8, p. 084005.

[59] Nozari K. and Asaiyan A., 2011, Class. Quantum. Grav., vol. 28, no. 12, p. 125017.

[60] Shiromizu T., Maeda K., and Sasaki M., 2000, Phys. Rev. D, vol. 62, no. 2, p. 024012.

[61] Maeda K-i., Mizuno Sh., and Torii T., 2003, Phys. Rev. D, vol. 68, no. 2, p. 024033.

[62] Do T. Q., 2016, Phys. Rev. D, vol. 93, no. 10, p. 104003.

[63] Dadhich N., Maartens R., Papadopoulos P., and Rezania V., 2000, Phys. Lett. B, vol. 487, no. 1-2, pp. 1-6.

[64] Kofinas G., Papantonopoulos E., and Zamarias V., 2002, Phys. Rev. D, vol. 66, no. 10, p. 104028.

[65] Middleton C. and Siopsis G., 2004, Mod. Phys. Lett. A, vol. 19, no. 30, p. 2259.

[66] Gregory R., 2009, Lect. Notes. Phys., vol. 769. pp. 259-298.

[67] Berezhiani L., Chkareuli G., de Rham C., Gabadadze G., and Tolley A. J., 2012, Phys. Rev. D, vol. 85, no. 4, p. 044024.

[68] Babichev E. and Brito R., 2015, Class. Quantum. Grav., vol. 32, no. 15, p. 154001.

[69] Tasinato G., Koyama K., and Niz G., 2013, Class. Quantum. Grav., vol. 30, no. 18, p. 184002.

[70] Babichev E. and Fabbri A., 2014, J. High. Energy. Phys., vol. 1407, p. 016.

[71] Babichev E. and Fabbri A. , 2014, Phys. Rev. D, vol. 90, no. 8, p. 084019.

[72] Cai R. G., Hu Y. P., Pan Q. Y., and Zhang Y. L., 2015, Phys. Rev. D, vol. 91, no. 2, p. 024032.

[73] Xu J., Cao L. M., and Hu Y. P., 2015, Phys. Rev. D, vol. 91, no. 12, p. 124033.

[74] Zhou Z., Wu J. P., and Ling Y., 2015, J. High. Energy. Phys., vol. 1508, p. 067.

[75] Hendi S. H., Eslam Panah B., and Panahiyan S., 2015, J. High. Energy. Phys., vol. 1511, p. 157.

[76] Ghosh S. G., Tannukij L., and Wongjun P., 2016, Eur. Phys. J. C, vol. 76, no. 3, p. 119.

[77] Do T. Q., 2016, Phys. Rev. D, vol. 94, no. 4, p. 044022.

[78] El-Menoufi B. K. and Sorbo L., 2015, Phys. Rev. D, vol. 91, no. 6, p. 064023.

13

Noether Symmetry in f(T) Theory at the anisotropic universe

A. Aghamohammadi

Sanandaj Branch, Islamic Azad University, Sanandaj, Iran; email:a.aqamohamadi@gmail.com; a.aghamohamadi@iausdj.ac.ir

Abstract. As it is well known, symmetry plays a crucial role in the theoretical physics. On other hand, the Noether symmetry is a useful procedure to select models motivated at a fundamental level, and to discover the exact solution to the given lagrangian. In this work, Noether symmetry in f(T) theory on a spatially homogeneous and anisotropic Bianchi type I universe is considered. We discuss the Lagrangian formalism of f(T) theory in anisotropic universe. The point-like Lagrangian is clearly constructed.The explicit form of f(T) theory and the corresponding exact solution are found by requirement of Noether symmetry and Noether charge. A power-law f(T), the same as the FRW universe, can satisfy the required Noether symmetry in the anisotropic universe with power- law scale factor. It is regarded that positive expansion is satisfied by a constrain between parameters.

Keywords: Noether symmetry, anisotropic Bianchi type I universe, modified theories of gravity, $f(T)$ gravity

1   Introduction

In this work, our aim is to study a Noether symmetry of scalar torsion gravity in anisotropic univers.

Recently, some astrophysical observations have shown that the Universe is undergoing an accelerated phase era. To justify this unexpected result, scientists have proposed some different models such as, scalar field models [1, 2, 3, 4] and modify theories of gravity [5, 6, 7, 8]. For the latter proposal, one can deal with telleparallel equivalent of general relativity [9, 10, 11, 12], in which the field equations are second order [13]. In addition, in this scenario the Levi-Civita connections are replaced by Weitzenböck connection where it has no curvature but only torsion [14].

A Bianchi type I (BI) universe, being the straightforward generalization of the flat FRW universe, is of interest because it is one of the simplest models of a non-isotropic universe exhibiting a homogeneity and spatial flatness. In this case, unlike the FRW universe which has the same scale factor for three spatial directions, a BI universe has a different scale factor for each direction. This fact introduces a non-isotropy to the system. The possible effects of anisotropy in the early universe have been investigated with BI models from different points of view [25, 26, 27, 28]. Some people [29, 30] have constructed cosmological models by using anisotropic fluid and BI universe. Recently, this model has been studied in the presence of binary mixture of the perfect fluid and the DE [31]. Further, there are some exact solutions for BI models in f (T) gravity [32]

The outline of this work is as follows. In the next section, a brief review of the general formulation of the field equations in a BI metric and $f(T)$ garavity are discussed, Sec. 3 is concerned with Lagrangian formalism of f(T) theory in anisotropic universe. Sec. 4, is related to Noether symmetry in f(T) theory in anisotropic universe. We summarize our results in last section.

2 General Framework

The telleparallel theory of gravity is defined in the Weitzenböck's space-time, with torsion and zero local Riemann tensor, in which we are working in a non-Riemannian manifold. The dynamics of the metric were determined using the scalar torsion T . The fundamental quantity in teleparallel theory is the vierbein (tetrad) basis $e^i{}_\mu$. This basis is an orthogonal, coordinate free basis defined by the following equation

$$g_{\mu\nu} = \eta_{ij}e^i{}_\mu e^j{}_\nu, \tag{1}$$

where $\eta_{ij} = diag[1,-1,-1,-1]$ and $e_i{}^\mu e^i{}_\nu = \delta^\mu_\nu$ or $e_i{}^\mu e^j{}_\mu = \delta^j_i$. and the matrix $e^a{}_\mu$ are called tetrads that indicate the dynamic fields of the theory, where Latin i,j are indices running over $0,1,2,3$ for the tangent space of the manifold, and Greek μ,ν are the coordinate indices on the manifold, also running over $0,1,2,3$. In the framework of $f(T)$ theory, Lagrangian density is extended from the torsion scalar T to a general function $f(T)$, similar to what happened in $f(R)$ theories. The action S of modified telleparallel gravity is given by [33, 34]

$$I = \int \mathrm{d}^4x|e|f(T) + L_m\Big], \tag{2}$$

where for convenience, we use the units $2k^2 = 16\pi G = 1$, $|e| = det(e^i_\mu) = \sqrt{-g}$ and e^i_μ forms the tangent vector of the manifold, which is used as a dynamical object in telleparallel gravity, L_M is the Lagrangian of matter. The components of the tensor torsion and the contorsion are defined respectively as

$$T^\rho{}_{\mu\nu} \equiv e_l{}^\rho \left(\partial_\mu e^l{}_\nu - \partial_\nu e^l{}_\mu\right), \tag{3}$$

$$K^{\mu\nu}{}_\rho \equiv -\frac{1}{2}\left(T^{\mu\nu}{}_\rho - T^{\nu\mu}{}_\rho - T_\rho{}^{\mu\nu}\right). \tag{4}$$

It was defined a new tensor $S_\rho{}^{\mu\nu}$ to obtain the scalar equivalent to the curvature scalar of general relativity i.e. Ricci scalar, that is as

$$S_\rho{}^{\mu\nu} \equiv \frac{1}{2}\left(K^{\mu\nu}{}_\rho + \delta^\mu_\rho T^{\alpha\nu}{}_\alpha - \delta^\nu_\rho T^{\alpha\mu}{}_\alpha\right). \tag{5}$$

Hence, the torsion scalar is defined by the following contraction

$$T \equiv S_\rho{}^{\mu\nu}T^\rho{}_{\mu\nu}. \tag{6}$$

By using the components (Eq.4,Eq.5), the torsion scaler (Eq. 6) gives

$$T \equiv -6H^2 + 2\sigma^2. \tag{7}$$

Bianchi cosmologies are spatially homogeneous but not necessarily isotropic. Here, we will consider BI cosmology. The metric of this model is given by

$$ds^2 = dt^2 - A^2(t)dx^2 - B^2(t)dy^2 - C^2(t)dz^2, \tag{8}$$

where the metric functions, A, B, C, are merely functions of time, t and related to scale factor by $a = (ABC)^{\frac{1}{3}}$. In this work for convenience, we assume $B = C = A^m$, where m is a constant. It is defined the shear tensor as describes the rate of distortion of the matter flow, that in a comoving coordinate system, from the metric (Eq.8), the components of the average Hubble parameter and the shear tensor are given by [38]

$$
\begin{aligned}
H &= \frac{1}{3}(\frac{\dot{A}}{B} + \frac{\dot{B}}{B} + \frac{\dot{C}}{C}), \\
\sigma^2 &= \frac{1}{2}[(\frac{\dot{A}}{A})^2 + (\frac{\dot{B}}{B})^2 + (\frac{\dot{C}}{C})^2] - \frac{3}{2}H^2.
\end{aligned}
\tag{9}
$$

3   Lagrangian Formalism of f(T) Theory in Anisotropic Universe

In this section, we discuss the Lagrangian formalism of f(T) theory in anisotropic universe .In the study of Noether symmetry,It is clear that the point-like Lagrangian plays a crucial role. From the action f(T) (Eq.2), and following [36, 35, 37], to derive the cosmological equations in the Bianchi I metric, (BIm), one can define a canonical Lagrangian $\mathcal{L} = \mathcal{L}(A, \dot{A}, T, \dot{T})$, whereas $Q = a, T$ is the configuration space, and $\mathcal{T}Q = [A, \dot{A}, T, \dot{T}]$ is the related tangent bundle on which L is defined. The factor A(t) and the torsion scalar T (t) are taken as independent dynamical variables. One can use the method of Lagrange multipliers to set T as a constraint of the dynamics (Eq. 6). Selecting the suitable Lagrange multiplier and integrating by parts, the Lagrangian $\mathcal{L}$ becomes canonical [36, 35] theory which is given by

$$
I = 2\pi^2 \int dt ABC \big[f(T) - \lambda(T + 6H^2 - 2\sigma^2) - \frac{\rho_{m0}}{ABC}\big],
\tag{10}
$$

where λ is a Lagrange multiplier. The variation with respect to T of this action gives

$$
\lambda = f_T.
\tag{11}
$$

So, the action (10) can be rewritten as

$$
I = 2\pi^2 \int dt ABC \big[f(T) - f_T(T + 6H^2 - 2\sigma^2) - \frac{\rho_{m0}}{ABC}\big],
\tag{12}
$$

and then the point-like Lagrangian reads (up to a constant factor $2\pi^2$)gives

$$
\mathcal{L}(A, \dot{A}, T, \dot{T}) = A^{1+2m}\big[f(T) - f_T(T + 6H^2 - 2\sigma^2)\big] - \rho_{m0},
\tag{13}
$$

where using from assume $B = C, a^3 = A^{1+2m}$. Writing (13) with respect (9) yield

$$
\mathcal{L}(A, \dot{A}, T, \dot{T}) = A^{1+2m}\big[f - f_T T + 2f_T(\frac{\dot{A}}{A})^2 c_0\big] - \rho_{m0},
\tag{14}
$$

where $c_0 = 1/3(2m+1) - (1+2m)^2/3 - m^2)$, and setting $m = 1$ reduce equation (Eq.14) to the same form of lagrangian equation in isotropic univerce, i.e. FRW metric [37], as well, the equation lagrangian form in the FRW metric constrained $c_0 \neq 0, -2/7$. As it is explicit for a dynamical system, the Euler-Lagrange equation is written

$$
\frac{d}{dt}(\frac{\partial \mathcal{L}}{\partial \dot{q}_i}) = \frac{\partial \mathcal{L}}{\partial q_i},
\tag{15}
$$

where q_i are A, T in this case. Substituting Eq. (14) into the Euler-Lagrange equation (Eq.15), we get the following equations with respect T, A respectively

$$A^{1+2m} f_{TT}\left[-T + 2(\frac{\dot{A}}{A})^2 c_0\right] = 0, \tag{16}$$

$$4f_{TT}\dot{T}\dot{A} + 4f_T\ddot{A}c_0 + 2f_T(\frac{\dot{A}}{A})^2(2m-1)c_0 - (1+2m)(f - f_T T) = 0. \tag{17}$$

From Eq. (16), it is easy to find that if $f_{TT} \neq 0$

$$T = 2(\frac{\dot{A}}{A})^2 c_0 = -6H^2 + 2\sigma^2. \tag{18}$$

That, setting $m = 1$ reduce equation (18) to the same form of torsion scaler from FRW metric, [37] . In addition, the relation (7) is recovered. Mainly, this is the Euler constraint of the dynamics. Substituting Eq. (18) into Eq. (17), we get

$$8f_{TT}c_0^2\frac{\dot{A}}{A}\left[\frac{2\ddot{A}\dot{A}}{A^2} - \frac{2\dot{A}^3}{A^3}\right] + 4f_T c_0\frac{\ddot{A}}{A} + 2f_T(\frac{\dot{A}}{A})^2 c_0(2m-1) - (1+2m)(f - 2f_T c_0(\frac{\dot{A}}{A})^2) = 0. \tag{19}$$

This is the modified Raychaudhuri equation, and by setting $m = 1$ in the c_0 parameter, the above equation is reduced to the same equation in isotropic universe, i.e. FRW metric [37] . By the way, it is explicit, that the corresponding Hamiltonian to Lagrangian $\mathcal{L}$ is given by

$$\mathcal{H} = \sum_i \frac{\partial \mathcal{L}}{\partial \dot{q}_i}\dot{q}_i - \mathcal{L}. \tag{20}$$

Replacing (Eq.14) into (Eq.20), one can rewrite the above Lagrangian density as follows

$$\mathcal{H} = 2f_T A^{2m-1} c_0 \dot{A}^2 - A^{1+2m}(f - f_T T) + \rho_{m_0}. \tag{21}$$

Using the zero energy condition, $\mathcal{H} = 0$, [36, 35, 39], we get

$$-2f_T A^{2m-1} c_0 \dot{A}^2 + A^{1+2m}(f - f_T T) = \rho_{m_0}, \tag{22}$$

where, it is clear again that by taking $m = 1$ in the c_0 parameter, the above equation end up to one in the modified Friedmann equation, i.e. the $f(T)$ gravity at FRW metric. As a result, we have found that the point-like Lagrangian obtained in (Eq.14) can yield all the correct equations of motion in anisotropic universe, that taken $m = 1$ in the c_0 parameter recovered what is in isotropic universe, i.e. the $f(T)$ gravity equation in FRW metric.

4   Noether Symmetry in f(T) Theory at Anisotropic Universe

As mentioned, one can find the exact solution to the given lagrangian by using Noether symmetry theorem. So in this section, we would like to investigate Noether symmetry in f(T) theory in anisotropic universe. Following references [36, 35, 37] , the generator of Noether symmetry is a killing vector

$$X = \alpha\frac{\partial}{\partial\alpha} + \beta\frac{\partial}{\partial\beta} + \dot{\alpha}\frac{\partial}{\partial\dot{\alpha}} + \dot{\beta}\frac{\partial}{\partial\dot{\beta}}, \tag{23}$$

where α, β, both are the function of the generalized coordinate of A, T. Requirement of Noether symmetry is that Lie differentiation with respect X to be zero. Hence we get

$$L_X \mathcal{L} = \alpha \frac{\partial \mathcal{L}}{\partial \alpha} + \beta \frac{\partial \mathcal{L}}{\partial \beta} + \dot{\alpha} \frac{\partial \mathcal{L}}{\partial \dot{\alpha}} + \dot{\beta} \frac{\partial \mathcal{L}}{\partial \dot{\beta}} = 0. \tag{24}$$

Therefore, based on Noether symmetry theorem, there should be a motion constant, so-called Noether charge [36, 35].

$$Q_0 = \sum_i \alpha_i \frac{\partial \mathcal{L}}{\partial \dot{\alpha}_i} = \alpha \frac{\partial \mathcal{L}}{\partial \dot{A}} + \beta \frac{\partial \mathcal{L}}{\partial T} = \alpha (4 f_T A^{2m-1} c_0 \dot{A}) = const, \tag{25}$$

where setting $m = 1, c_0 = -3$ in the above equation, recovered the same equation in [37] to isotropic universe. We know that $L_X \mathcal{L} = 0$, meaning $\mathcal{L}$ is constant along the flow generated by X, i.e. (Eq.24[39]. Therefore, evaluating (Eq.24) is a second degree function from $\dot{A}, \dot{T}$, whose coefficients are functions of a and T only. Hence, they have to be zero separately. So, replacing (Eq.13) into (Eq.24) and using the relations $\dot{\alpha} = \partial \alpha / \partial A \dot{A} + \partial \alpha / \partial T \dot{T}$ and $\dot{\beta} = \partial \beta / \partial A \dot{A} + \partial \beta / \partial T \dot{T}$ yield

$$\alpha(1 + 2m)(f - f_T T) + 2\alpha(2m - 1)f_T (\frac{\dot{A}}{A})^2 c_0 - \beta A f_{TT} T + 2\beta f_{TT} \dot{A}^2 A^{-1} c_0 + $$
$$4 \frac{\partial \alpha}{\partial A} f_T \dot{A}^2 A^{-1} c_0 + 4 \frac{\partial \alpha}{\partial T} \dot{T} f_T \dot{A} A^{-1} c_0 = 0. \tag{26}$$

As mentioned above, the coefficients $\dot{A}^2, \dot{T} \dot{A}$ should be zero, as a result, we get

$$4 \frac{\partial \alpha}{\partial T} f_T = 0, \tag{27}$$

$$2(2m - 1)f_T A^{-2} \alpha + 2\beta f_{TT} A^{-1} + 4 \frac{\partial \alpha}{\partial A} f_T A^{-1} = 0, \tag{28}$$

$$\alpha(1 + 2m)(f - f_T T) - \beta A f_{TT} T = 0. \tag{29}$$

It is explicit that solutions of (Eqs. 28, 27,29)are given if the explicit form of α, β are obtained, and if at least one of them is different from zero, then Noether symmetry exist[35]. From (Eq.27), it is clear that α is independent of T, so it is merely a function of A. In addition, from (Eq.29), we get

$$\alpha(1 + 2m)(f - f_T T) = \beta A f_{TT} T. \tag{30}$$

By substituting (Eq.30) in to (Eq.28), we get

$$2(2m - 1)f_T T A^{-2} \alpha + 2\alpha(1 + 2m)(f - f_T T)A^{-2} + 4 f_T A^{-1} T \frac{\partial \alpha}{\partial A} = 0. \tag{31}$$

By separation of variables, one can transform the above equation to two independent differential equations as follow

$$1 - \frac{A}{\alpha} \frac{\partial \alpha}{\partial A} = \frac{(1 + 2m)f}{2 f_T T}. \tag{32}$$

Since Right and left hand side are independent, hence, they must be equal to a same constant, that fore convenience, we set $\frac{1+2m}{n}$. As a result, (Eq.32) is separated into two ordinary differential equations as

$$1 - \frac{A}{\alpha}\frac{\partial \alpha}{\partial A} = \frac{1+2m}{2n}, \tag{33}$$

$$\frac{(1+2m)f}{2f_T T} = \frac{1+2m}{2n}. \tag{34}$$

It is readily obtained the solutions of these two above equation as follow

$$f = \mu T^n, \tag{35}$$

$$\alpha = \alpha_0 A^{\frac{2n-1-2m}{2n}}, \tag{36}$$

where, again the above equation is reduced into the same equation in isotropic universe by setting $m = 1$, i.e. [37]; hence it is the desired one and μ, α_0 are integral constants. Substituting (Eqs.35, 36) into (Eq.30) we get

$$\beta = -\frac{\alpha_0 \mu (1+2m)}{n} A^{-\frac{1+2m}{2n}} T. \tag{37}$$

Up till now, we obtained the non-zero solution of $f(T), \alpha, \beta$. Therefore, Noether symmetry exists in anisotropic universe on Bianchi type I. Now, we try to obtain a solution of scale factor for this $f(T)$ function. Hence substituting, the (Eqs.35,36,37) into (Eq.25) yields

$$\dot{A}A^{c_2} = (\frac{c_1}{c_0})^{\frac{1}{2}}, \tag{38}$$

where, $c_1 = (Q_0/4\mu n c_0 \alpha)^{1/n-1}$,$c_2 = \frac{4m^2+2m-4mn-1}{4m(n-1)}$. It is readily obtained the solution of (38) as follow

$$A = (1+c_2)^{\frac{1}{1+c_2}}\left[(\frac{c_1}{c_0})^{\frac{1}{2}}t - c_3\right]^{\frac{1}{1+c_2}}, \tag{39}$$

where c_3 is integral constant. From requirement $a(t=0) = 0$, it is easy to see that the constant c_3 is zero. As a result, $A \sim t^{\frac{1}{1+c_2}}$. Therefore, as mentioned, relation between scale factor a and component metric A in anisotropic Bianchi type I with assuming, $B = C = A^m$ is $a^3 = (ABC) = A^{1+2m}$, hence

$$a \sim t^{\frac{1+2m}{3(1+c_2)}}. \tag{40}$$

Note that, requirement of positive expansion, $\ddot{a} > 0$ requiring a constrained between the n and m, parameter as following

$$n - 1 < \frac{3m}{4+2m}. \tag{41}$$

It is clear, that one can readily obtained physical quantity corresponding to the exact solution a and $f(T)$ namely, $H, \dot{H}, \sigma^2$, and equation of state ω, that in this work, it is not our scope.

5 Discussion

As it is well known, symmetry plays a crucial role in the theoretical physics. On other hand, the Noether symmetry is a useful procedure to select models motivated at a fundamental level, and discover the exact solution to the given lagrangian. In this work, Noether symmetry in f(T) theory on A spatially homogeneous and anisotropic Bianchi type I universe have considered. We have addressed the Lagrangian formalism of f(T) theory in anisotropic universe, and a Lagrangian form was obtained. The point-like Lagrangian was clearly constructed. The explicit form of f(T) theory and the corresponding exact solution were found by requirement of Noether symmetry and Noether charge. A power-law f(T), have obtained in the anisotropic universe with power- law scale factor, that can satisfy the requirement of the Noether symmetry. It was regarded that positive expansion is satisfied. Our main conclusions can be summarized as follows

- A exact solution have been obtained to $f(T), a(t)$, that is reduced to those value in FRW metric with selecting the m parameter, equal one, i.e. $m = 1$ and $c_0 = -3$

- Requirement of positive acceleration have obtained a constrained between m and n parameter, i.e. $n - 1 < \frac{3m}{4+2m}$

- To regain the equation lagrangian form in the FRW metric, it was required $c_0 \neq 0, -2/7$.

- It was obtained a energy density matter form by the zero energy condition, $\mathcal{H} = 0$,

- We have obtained exact solution of scale factor a by Noether charge condition.

- At last,we have seen that the resulting f(T) theory from Noether symmetry can be study in anisotropic universe that may be, plays an important role at early universe.

References

[1] Brans. C., Dicke. R. H.,1996, Phys. Rev 124, 925–935

[2] Justin. Khoury., Amanda. Weltman., 2004, Annalen der Physik 69, 044026 .

[3] David F. Mota., John D. Barrow., 2004, Physics Letters B 581, 141–146 .

[4] Saaidi. Kh., Mohammadi. A., Sheikhahmadi. H., 2011, Phys. Rev. D 83, 104019.

[5] David. Wands., 1994, Classical and Quantum Gravity 11 269

[6] Nojiri. Shin'ichi., Odintsov. Sergei D.,2006, Phys. Rev. D 74, 086005 .

[7] Guarnizo. Alejandro., Castaeda. Leonardo., et al., 2010, General Relativity and Gravitation 42, 2713-2728.

[8] Saaidi. KH., Aghamohammadi. A., et al., 2012, International Journal of Modern Physics D 21, 1250057.

[9] Chao-Qiang Geng., Chung-Chi Lee., et al.,2011, Physics Letters B 704, 384 - 387.

[10] Linder. Eric V.,2010, Phys. Rev. D 81 127301.

[11] Bamba. Kazuharu., Geng. Chao-Qiang., et al.,2011, Journal of Cosmology and Astroparticle Physics 2011, 021.

[12] Aghamohammadi. A., 2014, Astrophysics and Space Science 352, 1–5 .

[13] Bengochea, Gabriel R.,2009, Ferraro, Rafael., Phys. Rev. D 79, 124019 5.

[14] Ferraro. Rafael., Fiorini. Franco.,2007, Phys. Rev. D 75, 084031.

[15] Saridakis.E.N.,2008, Phys. Lett. B 660, 138.

[16] Zhang.J., Zhang. X., et al., 2007, Eur. Phys. J. C 52, 693.

[17] Cai. R.G.,2007, Phys. Lett. B 657, 228.

[18] Wei. H., Cai.R.G., 2009, Eur. Phys. J. C 59, 99.

[19] Wei.H., Cai. R.G.,2008, Phys. Lett. B 660, 113.

[20] Wei. H., Cai.R.G.,2008, Phys. Lett. B 663, 1.

[21] Wu. J.-P., Ma. D.-Z., et al., 2008, Phys. Lett. B 663, 152.

[22] Neupane. I.P., 2009, Phys. Lett. B 673, 111.

[23] Zhang. J.,Zhang. X.,Liu. H., 2008, Eur. Phys. J. C 54, 303.

[24] Li.Y.H., Ma. J.Z., Cui.J.L.,et al, 2011, Sci. China Phys. Mech. Astron. 54, 1367.

[25] E. Komatsu., et al., 2009, Astrophys. J. Suppl. Ser. 180, 330.

[26] Bertschinger. E., 1994, Physica D 77, 354.

[27] Brevik. I., Pettersen.S.V., 1997, Phys. Rev. D 56, 3322.

[28] Khalatnikov.I.M., kamenshchik. A.Yu., 2003, Phys. Lett. B **553**, 119.

[29] Rodrigues, D.C.,2008, Phys. Rev. D 77, 023534.

[30] Koivisto, T., Mota, D.F.,2008b, Astrophys. J 679, 1.

[31] Yadav, A.K., Saha, B.,2012, Astrophys. Space Sci 337, 759

[32] Sharif, M., Rani, S., 2011, Mod. Phys. Lett. A 26, 1657.

[33] Ferraro, R., Fiorini, F.,2007, Phys. Rev. D 75, 084031. arXiv:gr-qc/ 0610067.

[34] Bengochea, G.R., Ferraro, R., 2009, Phys. Rev. D 79, 124019. arXiv: 0812.1205

[35] Vakili.B.,2008, Phys. Lett. B 664, 16 [arXiv:0804.3449].

[36] Capozziello.S., De Felice. A., 2008 JCAP 0808, 016 [arXiv:0804.2163]; Capozziello.S., Lambiase,G.,2000, Gen. Rel. Grav. 32, 673 [gr-qc/9912083].

[37] Wei. H., Guo. X.J., Wang, L.F., 2012, Phys. Lett. B , 707, 298304.

[38] Fayaz. V., Hossienkhani. H., et al., 2014, Astrophys. Space Sci. 353, 301.

[39] de Ritis.R., et al., 1990, Phys. Rev. D 42, 1091.

14

Pulsating red giant and supergiant stars in the Local Group dwarf galaxy Andromeda I

Elham Saremi[1] · Abbas Abedi[1] · Atefeh Javadi[2] · Jacco van Loon[3] · Habib Khosroshahi[2]

[1] Department of Physics, Faculty of Science, University of Birjand, Birjand, P.O.Box 97175-615, Iran; email: saremi@birjand.ac.ir
[2] School of Astronomy, Institute for Research in Fundamental Sciences (IPM), Tehran, P.O.Box 19395-5531, Iran;
[3] Lennard-Jones Laboratories, Keele University, ST5 5BG, UK

Abstract. We have conducted an optical long-term monitoring survey of the majority of dwarf galaxies in the Local Group, with the Isaac Newton Telescope (INT), to identify the long period variable (LPV) stars. LPV stars vary on timescales of months to years, and reach the largest amplitudes of their brightness variations at optical wavelengths, due to the changing temperature. They trace stellar populations as young as ~ 30 Myr to as old as ~ 10 Gyr whose identification is one of the best ways to reconstruct the star formation history.

The system of galactic satellites of the large Andromeda spiral galaxy (M 31) forms one of the key targets of our monitoring survey. In this first paper in the series, we present the first results from the survey in the form of a census of LPV stars in Andromeda I (And I) dwarf galaxy.

Photometry was obtained for 10585 stars in a 0.07 square degree field, of which 116 stars were found to be variable, most of which are Asymptotic Giant Branch (AGB) stars. Our data were matched to mid-infrared photometry from the *Spitzer* Space Telescope, and to optical catalogues of variable stars from the *Hubble* Space Telescope.

Keywords: stars: evolution, stars: red giants, supergiants, stars: mass-loss, stars: oscillations, galaxies: individual: Andromeda I, galaxies: stellar content

1 Introduction

Dwarf galaxies are the most common type of galaxies in the Universe, and the building blocks of more massive galaxies in hierarchical formation scenarios. The great variety in terms of stellar mass, luminosity, gas content, metallicity and surface brightness reflect the complex dynamical and astrophysical processes that drive galaxy evolution. Also, they represent the smallest scales on which astronomers are able to detect dark matter [1, 2, 3, 4, 5]. Therefore, their study is crucial for improving the cosmological models and our understanding of galaxy evolution. Naturally, our home, the Local Group (LG), is the best place to study dwarfs since their individual stars can be resolved and evolutionary histories can be derived in great detail.

The LG comprises dwarf galaxies of all diversity: dwarf spheroidals (dSphs), dwarf irregulars (dIrrs), and transition (dTrans) galaxies. DSphs are found in denser environments and have lower luminosities than dIrrs. They show no evidence of recent star formation (within the last 200 Myr). DTrans are located in similar environments occupied by dIrr galaxies; however, their luminosity and the star formation history are more comparable to dSphs [6, 7, 8, 9, 10, 11].

We have conducted an optical monitoring survey of the majority of dwarf galaxies in the LG, with the Isaac Newton Telescope (INT) for a duration of three years [12]. Our main objective is to identify all Long Period Variable stars (LPVs) in them. Then, we can determine the star formation histories from the mass function of LPVs, with a method that we successfully applied in some of the LG galaxies [13, 14, 15, 16, 17]. The most evolved stars with low to intermediate birth mass, $0.8 - 8 \, M_{\odot}$, at the tip of the Asymptotic Giant Branch (AGB), and somewhat more massive red supergiants (RSGs), show brightness variations on timescales of ≈ 100 to > 1000 days due to radial pulsations and this makes them powerful tools to trace stellar populations as young as ~ 30 Myr to as old as the oldest Globular Clusters.

In this paper, we present a first census of LPV stars in Andromeda I (And I) dwarf galaxy with data from our survey. And I is a bright dSph ($M_{\rm V} = -11.8 \pm 0.1$) mag [18], that was initially discovered on photographic plates by van den Bergh (1972). It lies some $3.3°$ from the center of M 31 at a position angle of $\sim 135°$ relative to the M 31 major axis [19]. Considerable efforts have been made to study the structure and properties of And I. For instance, McConnachie & Irwin (2006) have shown that And I is a strongly disrupted satellite of M 31. Also, they have derived some of the structural parameters such as position angle (22 ± 15 deg), ellipticity (0.22 ± 0.04), tidal radius (10.4 ± 0.9 arcmin) and half-light radius (2.8 arcmin) for this galaxy [18]. Kalirai et al. (2010) with spectroscopic data of And I have calculated the mean radial velocity 375.8 ± 1.4 km/s and intrinsic dispersion velocity 10.6 ± 1.1 km/s. They determined a metallicity of -1.45 ± 0.04 using theoretical isochrones [20]. Recent studies have shown that a large fraction of the M 31 dwarf galaxies such as And I have extended star formation histories, and appear inconsistent with an early truncation of their star formation histories [21].

The distance to And I has been determined *via* several methods. Da Costa et al. (1996) determined a distance of 810 ± 30 kpc by using Horizontal Branch stars, i.e. a distance modulus 24.55 ± 0.08 mag [19]. By using the tip of the red giant branch (RGB), McConnachie et al. (2004) found a distance of 735 ± 23 kpc, i.e. a distance modulus 24.33 ± 0.07 mag [22]. However, Conn et al. (2012) revised this to $\mu = 24.31 \pm 0.05$ mag. Martínez-Vázquez et al. (2017) calculated the distance based on the properties of the RR Lyrae stars and several independent techniques [23]: the reddening-free period-Wesenheit relations (24.49 ± 0.08), the luminosity-metallicity ($M_{\rm V}$ versus [Fe/H]) relation (24.54 ± 0.16), the first overtone blue edge relation (24.49 ± 0.10) and the RGB tip method (24.49 ± 0.12) [24]. In this paper, we adopted the last one, i.e. a distance modulus 24.49 ± 0.12 mag.

This paper is organized as follows: In Section 2, we describe the observations performed for this data set. Section 3 explains the data reduction and photometry method. The quality of data is discussed in section 4 and our method to detect LPVs is presented in Section 5. The discussion follows in Section 6.

2   Description of observations

Over the period June 2015 – October 2017, we used the Wide Field Camera (WFC) to conduct a survey of the majority of dwarf galaxies in the LG. The WFC is an optical mosaic camera at the 2.5m Isaac Newton Telescope (INT) of the Observatorio del Roque de los Muchachos (La Palma). It consists of four 2048×4096 CCDs, with a pixel size of 0.33 arcsec/pixel. The edge to edge limit of the mosaic, neglecting the $\sim 1'$ inter-chip spacing, is $34.2'$.

LPVs vary on timescales from ~ 100 days for low-mass AGB stars to ~ 1300 days for the dustiest massive AGB stars. Although we are not aiming to determine accurate periods,

Table 1: Log of WFC observations of And I dwarf galaxy.

Date (y m d)	Julian date	Epoch	Filter	t_{exp} (sec)	Airmass
2016 02 09	2457428.3494	2	i	60	1.475
2016 06 14	2457553.6758	3	i	60	1.612
2016 08 10	2457610.6268	4	i	60	1.086
2016 08 12	2457612.6985	4	V	80	1.014
2016 10 20	2457681.6113	5	i	60	1.189
2016 10 20	2457681.6286	5	V	80	1.263
2017 01 29	2457783.4433	6	i	60	2.393
2017 08 01	2457966.7014	7	i	60	1.018
2017 08 01	2457966.7174	7	V	80	1.014
2017 09 01	2457997.5625	8	i	60	1.087
2017 09 01	2457997.5799	8	V	80	1.054
2017 10 06	2458033.4097	9	i	60	1.286
2017 10 08	2458034.5389	9	V	80	1.014

to identify the LPVs and to determine their amplitude and mean brightness, we required monitoring over several epochs, spaced by a month or more. The first epoch was June 2015, and the last was completed in October 2017. Unfortunately, we could not obtain data in October 2015 due to bad weather conditions and we observed a total of 9 epochs.

Observations were taken in the WFC Sloan i and Harris V filters. We selected i band because the spectral energy distribution (SED) of cool evolved stars peaks around 1 μm, thus enhancing the contrast between the LPVs and other, warmer stars. Also, the bolometric correction needed to determine the luminosity in this band is smallest and least dependent on the colour and hence most accurate, and the effects of attenuation by dust are minimal. For monitoring the variations in temperature – and thus radius[1] – and to have more accurate SED modeling, we also observed in V band on several occasions to obtain colour information.

We chose exposure times that yield sufficient signal-to-noise (S/N) to detect small changes in magnitude at different epochs. i band amplitudes of pulsating AGB stars are > 0.1 mag. Therefore, we aimed for $S/N = 10$ for the faintest stars, equivalent to the tip of the RGB. In order to cover the chip gaps of WFC and increase the S/N, the galaxy is observed 9 times with offsets of $30''$ between the pointings each night. Additional observations were made of fields with photometric standard stars to be used for photometric calibration. The details of the observations used in this survey only for And I dwarf galaxy are listed in Table 1.

3   Data processing

The raw images can not be used for scientific analysis before they are combined, calibrated and corrected for artefacts. For this data reduction process, we used THELI (Transforming HEavenly Light into Image), an image processing pipeline for optical images taken by multi-chip (mosaic) CCD cameras [25]. It consists of a number of shell scripts that each perform a specific task and can be run in parallel on multi-chip CCDs. At first, image files are separated into frames of the individual chips (4 CCD chips in a WFC mosaic) and from then on, work is done on individual chips rather than whole images. Next, instrumental signatures

---

[1]Radius R and temperature T_{eff} are related via the luminosity L by the well-known formula $L = 4\pi R^2 \sigma T_{\text{eff}}^4$

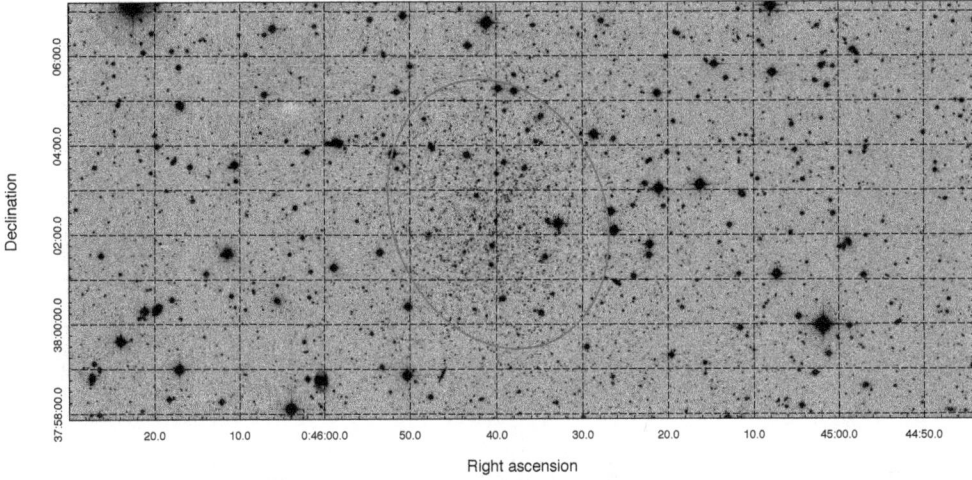

Figure 1: The master WFC image of And I dwarf galaxy. The half-light radii is marked with blue ellipse.

are removed from the data: the electronic offset (bias), pixel response and instrumental throughput variations (flatfield) and interference in the back-illuminated thinned detector chip, especially prevalent at the reddest wavelengths which more closely match the physical thickness of the chip (frigne pattern). While this process is repeated for all of the optical data, the subsequent step strongly depends on the scientific objectives and on the kind of data at hand. The data can be combined, optimised for astrometric accuracy, or optimised for automatised aperture photometry. Because of the crowded nature of the data in the direction of And I itself, it is more important at this stage to obtain accurate astrometry because the photometry will be done based on fitting and subtracting the individual stellar images.

Fortunately, in the THELI pipeline more emphasis is put on precise astrometry than precise photometry and it is perfectly suited to our goal. THELI uses the LDAC (Leiden Data Analysis Center) catalogue format, astronomical object catalogues created by Erik Deul and Emmanuel Bertin [26], and SCAMP package, astrometric tool developed in particular for multi-chip cameras [27], to create a full astrometric solution taking into account the gaps between the chips and overlapping objects. A useful feature of THELI is that it creates weights for individual frames. The weights are created based on the normalized flats. They mask defects such as cosmics and hot pixels in the images. The responsibility of the pipeline ends with the co-addition step. Sky background calculated with the SEXTRACTOR (software for source extraction) package [28] is subtracted from all frames and finally, the SWARP package [29] in THELI creates a co-added image using a weighted mean method [30].

To perform photometry in our crowded stellar field, we used the DAOPHOT/ALLSTAR software developed by Peter Stetson (1987). This package employs a Point Spread Function (PSF) method [31]. After identifying stars (from peaks above the noise) and obtaining aperture photometry for them, about 50–70 isolated stars were chosen in each frame to build an initial PSF model. By subtracting all but the PSF model stars from the frame, a final PSF model was build. Then ALLSTAR subtracts all the stars of image with using PSF fitting photometry along with the current best guesses of their positions and magnitudes.

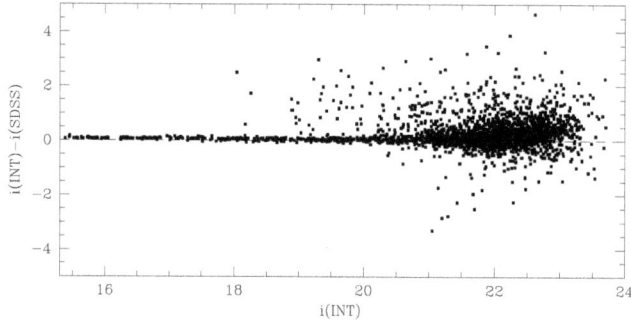

Figure 2: Magnitude differences between INT catalogue and SDSS of And I dwarf galaxy, plotted against i magnitude of our catalogue.

The individual images were aligned using the DAOMASTER routine, which computes the astrometric transformation equation coefficients from the ALLSTAR results. We combined the individual images using the MONTAGE2 routine [32] to create a master mosaic of And I (Fig. 1) and then a master catalogue of stars. This master catalogue was used as input for ALLFRAME, which simultaneously performed PSF-fitting photometry on these stars within each of the individual images. In this way, our final catalogue for And I was made.

To determine aperture corrections to the PSF-fitting photometry, i.e. the difference between the PSF-fitting and large-aperture magnitude of these stars, we used the DAOGROW routine [33]. It constructs growth-curves for each frame from which all stars had been subtracted except the PSF stars. We then applied the COLLECT routine [34] to calculate the aperture correction. We added this value to each of the PSF-fitting magnitudes using the NEWTRIAL routine [35].

Photometric calibration was then performed using the NEWTRIAL routine. Zero points were obtained for each frame based on the standard field observations. To obtain accurate zero points in the Sloan i filter, we applied transformation equations derived from comparing Stetson's compilation of the Landolt standard stars with the corresponding SDSS DR4 photometry [36]. For frames obtained on nights without standard star measurements, we adopted the average of zero points from other nights. Airmass-dependent atmospheric extinction corrections were applied adopting the extinction coefficients determined for La Palma [37].

In order to estimate the accuracy of calibration, we cross correlated the results with the SDSS. The matches were obtained by performing search iterations using growing search radii, in steps of $0.1''$ out to $1''$, on a first-encountered first-associated basis but after ordering the principal photometry in order of diminishing brightness (to avoid rare bright stars being erroneously associated with any of the much larger numbers of faint stars). As shown in Fig. 2, the result is consistent with good accuracy with SDSS catalogue within our desired range ($21 > i > 18$ mag, see below).

4   Quality assessment

To estimate the completeness of our catalogue, we used the ADDSTAR routine in the DAOPHOT package [31]. This task can add synthetic stars to an image, either placed at random by the computer or in accordance with positions and magnitudes specified by us. All photometric are then applied to the new images and so the star-finding efficiency and the photometric

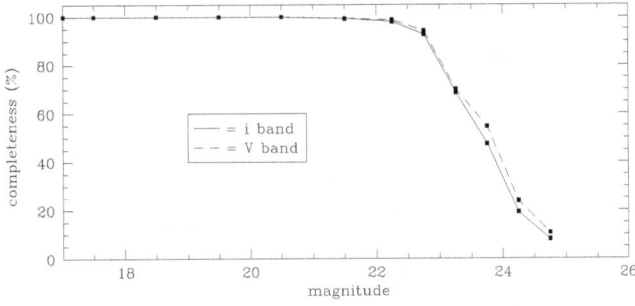

Figure 3: Completeness as a function of i-band (solid line) and V-band (dashed line) magnitude.

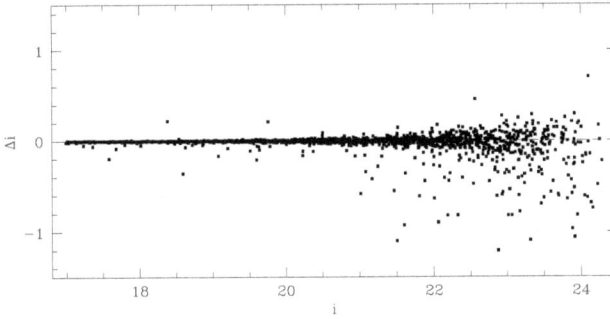

Figure 4: The difference between the input stellar magnitudes and the recovered stellar magnitudes from the artificial star tests.

accuracy can be estimated by comparing the output data for these stars to what was put in.

To avoid small number statistics in the artificial-star analysis without significantly changing the properties of the field (crowdedness), we added 300 artificial stars in each of 8 trials to the master mosaic and two of the individual frames in different bands in 1-mag bins starting from $i = 17$ mag until $i = 25$ mag (for V the same as i). Stars were positioned randomly in the image and Poisson noise was added. Then, we repeated the DAOPHOT/ALLSTAR/ALLFRAME procedure on the new frames as described before. Once the photometry was done and the final list created, we used DAOMASTER to evaluate what fraction of stars was recovered. As one can see in Fig. 3, our catalogue is essentially complete down to $i \sim 22$ mag (near the RGB–tip, see below), dropping to below 50% at $i = 23.6$ mag. The V-band reaches similar completeness levels but at ~ 0.1 magnitude fainter.

The result of estimating the accuracy of our photometry is shown in Fig. 4. The difference between the input stellar magnitudes and the recovered magnitudes down to $i \sim 22$ mag is very small, $|\Delta i| < 0.1$ mag, but it increases for fainter magnitudes. So, this photometry is deep enough to meet the scientific objectives of our project of studying AGB stars and RSGs, because they have $i < 21.5$ mag – except possibly for heavily dust-enshrouded cases, which however are rare.

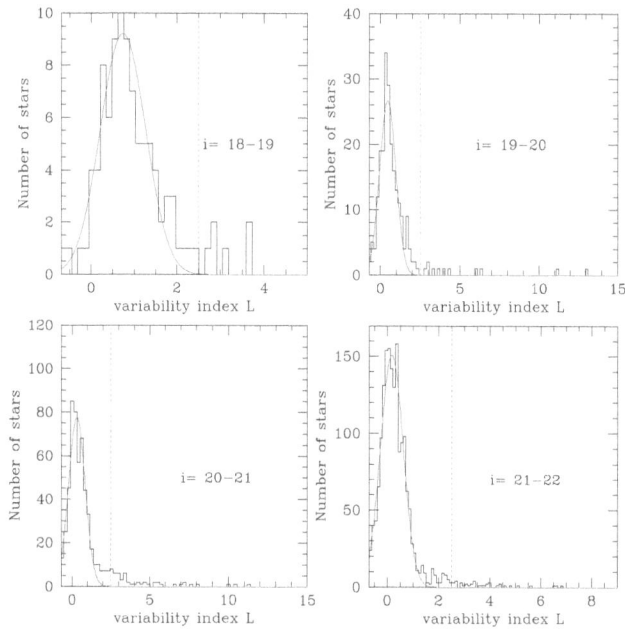

Figure 5: Histograms of the variability index L, for several i-band magnitude bins. The solid lines show Gaussian functions fitted to the histograms.

5   Variability analysis

For finding variable star candidates, we used the NEWTRIAL program [34]. This program was introduced by Welsh & Stetson (1993) and developed further by Stetson (1996). In this method, the observations are paired on the basis of timespan between observations such that the observations of each pair have a timespan less than the shortest period expected for the kind of the variable stars of interest (100 days or longer for LPVs). If within a pair of observations only one measurement is available for a particular star then the weight of the pair for that star is set to 0.5.

The NEWTRIAL program first calculates the J index:

$$J = \frac{\Sigma_{k=1}^{N} w_k \ \mathrm{sign}(P_k)\sqrt{|P_k|}}{\Sigma_{k=1}^{N} w_k}. \tag{1}$$

Here, observations i and j have been paired and each pair k has been given a weight w_k; the product of the normalized residuals, $P_k = (\delta_i \delta_j)_k$, where $\delta_i = (m_i - \langle m \rangle)/\epsilon_i$ is the deviation of measurement i from the mean, normalized by the error on the measurement, ϵ_i. Note that δ_i and δ_j may refer to measurements taken in different filters ($P_k = \delta^2 - 1$ if $i = j$). The J index has a large positive value for variable stars and tends to zero for data containing random noise only.

When we are dealing with a small number of observations or corrupt data we gain from also calculating the Kurtosis index:

$$K = \frac{\frac{1}{N}\Sigma_{i=1}^{N}|\delta_i|}{\sqrt{\frac{1}{N}\Sigma_{i=1}^{N}\delta_i^2}}. \tag{2}$$

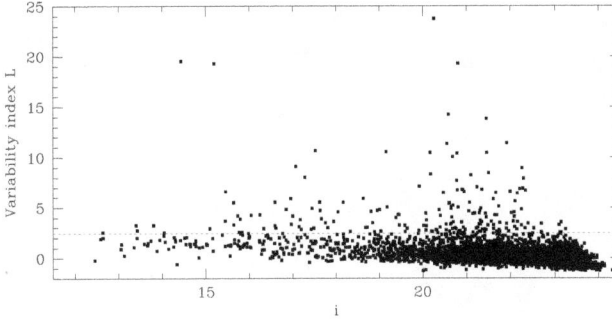

Figure 6: Variability index L vs. i-band magnitude. The dashed line indicates our threshold for identifying variable stars, at $L > 2.5$.

The value of K depends on the shape of the light-curve: $K = 0.9$ for a sinusoidal light variation, where the source spends most time near the extrema, $K = 0.798$ for a Gaussian distribution, which is concentrated towards the average brightness level (as would random noise), and $K \to 0$ for data affected by a single outlier (when $N \to \infty$).

Also, there is a variability index that depends on both the J and K indices [35]:

$$L = \frac{J \times K}{0.798} \frac{\Sigma_{i=1}^{N} w_i}{w_{\text{all}}},$$ (3)

where Σw is the total weight assigned to a given star and w_{all} is the total weight a star would have if observed in every single observation. We used L variability index for finding variable stars in this paper. One reason for using it is that the lightcurves of LPVs have shapes roughly between sinusoidal and triangular, and thus a kurtosis near 0.9-1. The L index is weighted by that kurtosis, making it larger for such variations compared to more random or erratic variations.

To determine the optimal variability threshold, we plotted histograms of the variability index for several i-band magnitude intervals in the range 18–22 mag along with a fitted Gaussian function to each of them. As shown in Fig. 5, while the Gaussian function is a near-perfect fit to the symmetrical distributions at low values for L, the distributions show a pronounced tail towards higher values for L. The departure from the Gaussian shape occurs typically around $L \approx 2.5$. Fig. 6 shows how the variability index L varies with i-band magnitude. The dashed line indicates our threshold for detected variability: $L > 2.5$.

The number of variable stars are suspect and we inspected all of the stars by eye. Some of these stars are located near the edge of the frame and some others located on the site of a saturated star in image, resulting in particularly poor photometry on several epochs with distorted stellar profiles and therefore, they are unreliable. We removed these stars and thus identified 116 variable stars in the And I. Two examples of likely long-period variability are shown in Fig. 7, with a non-variable star for comparison. One of the variable stars was also found to be variable at mid-IR wavelengths in the *Spitzer* monitoring survey DUSTiNGS [38] (cf. Section 6).

5.1 Amplitudes of variability

A measure for the amplitude of variability can be obtained by assuming a sinusoidal light-curve shape. It is done by comparing the standard deviation in the magnitudes to that

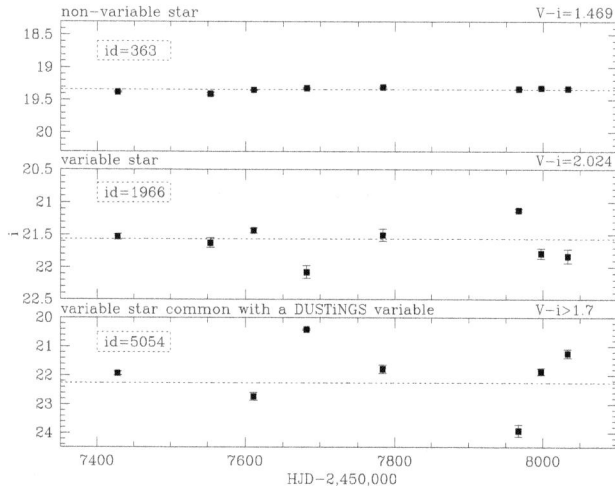

Figure 7: Example light-curves of two variable stars, with #5054 in common with the DUSTiNGS catalogue of variable *Spitzer* sources [38], with a non-variable star in the top panel for comparison.

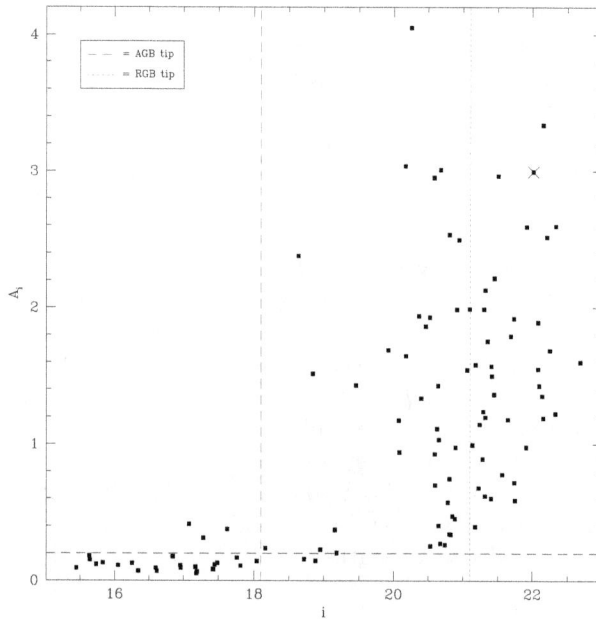

Figure 8: Estimated amplitude, A_i, of variability *vs.* i-band magnitude.

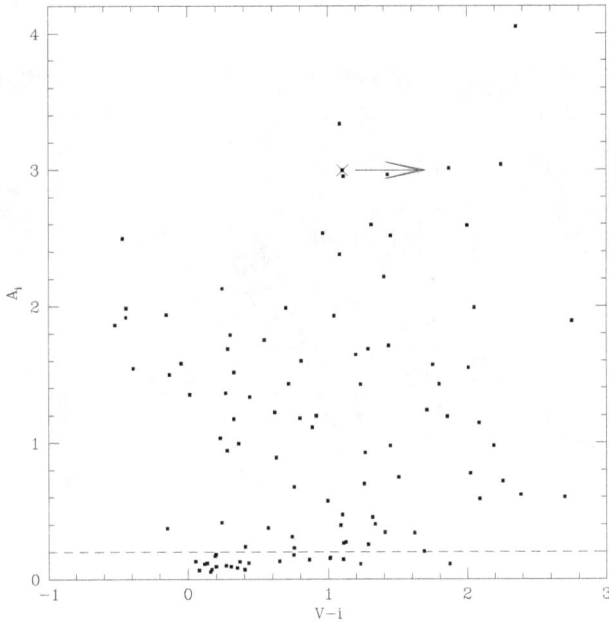

Figure 9: Estimated amplitude, A_i, of variability *vs.* colour.

expected for a completely random sampling of a sinusoidal variation. The estimated i-band amplitude of variability is plotted *vs.* i-band magnitude in Fig. 8. Variability could have been detected for $A_i > 0.2$ mag. There is a clear tendency for the amplitude to diminish with increasing brightness, which is a known [39, 40, 41] and to some extent understood [42] trend for AGB stars. The amplitudes stay below $A_i \sim 4.2$ mag and generally $A_i < 2.5$ mag. Very dusty AGB stars are known to reach such large amplitudes [39, 40, 41], but they are very rare. One example of such star, #5054 is highlighted in Fig. 8 with a cross; its lightcurve is displayed in Fig. 7 and it was also identified as a mid-IR variable by the DUSTiNGS survey [38]. The most interesting stars might be those three brighter than the tip of the AGB and with amplitude about 0.3–0.4 mag. They could be very massive AGB stars undergoing Hot Bottom Burning or RSGs and therefore, represent a "young" population (< 100 Myr).

The estimated i-band amplitude of variability is plotted *vs.* the $V - i$ colour in Fig. 9. Stars towards redder colours have larger amplitudes. This is not surprising as large-amplitude variability is known to be associated with abundant dust formation [39, 40, 41]. For the DUSTiNGS variable (#5054), we estimated $V > 23.7$ and thus $(V - i) > 1.7$ due to the completeness limit in V-band (23.7 mag).

6   Discussion

Our final catalogue contains 10585 stars in the region of CCD 4 of WFC (11.26×22.55 arcmin^2), with And I located near its center. Fig. 10 presents the spatial location of our variable candidates. The half-light radii is marked with a blue ellipse. The density of variable stars for the regions inside and outside the ellipse are 0.68 and 0.43 number/arcmin^2, respectively.

Colour–magnitude diagrams of this galaxy in i-band and V-band *vs.* $(V - i)$ colour are

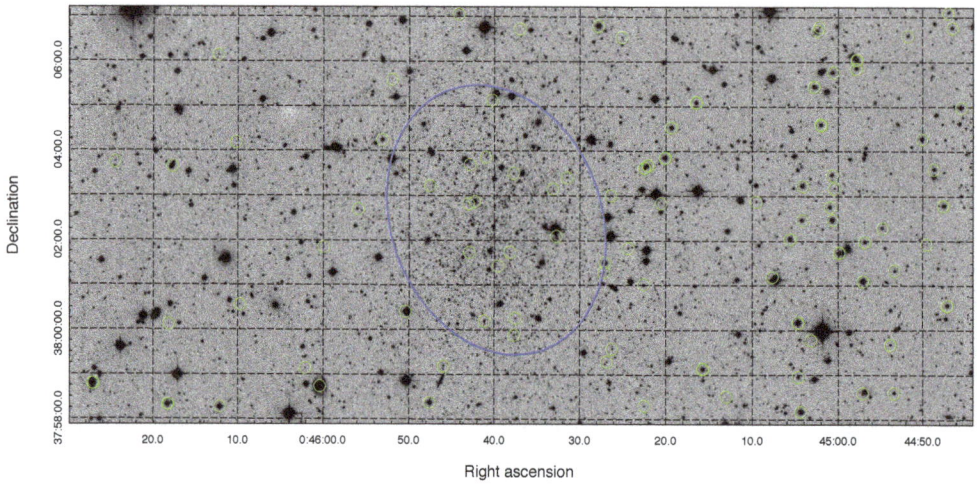

Figure 10: The master WFC image of And I dwarf galaxy with spatial location of our variable candidates. The half-light radii is marked with blue ellipse.

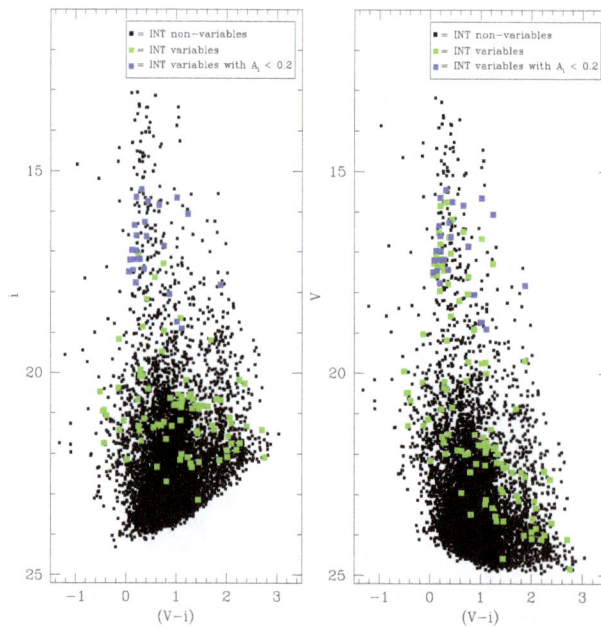

Figure 11: colour–magnitude diagrams showing the variable stars in green. The variable stars with $A_i < 0.2$ mag are highlighted in blue.

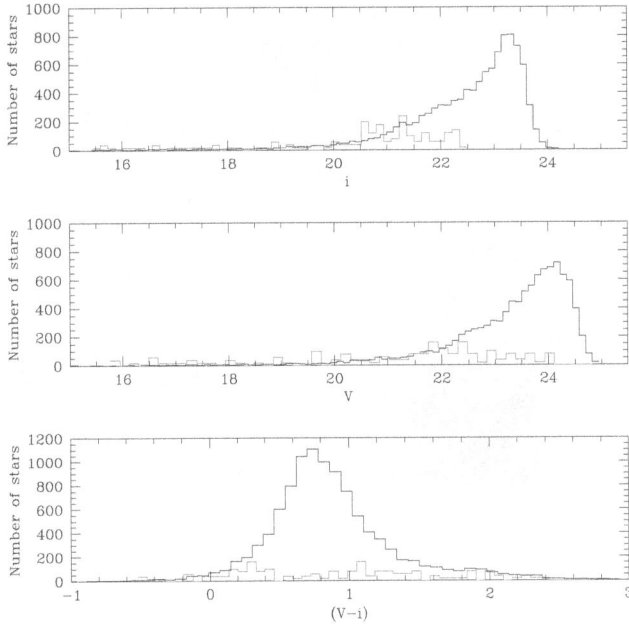

Figure 12: Distribution of all INT sources (black) and INT variable stars (red) as a function of brightness and colour. we multiplied the variable stars' histograms by 20.

shown in Fig. 11. The variable stars we identified are highlighted in green. These are mainly found between $i \sim 18$–22 mag, and as the stars get fainter, their number decreases. Some brighter variable RSGs are found (around $i \sim 16$–18 mag), too.

Fig. 12 presents histograms of the distributions over brightness and colour. The largest fraction of stars that are found to be variable occurs between $i \sim 18$–22 mag. As the frequency of stars increases from $i \sim 22$ mag, the frequency of variable stars decreases. This is probably because many stars have not yet reached the final phase of their evolution and they will still evolve to higher luminosities and lower temperatures before they develop large-amplitude variability.

To assess the level of contamination by foreground stars, we performed a simulation with the TRILEGAL tool [43]. We simulated two different sizes of the field, a 0.07 deg^2 (about the size of the entire CCD4 of WFC) and a 0.007 deg^2 (half the light of the galaxy), in the direction ($l = 121.68°$, $b = -24.82°$). As shown in Fig. 13 the foreground stars have contaminated our desired areas; therefore, the distribution of the population of AGB stars and then the star formation history can not be obtained from colour–magnitude diagrams alone. Because LPVs are relatively rare, and become rare still in directions away from the Galactic plane, one could instead use the distributions over LPVs to chart the star formation history, relatively free from foreground contamination.

The stellar population in the And I dwarf galaxy can be described using isochrones calculated by Marigo et al. (2008) (Fig. 14) [44]. The isochrones were calculated for And I metallicity, $Z = 0.00069$ [20]. The 100-Myr and 1-Gyr isochrones show consistency with the red branch of INT variables. The 10-Gyr isochrone defines the location of tip of the RGB, that here is 21.1 mag for i band. These isochrones are the most appropriate theoretical models for our purpose, for the following reasons:

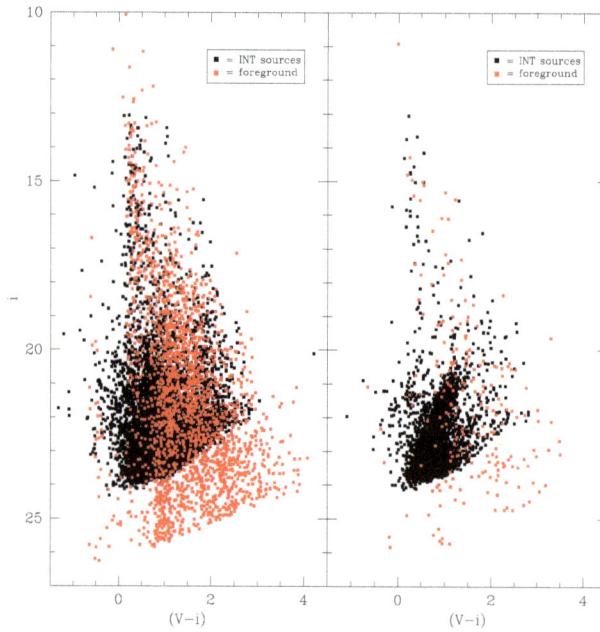

Figure 13: Estimated contamination by foreground stars (in red), from a simulation with TRILEGAL [43]. In the left panel, a 0.07 deg^2 field is considered while in the right panel, we have a 0.007 deg^2 field centered on And I (half the light of the galaxy).

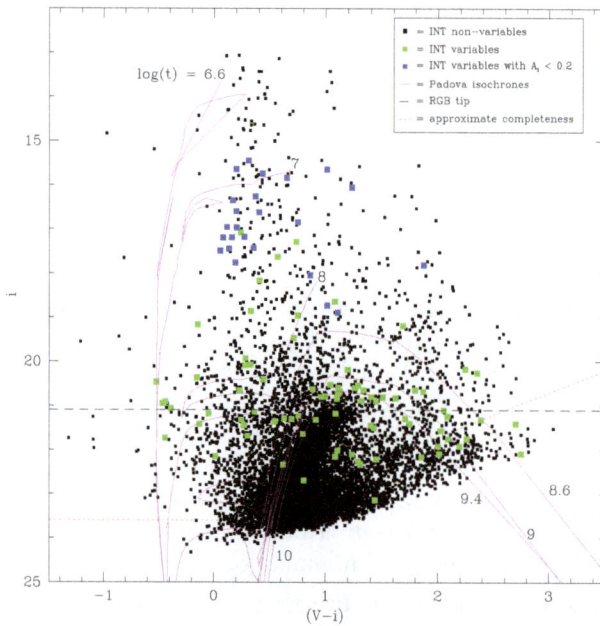

Figure 14: colour–magnitude diagram of $(V - i)$, with variable stars highlighted in green. The variable stars with $A_i < 0.2$ mag are highlighted in blue. Overplotted are isochrones from Marigo et al. (2008) for And I with a distance modulus of 24.49 mag [23].

- The star's evolution is followed all the way through the thermal pulsing AGB until the post-AGB phase. Crucially, two important phases of stellar evolution are included, viz. the third dredge-up mixing of the stellar mantle as a result of the helium-shell burning phase, and the enhanced luminosity of massive AGB stars undergoing hot bottom burning (HBB) [45];

- The molecular opacities which are important for the cool atmospheres of evolved stars have been considered in the models of stellar structure. The transformation from oxygen-dominated (M-type) AGB stars to carbon stars in the birth mass range $M \sim$ 1.5–4 $M_\odot$ is accounted for [46];

- The dust production in the winds of LPVs, and the associated reddening is included;

- The radial pulsation mode is predicted;

- Combination of their own models for intermediate-mass stars ($M < 7\ M_\odot$), with Padova models for more massive stars ($M > 7\ M_\odot$) [47]), gives a complete coverage in birth mass ($0.8 < M < 30\ M_\odot$);

- Magnitudes are calculated on a wide range of common optical and IR photometric systems;

- The isochrones are available via an internet-based form in a user-friendly format.

6.1   Cross-identifications in other catalogues

We cross-correlate our INT variability search results with the mid-IR variability search performed with the *Spitzer* Space Telescope [11] and also, with the variables catalogue obtained with the *Hubble* Space Telescope (HST) data [23]. The matches were obtained by search iterations using growing search radii, in steps of $0.1''$ out to $1''$, on a first-encountered first-associated basis after ordering the principal photometry in order of diminishing brightness (i-band/I-band for the optical catalogues, and 3.6-μm band for the *Spitzer* catalogue).

DUSTiNGS (DUST in Nearby Galaxies with *Spitzer*) was a 3.6 and 4.5 μm post-cryogen *Spitzer* Space Telescope imaging survey of 50 dwarf galaxies within 1.5 Mpc that was designed to identify dust-producing AGB stars and massive stars [11]. Using 2 epochs, spaced approximately 6 months apart, they identified a total of 4 variable stars for And I. As a result of cross-correlating, we obtained 5616 common stellar sources between our photometric catalogue and the DUSTiNGS survey in a 0.07 square degree field centered on And I. As shown in Fig. 15, all four of their variable stars are listed in our identified variables catalogue. We found a few variables among the brighter red giants, including a red one, at $[3.6] - [4.5] = 1$ mag. We did not find variables above the tip of the 10 million year isochrone, which is consistent with there being no LPVs among stars more massive than those which become RSGs. Also we did not find variables among the redder sources, $[3.6] - [4.5] > 0.4$ mag, around $[3.6] \sim 17 - 18$ mag. Is it possible that these sources have only been detected in our survey in one epoch, when they were near maximum brightness, so we could not determine any variability.

Martínez-Vázquez et al. (2017) presented the ISLAndS (Initial Star formation and Lifetimes of Andromeda Satellites) project, providing a census of variable stars in six M31 dSph satellites observed with the HST. They detected 296 RR Lyrae stars in And I [23]. There are 59 objects in common between their survey and ours, among which only two were identified by us as variable (Fig. 16). This illustrates that these surveys probe entirely different, complementary populations – the oldest populations (ISLAndS) and the intermediate-age and younger populations (our survey).

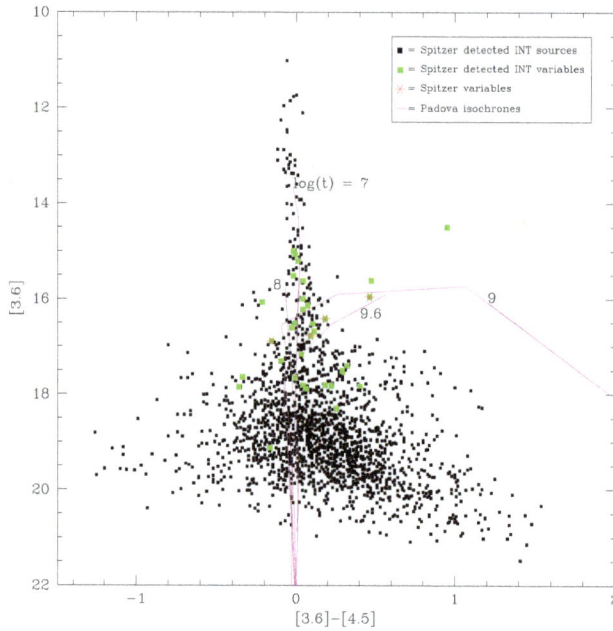

Figure 15: Mid-IR colour–magnitude diagram from DUSTiNGS survey, with INT variables highlighted in green and the variable stars of DUSTiNGS in red asterices. Isochrones from Marigo et al. (2008) for 10 Myr, 100 Myr 1 Gyr and 4 Gyr are drawn.

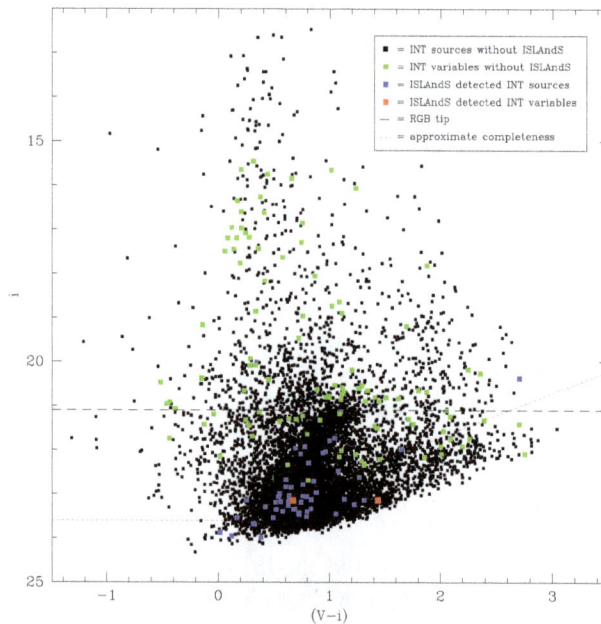

Figure 16: Optical colour–magnitude diagram showing the stars from the INT survey that were and were not detected in the ISLAndS variability survey [23]. Two red points represent common variable stars between the two surveys.

7 Conclusions

We have presented the preliminary, initial results from our long-term optical monitoring campaign at the Isaac Newton Telescope of a sample of the LG dwarf galaxies, namely for the And I satellite of M31. We found 116 variable stars, including a few among the brighter and redder sources, some of which have also been detected at mid-IR wavelengths. This indicates the presence of stars around a Gyr or younger in And I, which contribute to the replensihment of the interstellar medium within And I.

Acknowledgments

The observing time for this survey is provided by the Iranian National Observatory and the UK-PATT allocation of time to programmes I/2016B/09 and I/2017B/04 (PI: J. van Loon). The observers thank the Iranian National Observatory and the School of Astronomy (IPM) for the financial support of this project. The first author also thanks the School of Astronomy for the research grant. We thank the ING observers for service mode observations. Also, we thank Alireza Molaeinezhad, Arash Danesh, James Bamber and Iain McDonald for their efforts during the observations. We are grateful to Peter Stetson for sharing his photometry routines, Marta Boyer and Clara Martínez-Vázquez for sending us their full variability catalogue.

References

[1] Mateo M., Olszewski E. W., Vogt S. S., Keane M. J., 1998, AJ, 116, 2315

[2] Kleyna J. T., Wilkinson M. I., Evans N. W., Gilmore G., 2004, MNRAS, 354, L66

[3] Gilmore G., et al., 2007, ApJ, 663, 948

[4] Walker M. G., Mateo M., Olszewski E. W., Peñarrubia J., Evans N.W., Gilmore G. 2009, ApJ, 704, 1274

[5] Wolf J., et al., 2010, MNRAS, 406, 1220

[6] Dolphin A. E., Weisz D. R., Skillman E. D., Holtzman J. A., 2005, ASP Conference Series

[7] Read J. I., Pontzen A. P., Viel M., 2006, MNRAS, 371, 885

[8] Tolstoy E., Hill V., Tosi M., 2009, ARA&A, 47, 371

[9] Weisz D. R., et al., 2011, ApJ, 743, 8

[10] Kazantzidis S., Lokas E. L., Mayer L., Knebe A., Klimentowski J., 2011, ApJL, 740, L24

[11] Boyer M. L., et al., 2015, ApJ, 216, 10

[12] Saremi E., et al., 2017, J. Phys.: Conf. Ser, 869, 012068

[13] Javadi A., van Loon J. Th., Mirtorabi M. T., 2011a, MNRAS, 411, 263

[14] Rezaeikh S., Javadi A., Khosroshahi H., van Loon J. Th., 2014, MNRAS, 445, 2214

[15] Javadi A., et al., 2015, MNRAS, 447, 3973

[16] Javadi A., van Loon J. Th., Khosroshahi H., Tabatabaei F., Golshan R. H., 2017, MNRAS, 464, 2103

[17] Golshan R. H., Javadi A., van Loon J. Th., Khosroshahi H., Saremi E., 2017, MNRAS, 466, 1764

[18] McConnachie A. W., Irwin M. J., 2006, MNRAS, 365, 1263

[19] Da Costa G. S., Armandroff T. E., Caldwell N., Seitzer P., 1996, AJ, 112, 2576

[20] Kalirai J. S., et al., 2010, ApJ, 711, 671

[21] Martin N. F., et al., in press (arXiv:1704.01586)

[22] McConnachie A. W., et al., 2004, MNRAS, 350, 243

[23] Martínez-Vázquez., et al., in press (arXiv:1710.09038)

[24] Conn A. R., et al., 2012, ApJ, 758, 1

[25] Erben T., et al., 2005, AN, 326, 432

[26] E.Deul, ftp://ftp.strw.leidenuniv.nl/pub/ldac/software/

[27] Bertin E., 2006, ADASS XV: ASP Conf. Ser, 351, 112

[28] Bertin E., Arnouts S., 1996, A&A Supplement, 117, 393

[29] Bertin E., 2010, Astrophysics Source Code Library, 10068

[30] Schirmer M., 2013, ApJ Supplement, 209, 2

[31] Stetson P. B., 1987, PASP, 99, 191

[32] Stetson P. B., 1994, PASP, 106, 250

[33] Stetson P. B., 1990, PASP, 102, 932

[34] Stetson P. B., 1993, in: Stellar Photometry Current Techniques and Future Developments, IAU Coll. Ser. 136, 291

[35] Stetson P. B., 1996, PASP, 108, 851

[36] Jordi K., Grebel E. K., Ammon K., 2006, A&A, 460, 339

[37] García-Gil A., Muñoz-Tuñón C., Varela A. M., 2010, PASP, 122, 1109

[38] Boyer M. L., et al., 2015 ApJ, 800, 51

[39] Wood P. R., Whiteoak J. B., Hughes S. M. G., Bessell M. S., Gardner F. F., Hyland A. R., 1992, ApJ, 397, 552

[40] Wood P. R., 1998, A&A, 338, 592

[41] Whitelock P. A., Feast M. W., van Loon J. Th., Zijlstra A. A., 2003, MNRAS, 342, 86

[42] van Loon J. Th., et al., 2008, A&A, 487, 1055

[43] Girardi L., Groenewegen M. A. T., Hatziminaoglou E., da Costa L., 2005, A&A, 436, 895

[44] Marigo P., Girardi L., Bressan A., Groenewegen M. A. T., Silva L., Granato G. L., 2008, A&A, 482, 883

[45] Iben I. Jr., Renzini A., 1983, ARA&A, 21, 271

[46] Marigo P., Girardi L., 2007, A&A, 469, 239

[47] Bertelli G., Bressan A., Chiosi C., Fagotto F., Nasi E., 1994, A&A, 106, 275

15

A Simplified Solution for Advection Dominated Accretion Flows with Outflow

Seyede Tahere Kashfi · Shahram Abbassi

Department of Physics, Ferdowsi University of Mashhad, Mashhad, Iran

Abstract. The existence of outflow in the advection dominated accretion flows has been confirmed by both numerical simulations and observations. The outflow models for ADAF have been investigated by several groups with a simple self similar solution. But this solution is inaccurate at the inner regions and can not explain the emitted spectrum of the flow; so, it is necessary to obtain a global solution for ADAFs with outflow. In this paper, we use a simplified global solution to study the structure of ADAF in the presence of outflow. In this method which is proposed by Yuan et al (2008, hereafter YMN08), the radial momentum equation is replaced by a simple algebraic relation between angular velocity and Keplerian angular velocity to avoid the difficulty of the calculation of global solution. We consider the radial dependence for mass accretion rate $\dot{M} = \dot{M}_{out}(r/r_{out})^s$ where s is a constant and we do not change the other dynamical equations. We investigate the variation of physical quantity of accretion flow which is caused by outflow. The results that we obtained comply with our expectations from the influence of outflow on the structure of accretion flow.

Keywords: accretion; accretion disks; black hole; global solution; outflow

1 Introduction

Advection dominated accretion flow (ADAF) is an important model for black hole accretion. In this model which is introduced by Narayen & Yi (1994, Abramowicz et al. 1995, Chen et al. 1995), most of the energy is stored in the accretion gas and advected to the central black hole. The ADAFs have high temperature and thermal stability and can explain the hard X-ray emission from X-ray sources and AGNs. Also, because the advective nature of these flows, they are especially good to explaining low luminous systems.

On the one hand, as Narayan & Yi (1994, 1995a) noted, this type of accretion flow have an interesting feature that is the positive value for Bernoulli parameter. This parameter is the sum of the gravitational potential energy, kinetic energy and the enthalpy of the flow. When this parameter have a positive value, the gas can escape to infinity, and on the another hand, for a given density at a large distance from the black hole (e.g., measured by Chandra on 1″ scales), the density close to the black hole is much less than the ADAF or Bondi prediction (Yuan et al. 2003). Also, numerical simulations (Stone et al. 1999; Hawley & Balbus 2002; Igumenshchev et al. 2003) indicate that a large fraction of accretion material of the flow can not reach the central black hole and it is lost in the form of outflow. For these reasons, the existence of outflows should be considered in ADAF models.

Outflow models for ADAF have been investigated by several group (Xu & Chen 1997; Blandford & Begelman 1999; Beckert 2000; Xue & Wang 2005; Bu et al. 2009), in the frame of a self similar solution. Self similar solution is only valid for the regions away from the boundaries and does not match the boundary conditions of the flow; also, it is too simplified for calculating the emitted spectrum; so, we need to obtain the global solution.

Quataert & Narayan (1999) presented the global solution for ADAF with outflow for the first time. They use the form $\dot{M} \propto r^s$ to describe the accretion rate. In their investigation, all dynamical equations of the flow do not change, except the continuity equation. They calculated spectral models of ADAFs with outflow and compared their results with observations of several astrophysical systems. Also, Yuan et al. (2003; 2009) adopted the same approach for RIAFs and hot accretion flows.

However, the global solution is a two points boundary value problem and technically, it is difficult to solve. Because the difficulty of calculating the global solution, the effect of existence of outflow on the structure of ADAFs has not been investigated any more.

In this paper, we focus on the structure of ADAFs in the presence of outflow rather than the calculation of the emitted spectrum. By following the simplified global solution which is introduced by YMN08, we replace the radial momentum equation by a simple algebraic relation between the angular velocity and Keplerian angular velocity of the flow. In this way, we avoid the two points boundary value problem and the consequently, the approximate global solution for ADAF with outflow can be obtained.

We present the basic equations in section 2, and numerical results in section 3. Section 4 is devoted to a short summary. We should note that the effect of outflow on the angular momentum and energy equations is ignored here. Since the most important effect of outflow is to modify the density profile, we do not expect this treatment to cause a remarkable change on our results.

2 Basic Equations

We consider an advection dominated accretion flow which is axisymmetric and steady state ($\frac{\partial}{\partial t} = \frac{\partial}{\partial \varphi} = 0$). The Paczyński & Witta potential $\phi = -\frac{GM}{(r-r_g)}$ is adopted to mimic the geometry of a Schwarzschild black hole where M is the mass of black hole and $r_g = \frac{2GM}{c^2}$ is the Schwarzschild radius. Also, the hydrostatic equilibrium in the vertical direction is assumed; so, we have a heghit-integrated set of equations. In this formulation, all physical quantities are functions of only the cylindrical radius.

To take into account the role of outflow, the continuity equation must be modified. So, we consider the dependency of the mass accretion rate on radius as follows (Blandford & Begelman 1999)

$$\dot{M} = -4\pi r \rho H v_r = \dot{M}_{out}(\frac{r}{r_{out}})^s \tag{1}$$

where $\dot{M}_{out}$ is the mass accretion rate at the outer radius and s is a parameter that indicates the strength of outflow. All the other quantities have their usual meanings.

The other equations that describe the structure of flow are

$$v_r \frac{dv_r}{dr} = (\Omega^2 - \Omega_K^2)r - \frac{1}{\rho}\frac{dp}{dr}, \tag{2}$$

$$v_r(\Omega r^2 - j) = -\alpha r c_s^2, \tag{3}$$

$$\rho v_r(\frac{d\epsilon_i}{dr} - \frac{p_i}{\rho^2}\frac{d\rho}{dr}) = (1-\delta)q^+ - q_{ie}, \tag{4}$$

$$\rho v_r(\frac{d\epsilon_e}{dr} - \frac{p_e}{\rho^2}\frac{d\rho}{dr}) = \delta q^+ + q_{ie} - q^-. \tag{5}$$

In these equations, $\epsilon_i(\epsilon_e)$ is the internal energy of ions (electrons) per unit mass of gas and $p_i(p_e)$ is the pressure due to ions (electrons) which is defined as

$$\epsilon_e = \frac{1}{\gamma_e - 1}\frac{kT_e}{\mu_e m_H}, \qquad \epsilon_i = \frac{1}{\gamma_i - 1}\frac{kT_i}{\mu_i m_H}, \tag{6}$$

$$p_i = \frac{\rho}{\mu_i}\frac{k}{m_H}T_i, \qquad p_e = \frac{\rho}{\mu_e}\frac{k}{m_H}T_e \tag{7}$$

The q^+ is the net turbulent heating rate and the quantity δ determines the fraction heating rate that directly heats electrons. q_{ie} describes the energy transfer rate from ions to electrons by Coulomb collision and q^- is the radiative cooling rate that includes bremsstrahlung and synchrotron emission and their Comptonization (see Manmoto et al. 1997 for detail).

The set of dynamical equations includes two algebraic equations (1) and (3) and three first order differential equations (2), (4) and (5). To solve these differential equations we need to have the outer boundary conditions for v_r, T_i and T_e. Also, because the transonic nature of ADAF, the sonic point conditions have to be considered and the most difficult part of this solution is to apply these conditions in the radial momentum equation.

By following YMN08, we use a simple relation between angular velocity and Keplerian angular velocity;

$$\Omega = f\Omega_k \tag{8}$$

where

$$f = \begin{cases} f_0 & r > 3r_g \\ 3f_0\frac{r-r_g}{2r_g} & r < 3r_g. \end{cases}$$

In this relation, f_0 is an adjustable parameter that YMN08 defined it as $f_0 = 0.33$ for a wide rang of $\dot{M}$ and a large value of α.

With considering this simple relation, the radial velocity become

$$v_r = \frac{-\alpha r c_s^2}{(f\Omega_k r^2 - j)}. \tag{9}$$

By substituting the equations (1) and (9) into the energy equations and expressing the isothermal sound speed c_s in terms of ions and electrons temperature, we can obtain two first order differential equations for two unknown variables T_i and T_e. If we have T_i and T_e at the outer boundary, we can integrate these differential equations inward for a given value of α and $\dot{M}$. The other physical quantities such as ρ, v_r and c_s will be calculated from the relations which they have with temperature. We use the the standard fourth order Runge-Kutta method and solve the differential equations with initial conditions.

3   Results

In our calculations, we set $M = 10M_\odot$ and $\delta = 0.3$. We examine our approach for different values of s, α and $\dot{M}$. The conditions that we imposed at the outer boundary are the same as in YMN08.

Figure 1 shows the variation of the accretion rate as a function of radius for $s = 0$, $s = 0.2$, $s = 0.4$ and $s = 0.6$. In this figure, the solid line indicates the case without outflow. In the presence of outflow, as radius decreases, the accretion rate declines and for a powerful outflow $s = 0.6$, only a few of material can be accreted on the black hole.

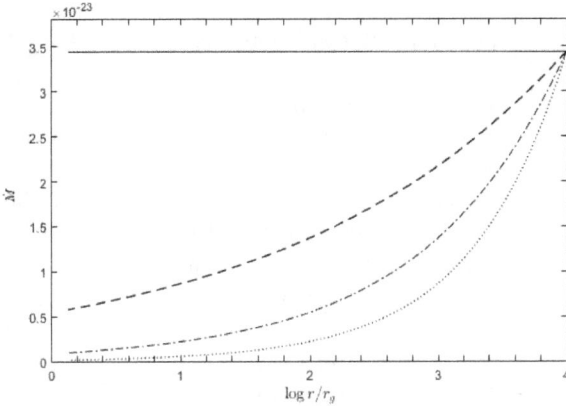

Figure 1: Accretion rate as a function of radius. The solid line, dashed line, dash dotted line and dotted line are related to $s = 0$, $s = 0.2$, $s = 0.4$ and $s = 0.6$ respectively.

Figure 2 indicates the radial variation of cooling rate. q^- includes three radiation mechanisms, bremsstrahlung radiation, synchrotron radiation and their Comptonization. In the outer regions of the flow, the bremsstrahlung emission is dominated while the synchrotron and Comptonization are the main emissions in the hot inner regions [8]. The bremsstrahlung emission is almost insensitive to the existence of outflow. As the figure 2 demonstrates, the cooling rate in the outer radii dose not change with the increasing the value of s because in these regions, the bremsstrahlung radiation is important and the effects of synchrotron and Comptonization emissions are negligible.

From the studies on the emitted spectrum [12], we know that the synchrotron emission decreases strongly with increasing s because when the value of s increases, the density and gas pressure decrease, so it is produced a weaker magnetic field. With increasing s the Compton power decreases strongly more than the synchrotron. In figure 2, we can see that the total cooling rate significantly reduces with large value for s.

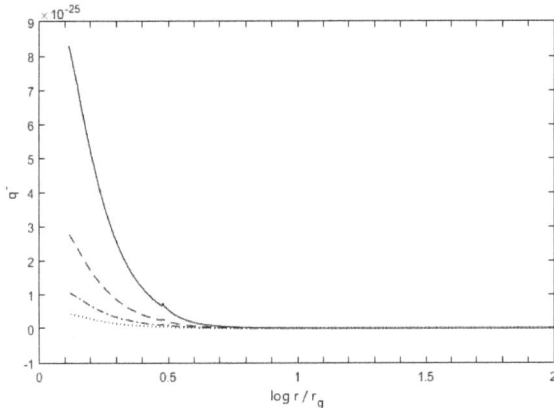

Figure 2

In figures 3 and 4, the profiles of ions and electrons temperature, radial velocity and density for different values of α and $\dot{M}$, are shown. In figure 3, $\alpha = 0.1$ and $\dot{M}_{out} =$

$10^{-3}\dot{M}_{Edd}$, and in figure 4, $\alpha = 0.3$ and $\dot{M}_{out} = 10^{-3}\dot{M}_{Edd}$. In figure 3, the radius of outer boundary is $r_{out} = 10^4 r_g$ and boundary conditions are $T_i = 0.2T_{vir}$, $T_e = 0.19T_{vir}$ where $T_{vir} = 3.6 \times 10^{12}(\frac{r_g}{r})$.

As we expected, the profile of density indicates an obvious decline in the presence of outflow. When $s = 0.6$, the value of density in the inner radii is almost 10^2 times less than the case without outflow. Also, the radial velocity decreases with increasing the value of s. In the profile of radial velocity, the variation of sound speed with radius is illustrated too. By comparing the profiles related to different value of s, we can realize that as s increases, the sonic point moves in and comes close to the black hole.

The plot of ions temperature shows that when s increases, the ions temperature decreases. From equation (4), we know there are two terms that affect the value of T_i, q^+ and q_{ie} and from their definitions, we know $q^+ \propto \dot{M}$ and $q_{ie} \propto \dot{M}^2$. Therefore, when the outflow exists, these two terms decrease and reduce the ions temperature.

In the outer regions of the electrons temperature profile, there is a decline in temperature when s increases. As we mentioned before, in these radii, the bremsstrahlung radiation dominates and this radiation is unaffected by the outflow. By return to the energy equation for electrons, we can see if q^- dose not change very much, the variation of two other terms q^+ and q_{ie} reduces the electrons temperature because they are the heating processes for electrons and with increasing s, these terms decrease. By contrast, in the inner radii, the cooling rate strongly decreases and this causes an increasing in electrons temperature. So, for the region close to the black hole, we have a larger T_e.

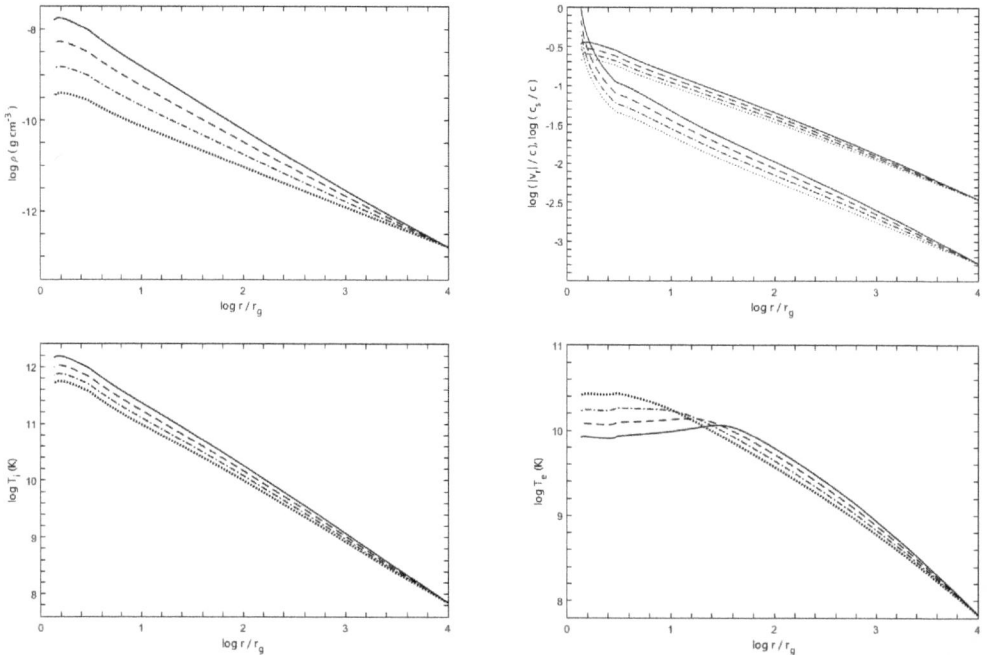

Figure 3: The variations with radii of density, radial velocity, ions temperature and electrons temperature for $s = 0$, $s = 0.2$, $s = 0.4$ and $s = 0.6$ are represented by solid, dashed, dash dotted and dotted lines respectively. The outer boundary is set at $10^4 r_g$ with the boundary conditions of $T_i = 0.2T_{vir}$ and $T_e = 0.19T_{vir}$ and $\alpha = 0.1$ and $\dot{M}_{out} = 10^{-3}\dot{M}_{edd}$ are fixed.

Figure 4 shows the variations of physical quantities in the presence of outflow for an another case. In this figure, the outer radius is $r_{out} = 10^2 r_g$ and boundary conditions are $T_i = 0.6T_{vir}$ and $T_e = 0.08T_{vir}$. The parameters α and $\dot{M}_{out}$ have a larger value in comparison with the case that was discussed in figure 3. So, the quantities q^+ and q_{ie} have larger values and the electrons temperature increases more clearly in the inner regions.

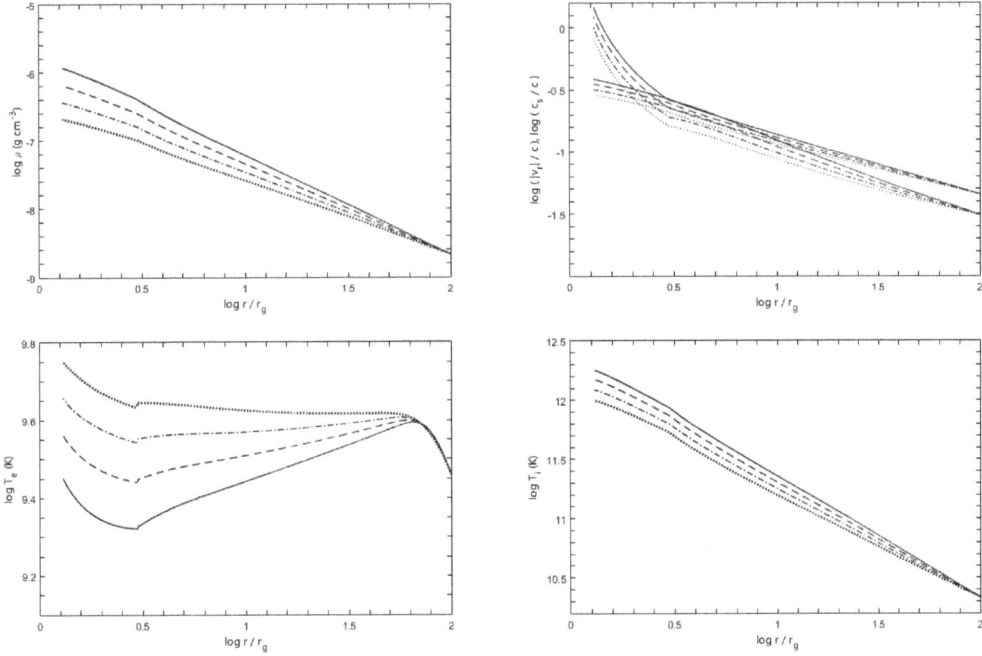

Figure 4: Radial dependency of density, radial velocity, ions temperature and electrons temperature for $s = 0$, $s = 0.2$, $s = 0.4$ and $s = 0.6$ are represented by solid, dashed, dash dotted and dotted lines respectively. The outer boundary radius is $10^2 r_g$ and the boundary conditions are $T_i = 0.6T_{vir}$ and $T_e = 0.08T_{vir}$. $\alpha = 0.3$ and $\dot{M}_{out} = 10^{-1}\dot{M}_{edd}$.

4 Summary

The observational evidence indicates the outflow exists in the accretion flow systems. For this reason, the study of structure of accretion flow in the presence of outflow is important. The spectral models for accretion flows with outflow have been investigated before in the frame of global solution. In these studies, a power law function of radius has been considered for the mass accretion rate and the other dynamical equations of the flow remain unchanged. Also, in these works, the emitted spectra has been taken into account.

In this paper, we focus on the structure of accretion flow in the presence of outflow. To overcome to difficulty of calculation the global solution, we use a simplified global solution to obtain the dynamical structure of accretion flow. We consider an advection dominated accretion flow and use a power law function for mass accretion rate to take into account the role of outflow. In spite of the simplifications that we imposed, our results have a acceptable agreement with the physical predictions.

Although if we want to have a full investigation, the other equations of the flow must be modified and the effects such as the angular momentum and energy transfer should be considered; as Xie & Yuan (2008) noted, the most important effect of outflow is on the density profile that is caused by the radial variation of mass accretion rate. If the strength of outflow that is determined by s is fixed, all other effects of outflow can only produce a difference for the density and temperature within a factor of ~ 2 [14]. So, we hope the simple method we used does not cause the significant error.

References

[1] Abramowicz, M. A., Chen, X., Kato, S., Lasota, J.-P., & Regev, O. 1995, ApJ, 438, L37

[2] Blandford, E. G., & Begelman, M. C. 1999, MNRAS, 303, L1

[3] Beckert, T. 2000, ApJ, 539, 223

[4] Bu D., Yuan F., & Xie F., 2009, MNRAS, 392, 325

[5] Chen, X., Abramowicz, M. A., Lasota, J.-P., Narayan, R., & Yi, I. 1995, ApJ, 443, L61

[6] Hawley, J. F., & Balbus, S. A. 2002, ApJ, 573, 738

[7] Igumenshchev, I. G., Narayan, R., & Abramowicz, M. A. 2003, ApJ, 592, 1042

[8] Manmoto, T., Mineshige, S., & Kusunose, M. 1997, ApJ, 489, 791

[9] Narayan, R., & Yi, I. 1994, ApJ, 428, L13

[10] Narayan, R., & Yi, I. 1995a, ApJ, 444, 231

[11] Paczyński, B., & Wiita, P. J. 1980, A& A, 88, 23

[12] Quataert, E., & Narayan, R. 1999, ApJ, 520, 298

[13] Stone, J. M., Pringle, J. E., & Begelman, M. C. 1999, MNRAS, 310, 1002

[14] Xie F. G., & Yuan F., 2008, ApJ, 681, 499

[15] Xu, G., & Chen, X. 1997, ApJ, 489, L29

[16] Xue, L., & Wang, J.-C. 2005, ApJ, 623, 372

[17] Yuan, F., Quataert, E., & Narayan, R. 2003, ApJ, 598, 301

[18] Yuan, F., Ma, R.Y., & Narayan, R., 2008, ApJ, 679, 984

[19] Yuan, F., Xie, F., & Ostriker, J. P. 2009, ApJ, 691, 98

Study of Solar Magnetic and Gravitational Energies through the Virial Theorem

Elham Bazyar[1] · Ali Ajabshirizadeh[2] · Zahra Fazel[3] · Jean-Pierre Rozelot[4]

[1] Research Institute for Astron. & Astrophys. of Maragha (RIAAM), 55134-441 Maraghe, Iran; email: bazyar_84@yahoo.com

[2] Research Institute for Astron. & Astrophys. of Maragha (RIAAM), 55134-441 Maraghe, Iran; email: a_adjab@tabrizu.ac.ir

[3] Astrophysics Depart., Physics Faculty, University of Tabriz, Tabriz, Iran; email: z_fazel@tabrizu.ac.ir

[4] Nice university, OCA-CNRS-Lagrange Depart., Av. Copernic, 06130 Grasse, France; email: rozelot@obs-azur.fr

Abstract. Virial theorem is important for understanding stellar structures. It produces an interesting connection between magnetic and gravitational energies. Using the general form of the virial theorem including the magnetic field (toroidal magnetic field), we may explain the solar dynamo model in relation to variations of the magnetic and gravitational energies. We emphasize the role of the gravitational energy in sub-surface layers which has been certainly minored up to now. We also consider two types of solar outer shape (spherical and spheroidal) to study the behavior of magnetic and gravitational energies. The magnetic energy affects by the solar shape, while the gravitational energy is not changed by the considered shapes of the Sun.

Keywords: ISM: virial theorem, ISM: magnetic field, rotation, gravitational energy

1 Introduction

There is a general agreement that the magnetic field of the Sun is generated by the dynamo process in which the kinetic energy is converted into the magnetic energy. The solar magnetic field may be divided into strong and weak components. Sunspots are the most prominent manifestation of the strong component. They show a regular cyclic behavior and they obey Hale's polarity rules. The strong field is assumed to result from dynamo action in the overshooting convection layer at the bottom of the convective zone. Differential rotation in the convective zone builds up a large toroidal magnetic field which is stored in the sub-adiabatically stratified medium. By contrast, the magnetic buoyancy instability is set up for producing strong fields of about $10^5 G$ [1, 2, 3, 4]. The weak field is irregular; it can be due to the turbulent dynamo action in the upper convective zone. In a weak regime, the instability provides a dynamo [5, 6], which regenerates the poloidal magnetic field and therefore close the dynamo cycle. A strong instability leads to the rise of the flux tubes through the convective zone, producing the bipolar active regions at the solar surface [7]. The azimuthal magnetic field that results from winding-up of an initially weak field is subject to instabilities.

In this paper, we study the importance of the virial theorem to understand the role of the gravitational energy in the sub-surface layers in comparison with magnetic energy. At first section, virial theorem is presented. In section 2, the solar shape is discussed. Section 3 gives the calculation used in this work. In section 4, results are presented. Finally, discussion and conclusion are followed in section 5.

2 Virial Theorem

Virial theorem can be important for understanding the structure and the evolution of stars. A general form of the virial theorem including the magnetic field was given by [8]:

$$\frac{1}{2}\frac{d^2 J}{dt^2} = 2K + \Omega + M + 2 \int P dV + S \tag{1}$$

$$J = \int \rho r^2 dV, M = \int \frac{B^2}{2\mu_0} dV, K = \frac{1}{2}\int \rho v^2 dV, \Omega = \frac{1}{2}\int \rho \Psi dV \tag{2}$$

where J is the inertial moment of the mass distribution in the region of radius R; K, the kinetic energy of (macroscopic) mass motion; Ω, the gravitational potential energy (with Ψ being the gravitational potential); M, the magnetic energy; P, the gas pressure and S, a surface integral over the boundary ∂R (with normal vector, n) of the region R, viz.

$$S = -\oint P(r \cdot n) ds + \frac{1}{\mu_0}\oint (r \cdot B)(B \cdot n) ds, \tag{3}$$

here, $P_{tot} = P_{gas} + \frac{B^2}{2\mu_0}$ is the total pressure (gas and magnetic). In the case of an ideal gas with constant ratio of specific heats, $\gamma = \frac{C_p}{C_v}$, the pressure integral over the region R can be related to the thermal (or internal) energy, U, that is

$$\int P dV = (\gamma - 1)U. \tag{4}$$

The region of integration can be the whole or a part of the studied star. If the stellar structure is in equilibrium, then it follows the virial theorem that

$$3(\gamma - 1)U + M + U = 0 \tag{5}$$

On the other hand, the total energy, E, is given by $E = U + M + \Omega$. Now by eliminating U:

$$E = -\frac{3\gamma - 4}{3(\gamma - 1)}(|\Omega| - M). \tag{6}$$

Considering Eq. (7), a necessary condition is obtained for the dynamical stability of an equilibrium state, that is

$$(3\gamma - 4)(|\Omega| - M) > 0, \tag{7}$$

which indicates a relation between the magnetic and the gravitational energies. Hence, the study of variations of the solar gravitational energy will be significant. It can be the key to identifying the seat region (the leptocline, [9]) of many phenomena; for example, an oscillation phase of the seismic radius together with a non-monotonic expansion of this radius with depth, a change in the turbulent pressure, an inversion in the radial gradient of the rotation velocity rate at about 50 in latitude, opacity changes and superadiabaticity. It could be also the cradle of hydrogen and helium ionization processes and probably the seat of in-situ magnetic fields and luminosity production. In addition, the solar radius changes as identified through space missions [10, 11, 12], or through temporal dependence of the f-modes are certainly induced both by magnetic and gravitational variations during the course of solar activity [13]. The mean equilibrium state of a star, at any given time, will follow the reduced magnetic virial theorem for equilibrium states:

$$2K + \Omega + M = 0 \tag{8}$$

where the radiant energy content is ignored, as being comparatively small. Moreover, the total energy must be conserved, for all states:

$$K + \Omega + M = C. \tag{9}$$

where C is constant. Note that the radiant energy losses at the stellar surface are not included, because the considered time intervals are short enough with respect to the star's cooling time (Kelvin-Helmholtz time, i.e. the time needed to radiate away a significant fraction of the thermal energy of a star). Following [14], when $M = 0$, hence $2K_0 + \Omega_0 = 0$ and $K_0 + \Omega_0 = C$. If the internal changes are not dynamically fast, then $K = K_0$ and $\Omega - \Omega_0 = -M$. This implies that the magnetic energy which decays into Joule heat ultimately gets converted into the gravitational potential energy. We differentiate Eqs. (9) and (10) for any change in the magnetic energy, δM, then

$$\delta K = 0, \delta M = -\delta \Omega. \tag{10}$$

Radiant energy losses at the surface can be neglected, as they are of second order. From $\Omega = -\int \frac{GM(r)}{r} dM(r)$, we deduce $\frac{\delta \Omega}{\Omega} = -\zeta \frac{\delta R}{R}$, where $0 < \zeta \leq 1$ (in the case of a homogenous structure of a star, $\zeta = 1$). Then

$$\frac{\delta M}{|\Omega|} = -\zeta \frac{\delta R}{R}, \tag{11}$$

where the sign is important. These equations are simply based on the virial theorem, which conserves the total energy. They predict shrinkage of the photospheric radius whenever the magnetic energy is increased, so that a minimum solar radius should occur in the maximum phase of the solar activity. If the magnetic flux tubes are produced near the base of the envelope, then $M \approx 10^{32}$ $Joules$ [15]. If it is generated in the super-adiabatic region near the surface, then $M \approx 10^{28}$ $Joules$ [16]. By comparison, we have $|\Omega| \approx 10^{42}$ $Joules$. Moreover, within the solar convective zone, ζ must be very small, but it is not yet clear how small it must be.

3   What Is the Shape of the Sun?

If the Sun is described as a sphere, then the gravitational and the pressure gradient forces are in hydrodynamic equilibrium. But due to the non-homogenous mass distribution and the differential rotation inside the Sun, its outer shape turns out to be distorted in latitude. The Sun has an extended atmosphere and it is not so simple to consider the upper limit of its photosphere. [17] have defined the free surface of the Sun as a level, where a given physical parameter such as the temperature, the density, the pressure, etc. is constant. This free surface does not exactly coincide with an ellipsoid of revolution. The true figure is a bit far more complex showing 'asphericities' (measured by 'shape coefficients'); to first order, it lies between the spherical and ellipsoid shape, which is a figure called 'spheroid'; this could be recognized by the flattening parameter, f (or by the oblateness, $\varepsilon = \frac{R_{eq} - R_{pol}}{R_{sol}}$). For the ellipsoidal shape of the Sun, the solar radius (limited to order 2) can be written (after some reductions) as,

$$r = R_{eq}(1 - \frac{1}{3}f - \frac{2}{3}fP_2 + O(f^2)), \tag{12}$$

where $f = \frac{R_{eq} - R_{pol}}{R_{sp}}$, is the flattening; $R_{sp} = (R_{eq}^2 R_{pol})^{\frac{1}{3}}$, the radius of the best sphere passing through the equatorial and the polar radii which is determined by applying $P_2 = 0$;

and $P_2 = P_2(\cos\theta)$, the Legendre Polynomial. Thus, as $R_{sp} = R_{eq}(1 - \frac{1}{3}f$, one gets

$$\frac{1}{r} = \frac{1}{R_{sp}}(1 + \frac{2}{3}fP_2) + O(f^2), \tag{13}$$

From this formalism, we will calculate in the next section, the gravitational and the magnetic energy for two cases, a spherical and an oblate Sun.

4   Calculation

By considering the variability of the magnetic energy, we derive the magnetic field and the magnetic energy by a suitable vector potential. The magnetic field in the tachocline (a layer between the radiation and the convection regions) is toroidal, so we have $A = (0, 0, A_\varphi)$ and $\mathbf{B} = \nabla\times \mathbf{A}_\varphi$ in a spherical coordinate system. The magnetic field is described by a polynomial function [18]:

$$B_r = \sum c_l(2l + 1)^{\frac{1}{2}} P_l(\cos\theta), \tag{14}$$

where

$$c_l = \frac{1}{2} \int_0^\pi B_r(2l + 1)^{\frac{1}{2}} P_l(\cos\theta) \sin(\theta)d\theta. \tag{15}$$

To derive a variability of the gravitational potential energy, we write:

$$\Phi_g(r, \theta, \varphi) = \sum_{n=0}^\infty \frac{1}{r^{n+1}} \sum_{m=0}^\infty [A_{nm}P_{nm}(\cos\theta)\cos(m\varphi) + B_{nm}P_{nm}(\cos\theta)\sin(m\varphi)], \tag{16}$$

where r is the radius and P_{nm} is the Legendre polynomial of degree n and order m. We suppose that the rotation is symmetric, so that

$$\Phi_g(r, \theta, \varphi) = -\frac{GM}{r}[1 - \sum_{n=1}^\infty (\frac{a}{r})^n J_n P_n(\cos\theta)]. \tag{17}$$

Here J_n is the gravitational moment of order n. Helioseismology data imply a significant temporal variation on both the angular momentum and the gravitational multipolar moments. Using such temporal variations would permit to constrain dynamical theories of the solar cycle. As J_n is not of enough magnitude for $n > 4$, we will use here the order 2 only, then [19]

$$\Phi_g = -\frac{GM}{r}[1 - (\frac{a}{r})^2 J_2 P2(\cos\theta)]. \tag{18}$$

From the above considerations, the two potentials can be written as:

$$\Phi_g = -\frac{GM}{R_{sp}}[1 + \frac{2}{3}f - J_2 P_2], \tag{19}$$

$$A_\varphi = -(\frac{9\sqrt{2}}{8}B_{cr})R_{sp}[1 - \frac{2}{3}fP_2]. \tag{20}$$

The magnetic energy, $M = \int \frac{B^2}{2\mu_0}dV$, and the gravitational energy, $\Omega = \frac{1}{2}\int \rho\Psi dV$, are calculated as followed: For a spherical Sun,

$$M_{spherical} = \int \frac{|(\nabla \times A_\varphi)^2|}{2\mu_0}dV = -\frac{9\sqrt{2}}{8}B_{cr}\int \frac{r}{2\mu_0}dV, \tag{21}$$

$$\Omega_{spherical} = \int \rho \frac{-GM}{r} dV = -GM \int \frac{\rho}{r} dV. \tag{22}$$

And for a spheroidal Sun,

$$M_{spheroid} = \int \frac{|(\nabla \times A_\varphi)^2|}{2\mu_0} dV = -\frac{9\sqrt{2}}{8} B_{cr} \int \frac{r_{sp}(1 - \frac{2}{3} f P_2)}{2\mu_0} dV, \tag{23}$$

$$\Omega_{spheroid} = \int \rho \frac{-GM}{r} [1 + \frac{2}{3} f - J_2 P_2] dV = -GM \int \frac{\rho}{r_{sp}} [1 + \frac{2}{3} f - J_2 P_2] dV. \tag{24}$$

5   Results

We compare the magnetic and the gravitational energies for the considered shapes of the Sun. Fig. 1 shows the variation of the magnetic energy as a function of the solar latitude for a spherical (upper) and a spheroidal (lower) shapes. For upper plot, the maximum occurs at the solar equator, and then decreases toward the pole (for the sake of clarity, only two curves are displayed, one for $r = 0.69R$ and the other one for $r = 0.73R$, as indicated in the left box). For lower plot, the variation of the magnetic energy has a local minimum at $28°$ and then increases to values less than the equator one (three curves are displayed, for $r = 0.69R$, $r = 0.71R$ and $r = 0.73R$, as indicated in the right box).

Fig. 2 displays the variation of magnetic energy as a function of solar latitude for $r = 0.72R$, i.e., at the center of the tachocline (the upper curve being the spherical shape and the lower one, the spheroidal shape). Finally, in Fig. 3 we indicate the variation of the magnetic energy as a function of radius, at the polar latitude ($90°$) for the considered solar shapes. For sake of clarity, the upper curve has been shifted one order of magnitude up.

Fig. 4 shows variation of the gravitational energy as a function of the solar radius, for three different heliographic latitudes ($0°$, $30°$ and $60°$, as shown in the right boxes) for two cases of solar shape (which are mentioned in figures), respectively.

In Fig. 4 (lower), a similar variation of the gravitational energy has been obtained for three considered latitudes from the base of the convective zone up to $0.95R$. By increasing the magnetic field in the tachocline layer, the magnetic energy is also increasing, while by rising from the tachocline to the surface, the magnetic energy decreases along with the magnetic field.

Comparing Figs. 3 and 4, the magnetic energy is decreasing from the tachocline to the solar surface, while the gravitational energy is gradually increasing. It seems that these opposite variations of magnetic and gravitational energies are related to the distribution of density in the solar interior: within the tachocline where the density is slightly higher than its general decrease trend, the gravitational energy increasing. By decreasing of density from the tachocline to the surface, the energy variation is smoother in magnitude (Fig. 4, lower). Moreover, the variation of the magnetic energy stored in the magnetic field (Fig. 3) depends on the magnetic field's behavior with the solar density (when the density decreases, so does the current and the magnetic field). Finally, Figs. 5 and 6 show the variation of the gravitational energy as a function of the solar radius (at $60°$ of latitude) and as a function of the solar latitude (at $r = 0.7R$) for two spherical and spheroid shapes, respectively. Here, again, for the sake of clarity, the upper curve has been shift up of one order of magnitude.

6   Discussion

In this paper, we considered radially variations of the magnetic field and we assumed that there exists a relation between distribution of the magnetic field and the differential rotation

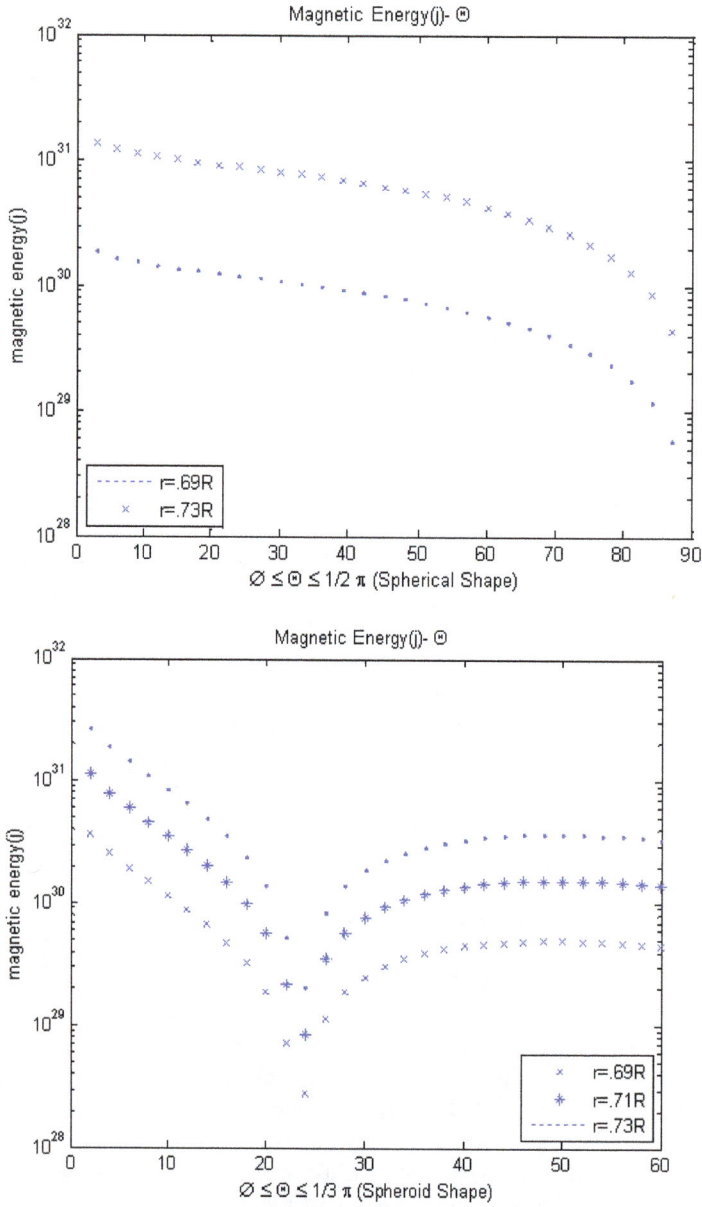

Figure 1: Variation of the magnetic energy with respect to the solar latitude for two different radii for spherical shape (upper) and spheroid shape (lower).

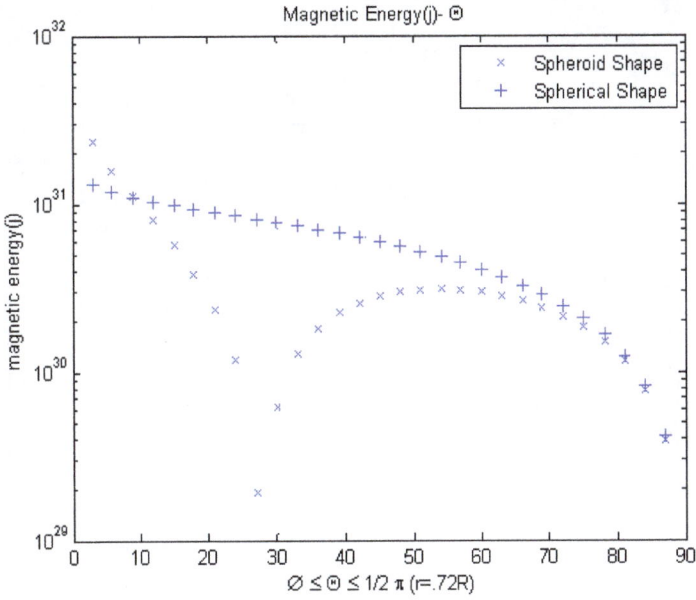

Figure 2: Variation of the magnetic energy as a function of solar latitude only for $r = 0.72R$ in two types of the solar shape.

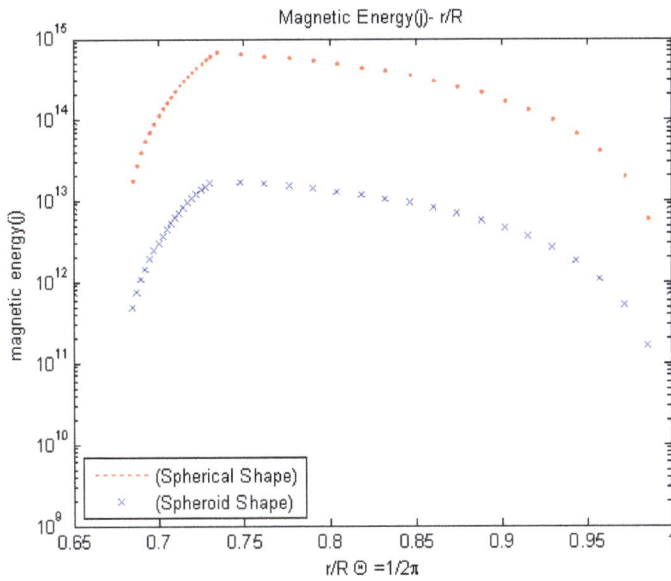

Figure 3: Variation of the magnetic energy as a function of solar radius at the polar latitude ($90°$) in two types of the solar shape.

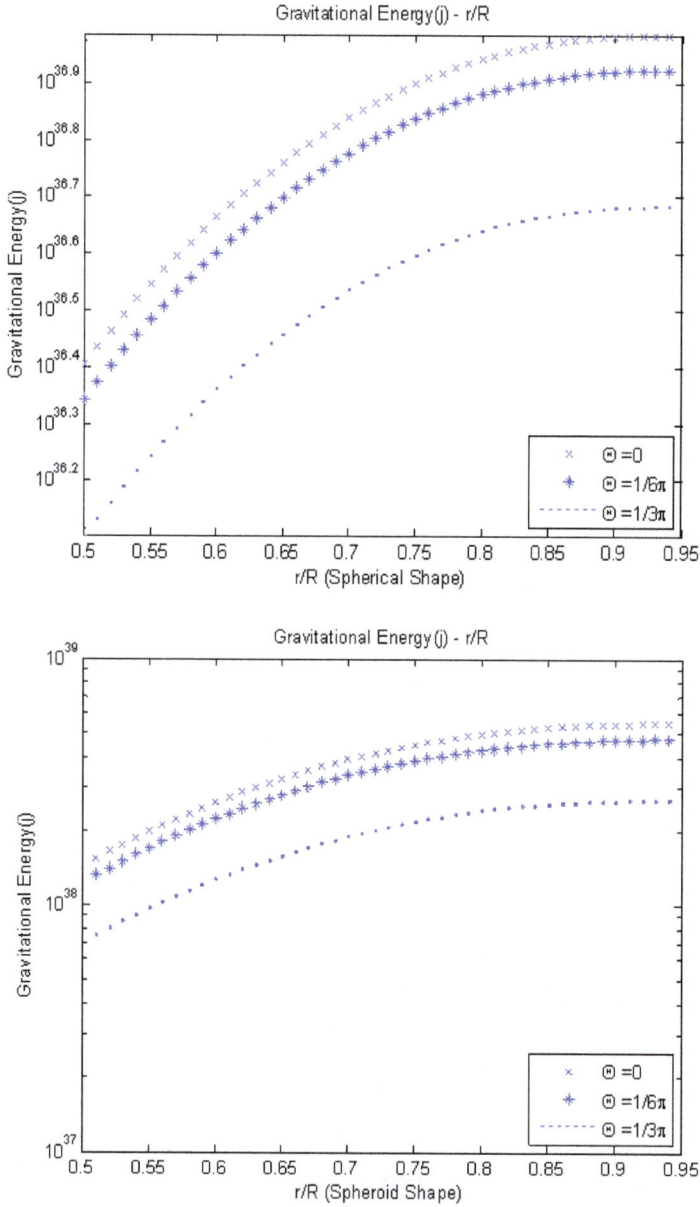

Figure 4: variation of the gravitational energy as a function of the solar radius, for three different solar latitudes in two cases of solar shape (which are mentioned in figures).

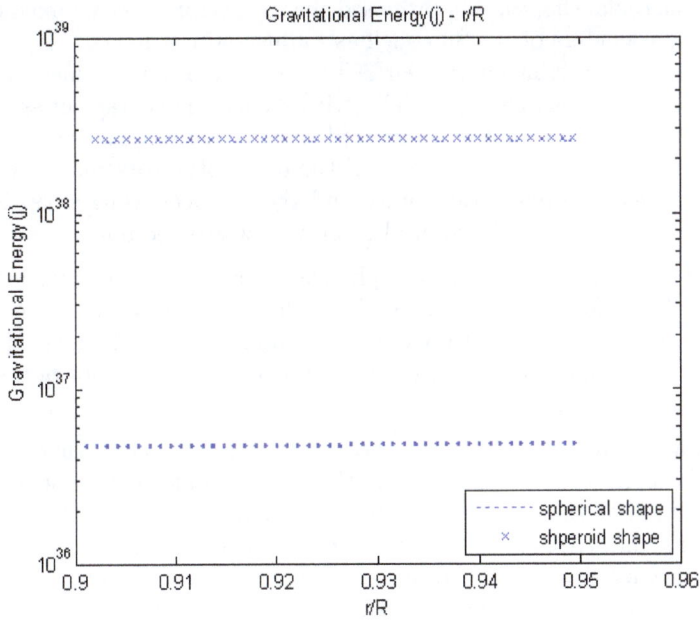

Figure 5: variation of the gravitational energy as a function of the solar radius at $60°$ of latitude.

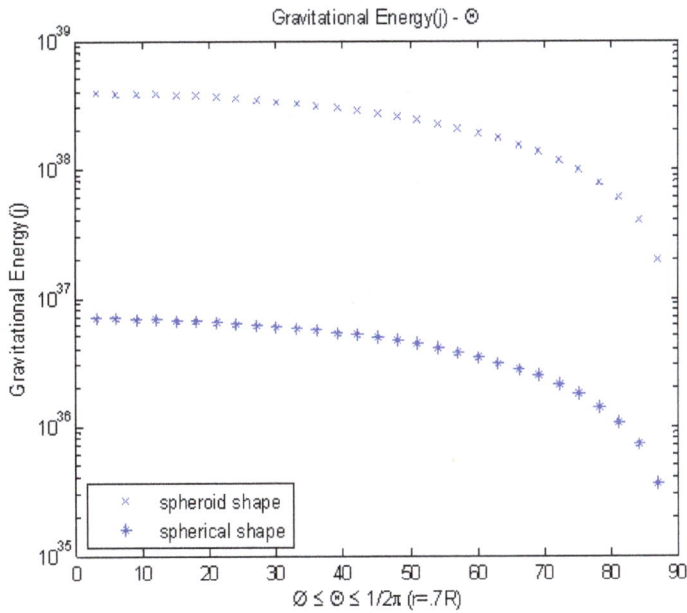

Figure 6: variation of the gravitational energy as a function of the solar latitude at $r = 0.7R$.

in the solar convective zone. Our results are investigated for two types of the solar shape: spherical and spheroidal shapes. [20] proposed that a variable internal magnetic field should affect all global parameters of the Sun such as radius and irradiance. [21] found distortions on the solar shape, i.e. a faint bump near 30° to 50°. Here, we obtained an increase of the magnetic energy from about 28° up to 50° (with a small percentage of error $\approx 3°$). So we argue that these distortions may be connected to this energy variation. But the magnetic energy variation is thwarted by a decrease of the differential rotation above the tachocline layer. We found that the magnetic energy undergoes a decreasing steep from 50° to 90° which agrees to the decreasing differential rotation towards the pole.

Variation of the magnetic field in the spherical shape is correlated with the distribution of the differential rotation (Fig. 1, upper). When differential rotation decreases at high latitudes, the magnetic field must be decreased. This diminishing undergoes until the magnetic energy becomes minimum at the pole, which is similar to the behavior of the magnetic field in the solar spheroidal shape.

Fig. 1 indicates that: a) for the spherical shape, the magnetic energy variation has a minimum at the poles and a maximum at the equator but, b) for the spheroidal shape, there is a maximum at the equator with a fast decreasing around 28°; another maximum takes place at 50° and then the magnetic energy decreases up to the pole. The maximum value of magnetic energy at both equator and pole reach to $10^{32} J$ which is in agreement with the observed solar irradiance variations in each solar cycle. By contrast, the gravitational energy shows the same variation for two types of the solar shapes (Fig. 4).

Since the maximum of magnetic energy variations is stored in the toroidal magnetic field at the equator at the bottom of the convection zone, it would be considered as a source for generation of the poloidal component of the magnetic field. This component is weak at the equator and it increases toward the poles. After the generation phase of the poloidal field, latitudinal differential rotation of the Sun begins to stretch the field lines beneath the faster-spinning equatorial regions. The poloidal field (north-south) is essentially changed into a toroidal (east-west) configuration by the meridional motions, in which the lines of force are near-circles around the solar axis. So the toroidal field must be produced at the poles and when reaching the equator, it becomes much stronger. The stretching process wraps the lines around the Sun several times causing them to intertwine and intensify. Finally they can be driven to the surface as magnetic loops [18]

Study of variations of the magnetic and the gravitational energies is important to develop the solar dynamo theory. We showed that regions of strong sub-surface rotational shear allow the toroidal magnetic flux to penetrate to the solar surface. In other words, the magnetic toroidal field generated in the sub-surface shear layer (the leptocline) does not contribute directly to the generation of the poloidal magnetic field; a fact that already recognized by [12], the subsurface-shear layer shaped the solar $\alpha\Omega$ dynamo.

Acknowledgment

This work has been supported financially by Research Institute for Astronomy & Astrophysics of Maragha (RIAAM) under research project No. 1/3720. The work of JPR was supported by the International Space Science Institute (ISSI) in Bern (Swisserland) under a Visiting Scientist grant.

References

[1] Moreno-Insertis, F. 1992, sto work, 385

[2] Schuessler, M., Caligari, P., Ferriz-Mas, A., Moreno-Insertis, F. 1994, A& A, 281, 69

[3] Ferriz-Mas, A., Schssler, M. 1993, GApFD, 72, 209

[4] Ferriz-Mas, A., Schssler, M. 1995, ApJ, 433, 852.

[5] Schmitt, D., Ferriz-Mas, A. 2004, Variable Solar and Stellar Activity by a Flux Tube Dynamo, p. 89-91

[6] Ferriz-Mas, A., Schmitt, D., Schuessler, M. 1994, A & A. 289, 949.

[7] Caligari, P., Moreno-Insertis, F., Schussler, M. 1995, ApJ, 441, 886

[8] Chandrasekhar, S., Fermi, E. 1953, ApJ, 118, 116

[9] Lefebvre, S., Nghiem, P.A.P., Turck-Chize, S. 2009, ApJ, 690, 1272

[10] Emilio, M., Kuhn, J.R., Bush, R.I., Scherrer, P. 2000, ApJ, 453, 1007

[11] Kuhn, J.R., Bush, R.I., Emilio, M., Scherrer, P.H. 2004, ApJ, 613, 1241

[12] Pipin, V.V., Kosovichev, A.G. 2011, ApJ, 727, 45

[13] Pap, J. M., Kuhn, J.R., Frhlich, C., Ulrich, R., Jones, A., Rozelot, J.P. 1998, ESA-SP 417, 267

[14] Stothers, R.B. 2006, ApJ, 653, L73-L75

[15] Spiegel, E.A., Weiss, N.O. 1980, Nature, 287, 616

[16] Dearborn, D.S.P., Blake, J.B. 1982, ApJ, 257, 896

[17] Rozelot, J.P., Lefebvre, S. 2003, in the Sun's surface and subsurface, ed. Rozelot, J.P. (LNP, 599, Springer), 4

[18] Durney, B.R. 1997, ApJ, 486, 1065

[19] Fazel, Z. 2007, Ph.D. thesis, University of Nice.

[20] Endal, A.S., Sofia, S., Twigg, L.W.H.C. 1985, ApJ, 290, 748

[21] Lefebvre, S., Rozelot, J.P. 2003, ESA-SP. 535, 53L

Generation of Alfvén Waves by Small-Scale Magnetic Reconnection in Solar Spicules

Zahra Fazel

Astrophysics Depart., Physics Faculty, University of Tabriz, Tabriz, Iran; email: z_fazel@tabrizu.ac.ir

Abstract. Alfvén waves dissipation is an extensively studied mechanism for the coronal heating problem. These waves can be generated by magnetic reconnection and propagated along the reconnected field lines. Here, we study the generation of Alfvén waves at the presence of both steady flow and sheared magnetic field in the longitudinally density stratified of solar spicules. The initial flow is assumed to be directed along the spicule axis, and the equilibrium magnetic field is taken 1-dimensional and divergence-free. We solve linearized MHD equations numerically and find that the perturbed velocity and magnetic field oscillate similarly which can be interpreted as generation and propagation of Alfvénic waves along spicule axis. The results of calculations give periods of around 25 and 70 s for these waves which are in good agreement with observations.

Keywords: ISM: solar spicules, ISM: Alfvén waves, ISM: magnetic reconnection

1   Introduction

The mechanism of coronal heating is one of the major problems in solar physics. The magnetic structure of the corona can play an important role on the problem of heating, so it should be necessary to study the converting of the magnetic energy to heat. A prime candidate for transferring energy up to coronal levels is a flux of Alfvén waves. Heyvaerts & Priest [1] proposed an idea for the behavior of Alfvén waves when the local Alfvén speed varies across the magnetic field lines. The propagation and damping of shear Alfvén waves in an inhomogeneous medium has been studied in more detail [2, 3, 4, 5, 6]. The damping of Alfvén waves is defined by various dissipative processes such as mode coupling [7], resonant absorption [8, 9, 10, 11], magnetohydrodynamic (MHD) turbulence [12], and phase mixing [1].

As an origin of Alfvén waves, Kudoh & Shibata [13] considered a photospheric random motion propagating along an open magnetic flux tube in the solar atmosphere, and performed MHD simulations for the solar spicule formation and the coronal heating. They have shown that Alfvén waves transport sufficient energy flux into the corona. Moriyasu et al. [7] performed $1.5D$ MHD simulations of the propagation of nonlinear Alfvén waves along a closed magnetic loop, including heat conduction and radiative cooling. They found that the corona is heated by fast- and slow-mode MHD shocks generated by nonlinear Alfvén waves via nonlinear mode-coupling.

He et al. [14] have shown the generation of kink waves due to chromospheric small-scale magnetic reconnection, by using Solar Optical Telescope (SOT) observations [15] in the Ca II H-line. The kink wave is identified by the upward propagation of a transverse-displacement oscillation along the spicule trace. This transverse oscillation appears to originate from the cusp position of an inverted Y-shaped magnetic structure, where a surge was taking place, including the occurrence of magnetic reconnection according to [16]. Cranmer & Van Ballegooijen [17] studied the generation, propagation and reflection of Alfvén waves in solar atmosphere. They modeled waves as thin-tube kink modes by assuming that all of the kink-mode wave energy is transformed into volume filling Alfvén waves above the merging height (a critical flux tube height which waves are modeled).

Alfvén waves can be generated by magnetic reconnection process. Magnetic reconnection is a fundamental dynamical process in highly conductive plasmas. Sweet, Parker, Petschek and Soward & Priest [18, 19, 20, 21] introduced magnetic reconnection as the central process allowing for efficient magnetic to kinetic energy conversion in solar flares and for interaction between the magnetized interplanetary medium and the magnetosphere of Earth. [22] modeled X-ray and EUV jets and surges observed with $H\alpha$ in the chromosphere by performing a resistive $2D$ MHD simulation of the magnetic reconnection occurring in the current sheet between emerging magnetic flux and overlying pre-existing coronal magnetic fields. *Hinode* observation revealed that jets are ubiquitous in the chromosphere [23]. De Pontieu et al. [24], from *Hinode* data estimated the energy flux carried by transversal oscillations generated by spicules. They indicated that the calculated energy flux is enough to heat the quiet corona.

Spicules, the grass-like, thin and elongated structures are one of the most pronounced features of the chromosphere. They are seen in spectral lines at the solar limb at speeds of about $20-25$ km s^{-1} propagating from the photosphere into the magnetized low atmosphere of the sun [25]. Their diameter and length varies from spicule to spicule having the values from 400 km to 1500 km and from 5000 km to 9000 km, respectively. Their typical lifetime is $5-15$ min. The typical electron density at heights where the spicules are observed is approximately $3.5\times10^{16}-2\times10^{17}$ m^{-3}, and their temperatures are estimated as $5000-8000$ K [26, 27]. Oscillations in spicules have been observed for a long time. [28] and [29] observed their transverse oscillations with the estimated period of $20-55$ and $75-110$ s by analyzing the height series of $H\alpha$ spectra in solar limb spicules.

Recently, Ebadi et al. [30] based on *Hinode*/SOT observations estimated the oscillation period of spicule axis around 180 s. They concluded that the energy flux stored in spicule axis oscillations is of order of coronal energy loss in quiet Sun.

In this paper, we are interested to study the generation of Alfvén waves in solar spicules by magnetic reconnection process. Section 2 gives the basic equations and theoretical model. In section 3, numerical results are presented and discussed, and a brief summary is followed in section 4.

2   Theoretical modeling

We consider effects of the stratification due to gravity in $2D$, $x-z$ plane in the presence of steady flow and shear magnetic field. The generation of Alfvén waves is studied in a region with nonuniform Alfvén velocity both along and across the spicule axis due to mass density and magnetic field inhomogeneities. The non-ideal MHD equations in the plasma dynamics are as follows:

$$\rho\frac{\partial \mathbf{v}}{\partial t} + \rho(\mathbf{v}\cdot\nabla)\mathbf{v} = -\nabla p + \rho\mathbf{g} + \frac{1}{\mu_0}(\nabla\times\mathbf{B})\times\mathbf{B}, \tag{1}$$

$$\frac{\partial \mathbf{B}}{\partial t} = \nabla\times(\mathbf{v}\times\mathbf{B}) + \eta\nabla^2\mathbf{B}, \tag{2}$$

$$\nabla\cdot\mathbf{B} = 0, \tag{3}$$

$$p = \frac{\rho RT}{\mu}. \tag{4}$$

where η is constant resistivity coefficient, which its typical value in the solar chromosphere is $8 \times 10^8 T^{-3/2}$ m^2 s^{-1} [31]. μ_0 is the vacuum permeability, μ is the mean molecular weight. We assume that spicules are highly dynamic with speeds that are significant fractions of the Alfvén speed. Perturbations are assumed to be independent of y, i.e.:

$$\mathbf{v} = v_0 \hat{k} + v_y(x, z, t)\hat{j},$$
$$\mathbf{B} = B_{0z}(x)\hat{k} + b_y(x, z, t)\hat{j}, \tag{5}$$

and the equilibrium sheared magnetic field is one-dimensional and divergence-free as:

$$B_{0z}(x) = B_0 \tanh(\frac{x - z_t}{z_w}). \tag{6}$$

where z_w is the thickness of the initial current sheet and z_t is the position of the middle of a spicule. The equilibrium magnetic field is assumed force-free and pressure is balanced by gravity force. The longitudinally stratified pressure and density are given by [32, 33]:

$$p_0(z) = p_0 \exp(z/H), \tag{7}$$

where H is the pressure scale height. The linearized dimensionless MHD equations with these assumptions are:

$$\frac{\partial v_y}{\partial t} + v_0 \frac{\partial v_y}{\partial z} = \frac{1}{\rho_0(z)} \left[B_{0z}(x) \frac{\partial b_y}{\partial z} \right], \tag{8}$$

$$\frac{\partial b_y}{\partial t} + v_0 \frac{\partial b_y}{\partial z} = B_{0z}(x) \frac{\partial v_y}{\partial z} + \eta \nabla^2 b_y. \tag{9}$$

where density, velocity, magnetic field, time and space coordinates are normalized to ρ_0 (the plasma density at dimensionless $z = 6$), V_{A0}, B_0, τ, and a (spicule radius), respectively. Also the gravity acceleration is normalized to a^2/τ. The second terms in the left hand side of Eqs. (8) and (9) present the effect of steady flows. Eqs. (8) and (9) should be solved under following initial and boundary conditions:

$$v_y(x, z, t = 0) = 0,$$
$$b_y(x, z, t = 0) = 0. \tag{10}$$

and

$$v_y(x = 0, z, t) = v_y(x = 4, z, t) = 0,$$
$$b_y(x = 0, z, t) = b_y(x = 4, z, t) = 0,$$
$$v_y(x, z = 0, t) = v_y(x, z = 16, t) = 0,$$
$$b_y(x, z = 0, t) = b_y(x, z = 16, t) = 0. \tag{11}$$

3 Results

We use the finite difference and the Fourth-Order Runge-Kutta methods to take the space and time derivatives in the coupled Eqs. (8) and (9). The implemented numerical scheme is used by the forward finite difference method to take the first spatial derivatives with the truncation error of (Δx), which is the spatial resolution in the x direction. The order of approximation for the second spatial derivative in the finite difference method is $O((\Delta x)^2)$. On the other hand, the Fourth-order Runge-Kutta method takes the time derivatives in the questions. The computational output data are given in 17 decimal digits of accuracy.

We set the number of mesh-grid points as 256×256. In addition, the time step is chosen as 0.0005, and the system length in the x and z dimensions (simulation box sizes) are set to be (0,4) and (0,16).

The parameters in spicule environment are as follows: $a = 250$ km(spicule radius), $L = 6000$ km (Spicule length), $v_0 = 25$ km s^{-1}, $n_e = 11.5 \times 10^{16}$ m^{-3}, $B_0 = 1.2 \times 10^{-3}$ Tesla, $T_0 = 14\,000$ K, $g = 272$ m s^{-2}, $R = 8300$ m^{2}s^{-1}k^{-1} (universal gas constant), $V_{A0} = 75$ km/s, $\mu = 0.6$, $\tau = 3.5$ s, $\rho_0 = 1.9 \times 10^{-10}$ kg m^{-3}, $p_0 = 3.7 \times 10^{-2}$ N m^{-2}, $\mu_0 = 4\pi \times 10^{-7}$ Tesla m A^{-1}, $z_w = 0.1$ and $z_t = 2$ (in our dimensionless units), $H = 750$ km, $\eta = 10^3$ m^2 s^{-1}, and $k = \pi/8$ (dimensionless wavenumber normalized to a).

Fig. 1 shows perturbed velocity variations with respect to time in $x = 250$ km, $z = 875$ km and $x = 250$ km, $z = 2500$ km, respectively. In Fig. 2, perturbed magnetic field variations are presented for $x = 250$ km, $z = 875$ km and $x = 250$ km, $z = 2500$ km, respectively.

At the first height ($z = 875$ km), total amplitude of both velocity and magnetic field oscillations have values near to the initial ones. As height increases, these amplitudes does increase. This means that with an increase in height, amplitude of velocity oscillations is expanded due to significant decrease in density, which acts as inertia against oscillations. Similar results are observed by time-distance analysis of solar spicule oscillations [30]. It is worth to note that the density stratification effect on the magnetic field is weak, which is in agreement with Solar Optical Telescope observations of solar spicules [34].

Behavior of both perturbed velocity and magnetic field with time is similar in Figs. 1 and 2. In these figures, we can see two types of periodic oscillations: one is related to the generated Alfvén waves at the first height; and the other one which occurs at higher heights shows a periodic standing oscillations. Figs. 3 and 4 illustrate the $3D$ plots of the perturbed velocity and magnetic field with respect to x, z for $t = 5\tau$ s, $t = 50\tau$ s, and $t = 80\tau$ s.

4 Discussion

The generation of Alfvén waves is investigated in a medium with steady flow and sheared magnetic field along spicule axis. I take into account the density stratification and an initial antiparallel magnetic field. This component of the magnetic field indicates a simple magnetic reconnection at the base of a spicule in solar chromosphere. The initially generated Alfvénic waves are propagated in the medium and led to packets in higher heights. The period of transverse oscillations that are included in the medium due to the propagation of Alfvén waves are in agreement with those observed in spicules. Two types of periods are seen at the perturbed velocity and magnetic field variations. At the first height, Alfvén waves oscillate with the period of $\simeq 25$ s which may correspond to the transverse oscillations observed in spicules. Moreover, at the higher heights, Alfvénic wave packets are formed with the period of $\simeq 70$ s [29]. These periods depend on the plasma density and the initial magnetic field in the medium. The perturbed velocity and magnetic field oscillate with the same periods and amplitudes which may be interpreted as generation and propagation of Alfvénic waves along spicule axis.

References

[1] Heyvaerts, J., Priest, E.R., 1986, AAP 117, 220

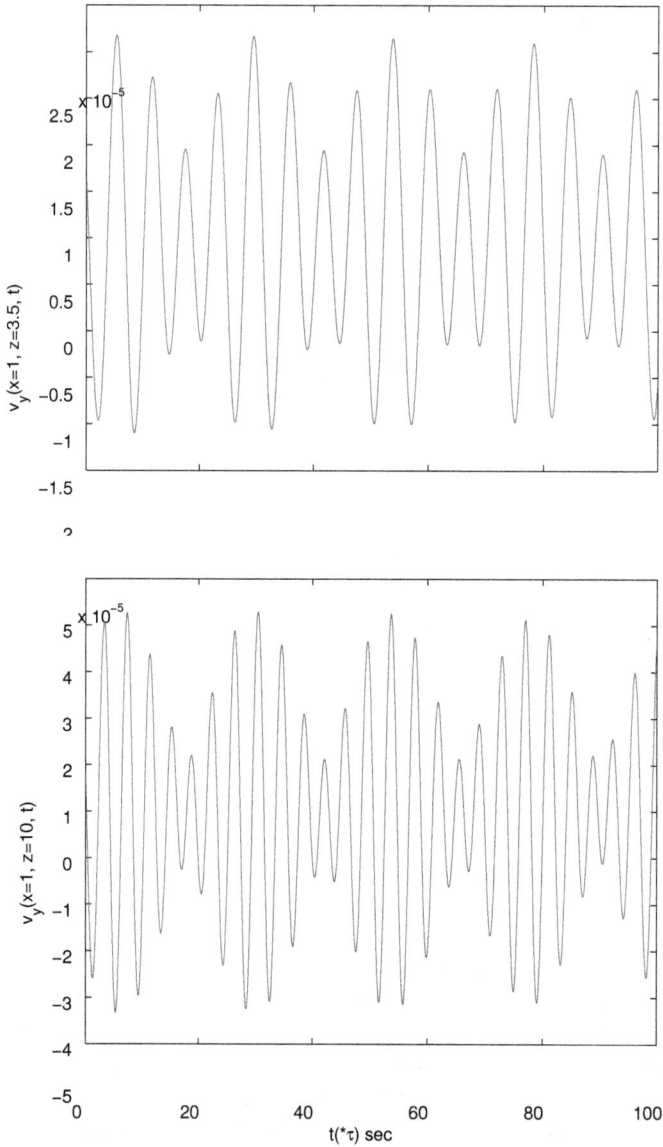

Figure 1: The perturbed velocity variations are shown with respect to time and $x = 250$ km for two values of $z = 875$ km and $z = 2500$ km from top to bottom. The perturbed velocity is normalized to V_{A0}.

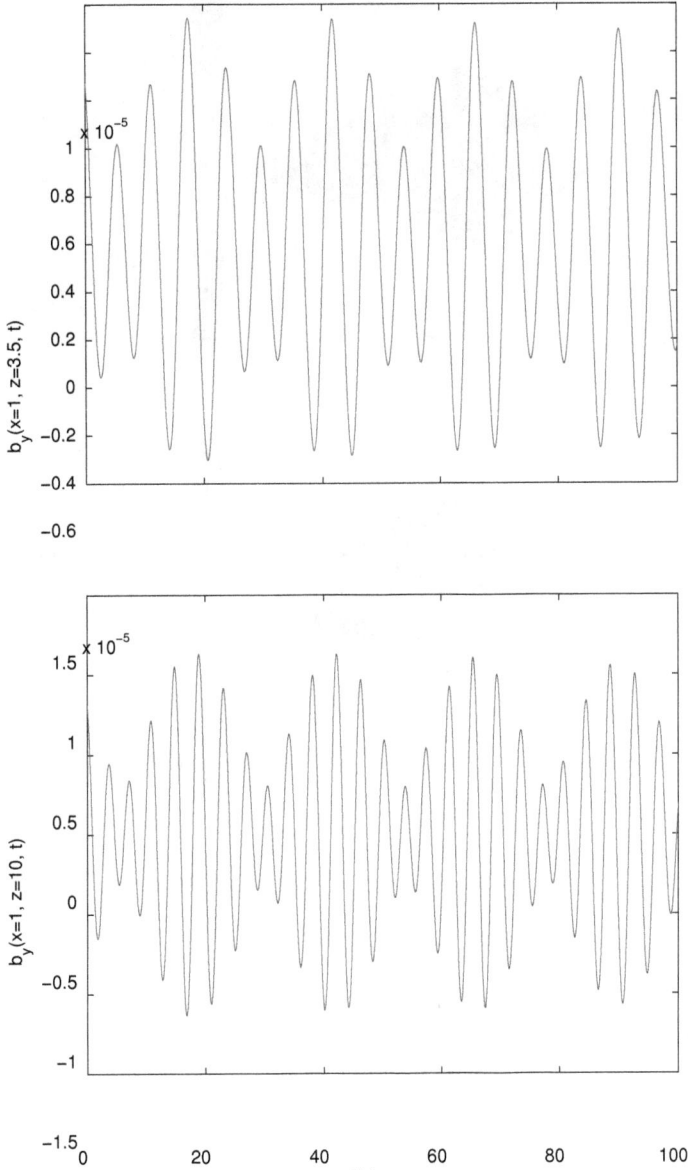

Figure 2: The perturbed magnetic field variations are shown with the same coordinates as inferred in figure 1. The perturbed magnetic field is normalized to B_0.

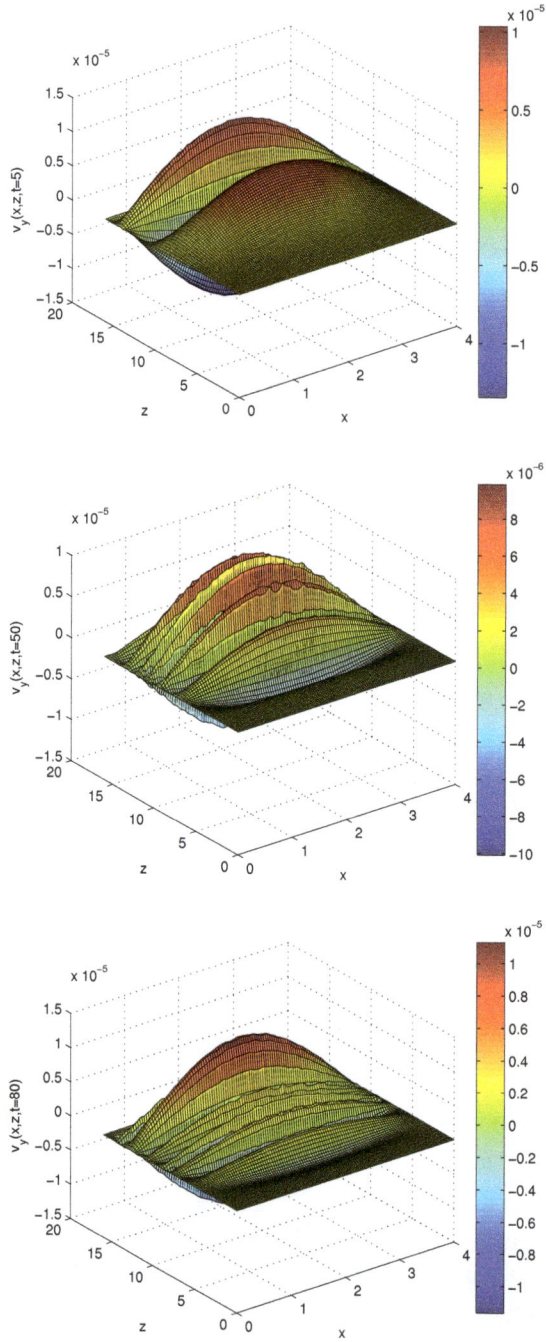

Figure 3: The $3D$ plots of the transversal component of the perturbed velocity with respect to x, z in $t = 5\tau$ s, $t = 50\tau$ s, and $t = 80\tau$ s.

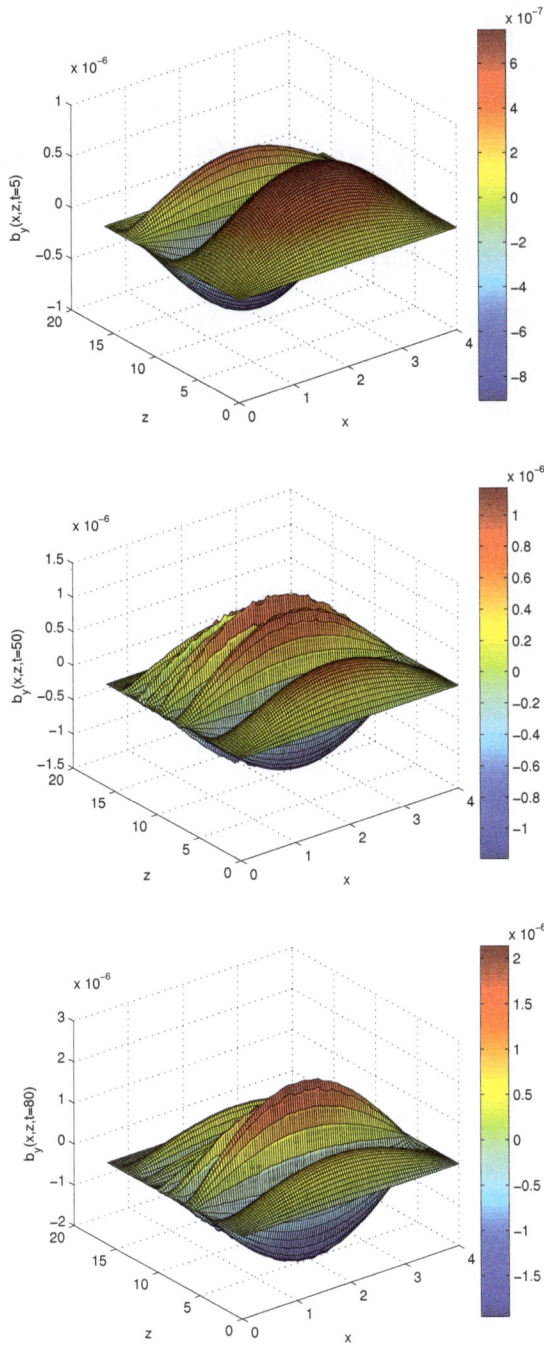

Figure 4: The same as in Fig. 3 for the perturbed magnetic field.

[2] Ireland, J., Priest, E.R., 1997, SolPhys 173, 31

[3] Browning, P.K., Priest, E.R., 1984, AAP 131, 283

[4] Hollweg, J.V., 1986, J. Geophys. Res. 91, 4111

[5] McKenzie, J.F., Banaszkiewicz, M., Axford, W.I., 1995, AAP 303, 45

[6] Karami, K., Ebrahimi, Z., 2009, Publ. Astron. Soc. Aust. 26, 448

[7] Moriyasu, S., Kudoh, T., Yokoyama, T., Shibata, K., 2004, APJ 601, 107

[8] Ionson, J.A., 1978, APJ 226, 650

[9] Ruderman, M.S., Berghmans, D., Goossens, M., Poedts, S., 1997, AAP 320, 305

[10] Andries, J., Goossens, M., Hollweg, J.V., Arregui, I., Van Doorsselaere, T., 2005, A & A 430, 1109A

[11] Safari, H., Nasiri, S., Karami, K., Sobouti, Y., 2006, A & A 448, 375S

[12] Matthaeus, W.H., Zank, G.P., Oughton, S., Mullan, D.J., Dmitruk, P., 1999, APJL 523, 93

[13] Kudoh, T., Shibata, K., 1999, APJ 514, 493

[14] He, J., Marsch, E., Tu, Ch., Tian, H., 2009, APJ 705, L217

[15] Tsuneta, S. et al., 1008, SolPhys 249, 167

[16] Shibata, K., et al., 2007, Science 318, 1591

[17] Cranmer, S.R., Van Ballegooijen, A.A., 2005, Astrophysical J. Supplementary Series 156, 265

[18] Sweet, P.A., 1958, in IAU Symp. 6, Electromagnetic Phenomena in Cosmical Phys., ed. Lehnert, B. (Cambridge Uni. Press) 123

[19] Parker, E.N., 1963, APJS 8, 177

[20] Petschek, H.E., 1964, in Proc. AAS-NASA Symp. on Physics of Solar Flares, ed. W.N. Hess (NASA SP-50), 425

[21] Soward, A.M., Priest, E.R., 1982, J. Plasma Phys. 28, 335

[22] Yokoyama, T., Shibata,K., 1995, Nature 375, 42

[23] Kosugi, T. et al., 2007, SolPhys 243, 3

[24] De Pontieu, B., McIntosh, S.W., Carlsson, M., et al., 2007, Science 318, 1574

[25] Zaqarashvili, T.V., Erdélyi, R., 2009, SSR 149, 335

[26] Beckers, J.M., 1968, SolPhys 3, 367

[27] Sterling, A.C., 2000, SolPhys 196, 79

[28] Kukhianidze, V., Zaqarashvili, T. V., Khutsishvili, E., 2006, AAP 449, 35

[29] Zaqarashvili, T. V., Khutsishvili, E., Kukhianidze, V., Ramishvili, G., 2007, AAP 474, 627

[30] Ebadi, H., Zaqarashvili, T.V., Zhelyazkov, I., 2012, ApSS 337, 33

[31] Priest, E.R., 1982, Solar magnetohydrodynamics. p. 79, Reidel, Dordrecht.

[32] Safari, H., Nasiri, S., Sobouti, Y., 2007, A & A 470, 1111S

[33] Fathalian, N., Safari, H., 2010, APJ 724, 411F

[34] Verth, G., Goossens, M., He, J.S., 2011, APJL 733, 15

18

AGN Zoo and Classifications of Active Galaxies

Areg M. Mickaelian

Byurakan Astrophysical Observatory (BAO), Byurakan 0213, Aragatzotn Province, Armenia;
email: aregmick@yahoo.com

Abstract. We review the variety of Active Galactic Nuclei (AGN) classes (so-called "AGN zoo") and classification schemes of galaxies by activity types based on their optical emission-line spectrum, as well as other parameters and other than optical wavelength ranges. A historical overview of discoveries of various types of active galaxies is given, including Seyfert galaxies, radio galaxies, QSOs, BL Lacertae objects, Starbursts, LINERs, etc. Various kinds of AGN diagnostics are discussed. All known AGN types and subtypes are presented and described to have a homogeneous classification scheme based on the optical emission-line spectra and in many cases, also other parameters. Problems connected with accurate classifications and open questions related to AGN and their classes are discussed and summarized.

Keywords: AGN, quasars, Seyfert galaxies, LINERs, Starburst galaxies, radio galaxies, jets

1   Introduction

Active Galactic Nuclei (AGN) are the most interesting and crucial topic in extragalactic astronomy. Their studies are connected to galaxy evolution, understanding of energy sources, galaxy morphology, interactions and merging, binary and multiple structure and clustering. AGN are the most luminous persistent sources of the Universe, and as such can be used as means for discovery of distant objects. On the other hand, their evolution as a function of cosmic time also puts constraints on cosmological models of the Universe. According to modern views, AGN is a compact region at the centre of a galaxy having much higher than normal luminosity over the whole or at least some part of the electromagnetic spectrum. Such excess emission has been observed in radio, microwave, IR, optical, UV, X-ray and gamma-ray wavelengths. That is why it is so important to have an overall picture of radiation, multi-wavelength (MW) Spectral Energy Distribution (SED). AGN hosting galaxies are called active galaxies, however galaxies show **Starburst (SB)** activity as well (having high star-formation rate, SFR). So active galaxies may have either nuclear or starburst activity or both. The radiation from AGN is believed to be the result of accretion of mass by a **Super-Massive Black Hole (SMBH)** at the centre of its host galaxy having 10^6-10^{10} M$\odot$. Accretion can potentially give very efficient conversion of potential and kinetic energy to radiation, and SMBH has a high Eddington luminosity (the maximum luminosity of a body when there is balance between radiation and gravitation), and as a result, it can provide the observed high persistent luminosity. SMBHs are now believed to exist in the centres of the most if not all massive galaxies since the mass of the black hole correlates well with the velocity dispersion of the galactic bulge or with bulge luminosity. Thus AGN-like spectrum is expected whenever a supply of material for accretion comes within the sphere of influence of the central BH.

There is a number of **observational signatures to distinguish an AGN**. Here we list some important features that allow to identify an AGN:

- **Optical emission lines.** Depending on the physical conditions of the regions of their origin, they can be broad or narrow corresponding to Broad Line Regions (BLR, closer to the nucleus) or Narrow Line Region (NLR, relatively farther from the nucleus), respectively.

- **Optical continuum emission** comes from the nucleus and is visible whenever there is a direct view of the accretion disc. Jets can also contribute to this component. The optical emission has a roughly power-law dependence on wavelengths.

- **Radio continuum emission** always comes from a jet. It shows a spectrum characteristic of synchrotron radiation. However, radio structures show a big variety; from central ones to large radio lobes.

- **Nuclear IR emission** is detectable if the accretion disc and its environment are obscured by gas and dust close to the nucleus. They re-emit UV and optical radiation into IR. As it is thermal emission, it can be distinguished from any jet or disc emission by its distribution.

- **X-ray continuum emission** comes both from the jet and from the hot corona of the accretion disc through the scattering process and show power-law spectrum. In some radio-quiet AGN there is an excess of soft X-ray in addition to the power-law component.

- **X-ray line emission.** The illumination of cold heavy elements by X-ray continuum causes fluorescence of X-ray emission lines. Such well-known feature is Fe-K line centered at 6.4 keV. It may be narrow or broad. Using relativistically broadened Fe lines, one can study the dynamics of the accretion disc very close to the nucleus, hence the nature of the central SMBH.

- **Other manifestations of activity.** These may be jets, compact components, interactions, merging, etc., including those not directly related to the nucleus, but somehow resulting from its activity, so-called AGN feedback.

The variety of observational manifestations has led to the **variety of AGN types**, especially taking into account historical classifications, when any classification was made based on the not complete knowledge of the given epoch. And even though **unified models** (or unified schemes; Antonucci & Miller 1985; Antonucci 1993; Urry & Padovani 1995) propose that different observational classes of AGN are a single type of physical object observed under different conditions (in fact, their different orientations in the space and hence, angles of view to the observer), anyway one should carefully study and understand all observable features, and most importantly the optical emission-line spectrum of AGN. We focus on the observable characteristics of AGN, including optical spectrum and other wavelengths features that help classifying and understanding these objects among the zoo of various types.

Classification is especially important for further studies of any object, both to confirm or reject any observable (astronomical) or physical relation based on definite types of objects. In case of AGN, we study and set up their characteristics for different types based on accurate classifications, otherwise any study will involve errors and uncertainties. Hundreds of thousands of AGN have been discovered and catalogued; here we list some of the biggest and/or most important ones that serve as bases for homogeneous classifications:

- **Catalogue of Quasars and Active Nuclei**, 13th version, hereafter VCV-13, 168,940 objects (Véron-Cetty & Véron 2010).

- **SDSS-DR7 5th Quasar Catalog**, 105,783 quasars (Schneider et al. 2010).

- **SDSS-DR9 Quasar Catalog**, 87,822 quasars, mostly new ones (Pâris et al. 2012).

- **Large Quasar Astrometric Catalogue (LQAC-2)**, 187,504 quasars (Souchay et al. 2012).

- **Roma Multi-frequency Catalogue of Blazars (BZCAT)** 5th version, 3,561 blazars (Massaro et al. 2015; http://www.asdc.asi.it/bzcat/).

- **Low-frequency radio catalog of flat-spectrum sources (FSS)**, 28,358 sources (Massaro et al. 2014).

- **Catalogue of X-ray selected AGN (HRC/BHRC)**, 4,253 ROSAT objects identified as AGN (Paronyan & Mickaelian 2015).

- Some other AGN lists from SDSS and elsewhere: **Type II QSOs** (Zakamska et al. 2003), **Seyferts and LINERs** (Hao et al. 2005), **BL Lacs** (Collinge et al. 2005), etc.

However, none of these samples give accurate and detailed classifications. This is because VCV-13 uses classifications from source papers independent of their accuracy. Quasars in fact do not have any regular classification, and on the other hand, SDSS automatic procedures lead to a lot of erroneous classifications, as well as to misidentification of spectral lines and hence, redshift measurements. Moreover, there is no unique classification scheme that could be used in all these and other catalogues.

There are a number of issues related to AGN and other active galaxies classifications. Here we give some of them:

- **Physics.** Of course the main task is to understand what physics is underlying the observed spectrum and other observable features. What physical phenomena cause the given emission and absorption lines, and the continuum? How it works to produce the summarizing spectrum? Both observational (empirical) and theoretical approaches are useful that may solve direct or inverse problems.

- **Technical issues.** There are many technical problems to be solved for accurate reduction, measurements and classifications. Many parameters may play role: the slit/fiber size, distance of the observed objects and hence angular size, wavelength range, signal-to-noise (S/N) ratio, spectral resolution, variability, reduction/measurements effects, etc.

- **Methodology.** Reduction, measurement and classification methodology is also rather important; what we are distinguishing. Do we give importance only to optical spectra and hence lose hidden AGN (IR, X-ray)? Do we combine optical spectral data with those of other wavelength ranges?

- **Philosophy.** The philosophy of classifications is that one needs to have definite types of the studied samples and/or objects for making any conclusion. During the whole history of science, classifications have played important role in different science disciplines (astronomy, particle physics, chemistry, biology, etc.). Only after the classification one can study definite objects. However, they also introduce discontinuity, as intermediate objects are artificially lost or neglected. Scientists try to find regular relations for all objects and properties, though this might not always be the case in the nature. This is the reason why most of the astronomers call most of the classifications as historical ones and different types of objects as historical types. One solution is to turn qualitative nomenclature to quantitative data and study everything quantitatively, which is much harder but much safer. An example could be the stellar classification into O-B-A-F-G-K-M-L-T. When studying properties of these definite types, one can for example compare those for A and F types. However we know that there are also subtypes (A0, A1, A2, ..., A8, A9, F0, F1, etc.) and e. g. A9 is much closer to F0 than to A1. However we combine all A subtypes together and may lose some fine characteristics. Therefore, quantitative relations (e. g. using the temperature or colour index rather than spectral type or even subtype) are much better. The same was with Seyfert galaxies. Until now we try to compare Sy1 and Sy2 types and find differences, because they have been classified differently, though intermediate subtypes were introduced later. Similarly, radio galaxies have a wide range of radio luminosities and one should not study all galaxies as of two types (radio loud and radio quiet) not to lose numerous characteristics of intermediate objects. The same is with FR I and FR II types sources.

Anyway, we give high importance to **accurate spectral reduction, line measurements and classifications**, as many parameters that we define and use for calculation and derivation of the physical properties and hence understanding the nature of objects depend on this; distribution of physical parameters, dependencies, flux ratios, etc.

2 Historical Classifications and AGN Zoo

Since the end of the XIX century, astronomers have noticed that many nebulae had emission lines; these were planetary nebulae, a number of diffuse ones, as well as since the beginning of the XX century some spiral nebulae also showed emission lines. The previous experience of the study of emission nebulae would play an important role for further investigations of galaxies.

In 1908, E. A. Fath observed first emission lines in spectrum of NGC 1068 (at that time known as nebula; Fath 1908). Later on, in 1917-1934, Slipher, Cambell, Moore, Hubble, Humason and Mayall confirmed the emission lines for NGC 1068 and observed such in NGC 5236, NGC 4151, NGC 4051 and NGC 1275 (Slipher 1917; Cambell & Moore 1918; Hubble 1926; Humason 1932; Mayall 1934). However, these papers were left without enough importance.

On the other hand, historically the first discovery of any kind of activity (from modern point of view) in galaxies was reported by Grote Reber in 1939 (Reber 1940), when he discovered Cygnus A as a radio source. Anyway, it was also not known as an extragalactic object.

Later on, Carl Seyfert (Seyfert 1943) observed emission-lines in the spectra of some spiral galaxies ("extragalactic nebulae"), including presently well-known AGN NGC 4151, 4051, 1068, 1275, 3516, 5548, 7469. Especially surprising were broad emission lines (or broad wings of lines) that were not observed in the spectra of the galactic nebulae. Out of Seyfert's 12 emission-line galaxies (NGC 1068, 1275, 2782, 3077, 3227, 3516, 4051, 4151, 4258, 5548, 6814, 7469), now only 8 are considered as genuine Seyferts.

In 1940s, after observations of discrete radio sources, some of them were for the first time identified with extragalactic objects (galaxies) and these galaxies were named **radio**

galaxies. Hey, Parsons and Phillips (Hey et al. 1946) discovered variations in the intensity of galactic noise from the direction of the constellation of Cygnus, with a period of about one minute-suggesting that this particular radiation has its origin in a discrete source, Cygnus A (later named as 3C 405). Several other radio galaxies (Per A = NGC 1275, etc.) were detected that seemed to have double structure (Bolton et al. 1949). These objects emitted huge amounts of energy attributed to the collisions of two galaxies (Baade & Minkowski 1954).

In the optical wavelengths, using three-color filter technique, 44 blue galaxies were found by Haro (1956), some of them showing emission lines in their spectra ([OII] 3727 doublet, etc.). However, very few observational facts were known in the middle of 1950s related to peculiar radiation from galaxies.

Since the beginning of 1950s, Ambartsumian carefully analyzed all accumulated data on emission-line galaxies (Seyfert 1943 and other papers), radio galaxies, blue components around giant galaxies, some other multiple galaxies (Zwicky 1956), Haro's blue galaxies (Haro 1956), etc. and came to a conclusion that all these various manifestations of activity related to the same physical phenomenon, namely activity of the galactic nuclei (Ambartsumian 1955; 1956). It was not straightforward and obvious, as the data were very few and each seemed to have independent explanation. Moreover, blue-UV emission of some nearby galaxies obviously came from their spiral arms and was explained by a large number of hot stars. Ambartsumian rejected the collisional model by Baade & Minkowski (1954) based on calculations of probabilities of collisions, which appeared to be almost impossible. According to Ambartsumian, manifestations or forms of activity could be rather different:

- emission or outflow of ordinary gas matter from the central part of the galaxy having velocities up to several hundreds of km/sec,

- continuous emission of fluxes of relativistic particles originating high-energy particles (forming radio halos around the nuclei),

- eruptive ejections/outbursts of gas matter (M82),

- eruptive ejections/outbursts of relativistic plasma (NGC 4486, NGC 5128),

- ejections/outbursts of blue condensations having absolute luminosities typical of dwarf galaxies, etc.

These various forms of activity were presented as different manifestations of the same phenomenon of the activity of galactic nuclei. The evolutionary significance of the activity in the galactic nuclei was emphasized and a further hypothesis was suggested on the ejection of new galaxies from AGN. Thus, a **hypothesis on the activity of galactic nuclei** was proclaimed by Ambartsumian (1955; 1956). Similar discussions and direct indication on massive nuclei were given by Woltjer (1959).

A comparative analysis of all these observational data shows that independent on their apparent differences, all these phenomena have a common physical nature. Ambartsumian came to such conclusion at the very beginning of investigations and reported this in important papers (Ambartsumian 1958; 1961). From the modern point of view, these ideas could be regarded as the same as a unified model for all types of AGN.

However, during many years in 1960s–1980s, all types of the revealed AGN were regarded as different kinds of objects, probably with different mechanisms of radiation. Moreover, all historical classifications (Seyfert 1 and 2, radio galaxy, QSO, LINER, BL Lac objects, etc.) supported an idea to explain them individually and then (if possible) try to find similarities or links between these classes.

The theoretical study of the numerous observational evidences of various sorts of physical instability in galaxies led Ambartsumian to a fundamental conclusion that in processes of origin and evolution of galaxies, the role of the central small in their sizes condensations, the nuclei of galaxies, is huge. He justified an essentially new understanding that all observational evidences of the instability of galaxies are a consequence of activity of the galactic nuclei. The hypothesis on the superdense protostellar matter was engaged to explain the observational data, though later on not accepted. Anyway, modern understanding on SMBH in the centre of AGN very much resembles Ambartsumian's ideas.

In 1959, after the completion of the 3rd Cambridge radio survey (3C Catalogue; Edge et al. 1959), many new radio sources were found and many of them were identified with extragalactic objects. Optical spectra of a number of the identified point-like objects showed peculiar emission lines that could not be identified. Using an optical spectrum obtained with the 200-inch Hale Telescope on Mt. Palomar, Maarten Schmidt was the first (Schmidt 1963) to interpret the spectrum of 3C 273 as having very largely redshifted (z=0.158) broad emission Balmer lines corresponding to recession velocity of 47,000 km/s. Immediately, Greenstein & Matthews (1963) identified spectral lines in the optical spectrum of the radio source 3C 48 (z=0.37). This discovery allowed other astronomers to find redshifts from the emission lines from other radio sources thus extending our knowledge to much farther extragalactic universe. These point-like extragalactic radio sources were called **quasi-stellar radio sources (quasars) or quasi-stellar objects (QSOs)**.

In 1963, B. E. Markarian published a list of 73 galaxies having peculiar colours compared to their spectral types and showed that there should be some additional radiation (Markarian 1963). Based on this consideration and powered by Ambartsumian's ideas, in 1965 Markarian initiated a large objective prism survey for detection of galaxies with UV-excess (UVX) (Markarian 1967). 17,000 sq. degrees were covered in the Northern and some part of the Southern extragalactic skies and 2,000 plates were obtained. **Markarian Survey** (also known as the First Byurakan Survey, FBS) played an important role in the discovery of many new AGN and understanding their types and cosmic abundance. Weedman and Khachikian obtained many spectra for **Markarian galaxies** and due to more statistics classified Seyferts into Sy1 and Sy2 types (Weedman & Khachikian 1968; Khachikian & Weedman 1971; 1974); Sy1s having both broad and narrow emission lines and Sy2s having only narrow emission lines. Later on, Osterbrock introduced more subtypes: Sy1.0, Sy1.2, Sy1.5, Sy1.8, Sy1.9, and Sy2.0 (Osterbrock 1981). However, even Sy1 and Sy2 were regarded as different types of objects with different physical nature and radiation mechanisms. Based on Markarian galaxies, Weedman also introduced the class of the Starburst galaxy (Weedman 1977). Catalogues of Markarian galaxies were published by Mazzarella & Balzano (1986), Markarian et al. (1989; 1997) and Petrosian et al. (2007). Some important features of Markarian survey are:

- Markarian Survey was the first systematic objective-prism survey in the world

- Until now it is the largest objective-prism survey of the Northern sky (17,000 sq. deg)

- It was a new method of search for active galaxies

- 1515 UVX galaxies were discovered, including 181 Seyferts, 17 LINERs, 13 QSOs, 3 BLLs, 95 Starburst, 26 HII galaxies (Markarian et al. 1989)

- Classification of Seyferts was carried out into Sy1 and Sy2 (Khachikian & Weedman 1974)

- The definition of Starburst galaxies was introduced (Weedman 1977; Balzano 1983)

- Many new Blue Compact Dwarf Galaxies (BCDG) were discovered

- Similar surveys were conducted and many new AGN were discovered; Second Byurakan Survey (SBS, Markarian et al. 1983, Stepanian 2005) and others

- Other projects were carried out based on FBS plates: FBS Blue Stellar Objects (BSOs; Mickaelian 2008), FBS Late-Type Stars (Gigoyan & Mickaelian 2012), optical identifications of IRAS points sources; Byurakan-IRAS Stars (BIS; Mickaelian & Gigoyan 2006) and Byurakan-IRAS Galaxies (BIG; Mickaelian & Sargsyan 2004), including many new AGN and ULIRGs

More details on Markarian Survey are given by Mickaelian (2014).

In addition to Seyfert, Haro and radio galaxies known since 1940s-1950s, during 1960s-1980s, quasars (Schmidt 1963), peculiar galaxies (Apr 1966), Markarian galaxies (Markarian 1967), BL Lacertae objects (as extragalactic sources similar to quasars; Strittmatter et al. 1972), High surface brightness (Arakelian) galaxies (Arakelian 1975), Starburst galaxies (Weedman 1977), blazars (as a new class unifying BL Lacs and OVV/HPQ quasars; Spiegel 1978), LINERs (Heckman 1980), IR galaxies (after IRAS mission in 1983, the first big list was given by Soifer et al. (1989) and revised by Sanders et al. (2003)) were discovered either introduced. AGN zoo appeared with a big mixture of properties and confusion in definitions and classifications.

In 1985, Antonucci and Miller published a paper *"Spectropolarimetry and the nature of NGC 1068"* (a classical Seyfert 2 type showing only narrow emission lines) (Antonucci & Miller 1985). The polarized flux plot revealed the presence of very highly polarized, very broad symmetric Balmer lines and also permitted Fe II closely resembling the flux spectra of Seyfert type I nuclei. This line emission indicated that both polarizations were due to scattering, probably by free electrons which must be cooler than a million K. A model was suggested in which the continuum source and broad line clouds were located inside a thick disk, with electrons above and below the disk scattering continuum and broad-line photons into the line of sight. All of the narrow lines, including the narrow Balmer lines, had similar low polarizations, unrelated to that of the continuum. Further studies strengthened such a geometrical understanding of the difference between the AGN, so that each type (the classification) depended on the observed angle. So the major breakthrough in understanding the connection between Sy1s and Sy2s was the discovery of a "hidden" broad line region (BLR) in the Sy2 NGC 1068.

According to this understanding, Sy1s have nuclear emission line spectrum characterized by broad (a few thousands km s^{-1}) permitted lines and narrow (a few hundreds km s^{-1}) high excitation lines. Seyfert 2s have narrow emission line spectrum like the Seyfert 1s, but they are lacking both the compact nucleus and the broad emission lines. QSOs are high luminosity Sy1s. Seyferts and QSOs contain a compact nuclear continuum source ionizing a broad line region, surrounded by an optically thick torus of dust. Depending on the orientation of this torus with respect to the line of sight, the central object is seen or hidden; when it is hidden, we see only the narrow, extended emission line region; the galaxy is a Sy2. This is, very schematically, the generally accepted **Unified Scheme of AGN**. Thus, all kinds of AGN now are put in the same scheme and are regarded as a common phenomenon; an approach developed by Ambartsumian since the middle of 1950s.

Later on, new classes and subclasses of active galaxies were discovered and introduced, such as BCDGs (Thuan & Martin 1981), Narrow-Line Seyfert Galaxies (Osterbrock & Pogge 1985), composite spectrum objects (Véron et al. 1997), subtypes of Starbursts (Terlevich 1997; 2000), etc.

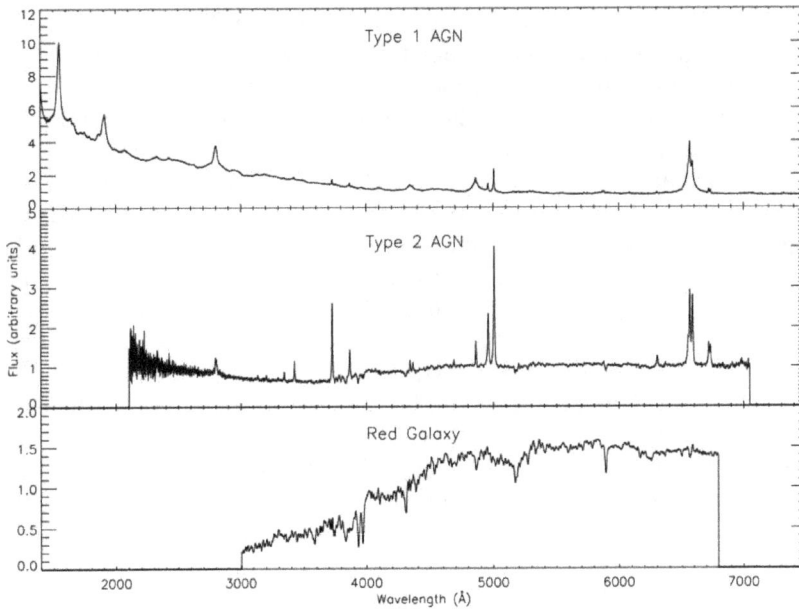

Figure 1: Three main types of optical spectra of galaxies: Type 1 AGN with both broad and narrow emission lines, Type 2 AGN with narrow emission lines, and normal galaxies with absorption lines.

3   Optical Emission Line Spectrum of Active Galaxies

The optical emission line spectrum of active galaxies contains narrow (line widths up to 300 km/s) and broad (line widths – several thousand km/s) lines. They have been crucial for understanding the physical processes in these galaxies and for classifications, as they are the most significant features and provide a big variety of information. First of all, one should distinguish normal (typically red) galaxies from active ones by their absorption spectrum. Active galaxies (both AGN and SB) show many emission lines, including H Balmer series, He, O at different ionization level, N, S, metallic lines, etc. Two big groups that should be distinguished immediately, are Type 1 and Type 2 AGN. Type 1 AGN have both broad and narrow emission lines, Type 2 AGN, as well as SB have only narrow emission lines, and normal galaxies have no emission lines (Fig. 1). Along with the spectral lines, an important role also plays the continuum; stronger the continuum, relatively weaker the lines and less accurate is the classification. Therefore, in many cases, templates for red elliptical galaxies are being built, fitted to the studied spectra and extracted to have pure emission line spectrum for accurate measurements and classifications. The presence of the underlying host galaxy is displayed by the level of the continuum with typical absorption galaxy shape and absorption lines (NaI 5890/5896 doublet, MgI, Hydrogen Balmer lines, etc.). Among the emission lines, in the optical range most prominent are Hydrogen Balmer series lines (Hα 6363, Hβ 4861, Hγ 4340, etc.), Oxygen lines ([OIII] 4959 and 5007, [OII] 3727 and [OI] 6300), Nitrogen lines ([NII] 6548 and 6484), Sulfur doublet ([SII] 6716/6731), Helium lines HeI 5876 and HeII 4686, etc.

There are two kinds of emission lines: 1) Recombination lines (e.g. Balmer H lines) and 2) Collisionally excited lines. The first type lines are due to electric dipole transitions, so they occur easily, they have very short lifetimes, and are called "permitted" transitions and hence, **"permitted lines"**. Collisionally excited lines, ground state often split by small energies $E \sim kT$ thermal collisions can populate these low lying levels. De-excitation occurs either by collisions or radiatively; which dominates depends on which occurs fastest. Often, the radiative lifetimes are long, because the transitions are "forbidden" (they only occur via electric quadrupole or magnetic dipole transitions). The lines hence are called **"forbidden lines"** (usually taken in square brackets, e. g. [OIII]). They tend to be suppressed at high densities, when collision times are fast. At the critical density, the radiative and collisional rates are equal. In gas with density above the critical density, the line is not strongly produced (it is collisionally suppressed). For each forbidden line, there is its critical density. For some of the most important forbidden lines that appear in AGN and SB spectra, these critical densities in cm^{-3} (log) are: CIII] 1909 – 9.0, [OII] 3726.1 – 3.5 and [OII] 3728.8 – 2.8, [OIII] 5006.9 – 5.8, [OI] 6300.3 – 6.3, [NII] 6583.4 – 4.9, [SII] 6716.4 – 3.2, [SII] 6730.8 – 3.6. Permitted and semi-forbidden lines can be both broad and narrow, whereas forbidden lines never appear as broad ones.

Because there are numerous lines in the optical part of the spectrum, their identification and classifications are not easy tasks. There also are a number of technical problems with emission lines that should be taken into account to consider further classifications. Here are some of them:

- **Reduction problems.** These cause different numbers in line measurements, slope of the continuum, etc. Even the SDSS spectra that are relatively good by their spectral resolution and signal-to-noise (S/N) ratio, very often are measured differently.

- **Host galaxies.** These produce significant continuum and absorption spectrum that strongly affect emission line measurements. Very often Balmer lines have both emission and absorption components superposed and it is not easy to define what the true values are for each. Better emission-line spectrum is obtained if only the central part of the galaxy is observed.

- **Fe lines.** AGN spectra contain numerous iron (FeI, FeII and FeIII) lines. They appear around $H\beta$ (from both sides) and elsewhere and interfere accurate line identification and measurements. Fe templates have been built to be fitted and subtracted from a given spectrum. The best one is IZw1 FeII template obtained with MIDAS (Fig. 2).

- **Line identification.** Due to the presence of numerous lines, their identification becomes rather tricky. Most of the models use emission lines well defined from nebular spectra, as well as from emission-line stars. However, very often one needs to make own choice between many possibilities (Fig 3).

- **Multiple profiles.** Emission line systems are not unique. Depending on the physical properties of the regions producing these lines, several broad and narrow lines systems may appear. In this case the summarizing profiles have no physical sense and one should be careful when measuring any line parameter (Fig. 4).

- **Variability.** This causes change of both continuum and line parameters. AGN typically are optically variable (e. g. blazars, QSOs; Véron & Hawkins 1995). In this case typical (average) line parameters are given with possible limits of changes. However, one should carefully measure all parameters and in case of multiple observations, follow the changes in individual lines; both positions, intensities and widths (e. g. Goodrich 1995; Edelson et al. 2001, for NGC 4151). In Fig. 5 we give two examples of variable AGN spectra superposed on each other to show the changes in both continuum and emission lines.

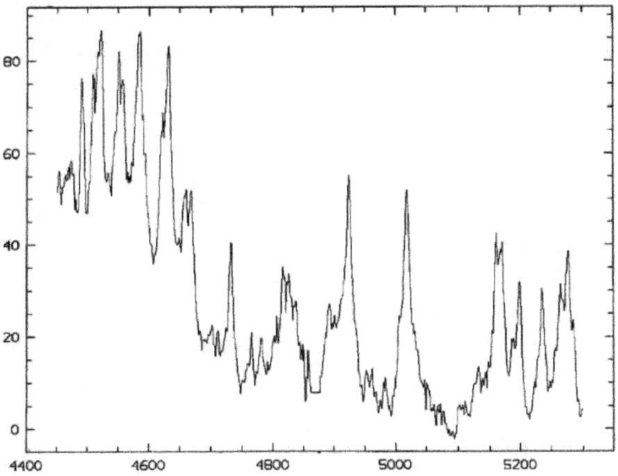

Figure 2: FeII template obtained from IZw1 spectrum and especially important around Hβ.

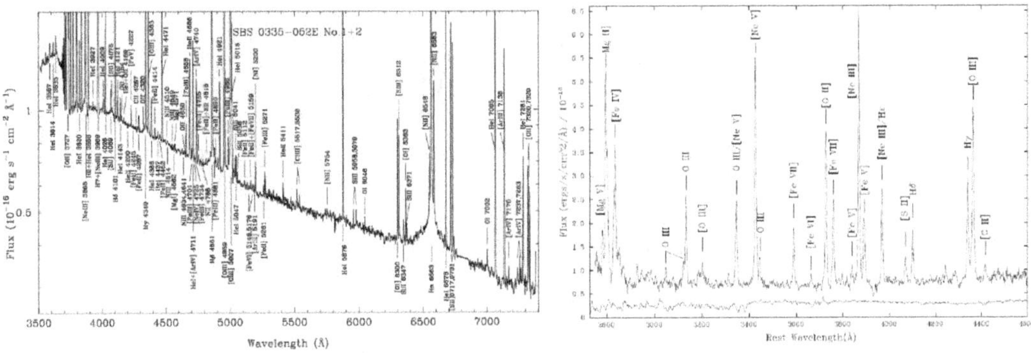

Figure 3: Numerous emission lines typical for optical spectra of active galaxies.

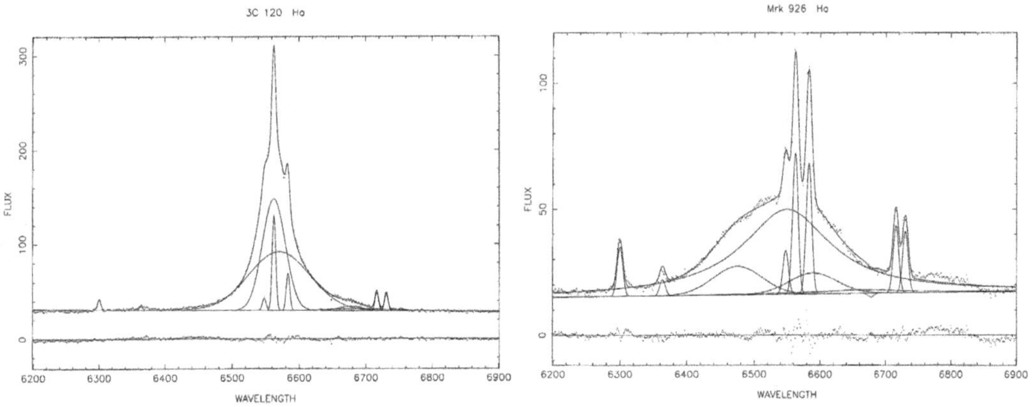

Figure 4: Individual line profiles and summarizing observed lines in spectra of 3C 120 and Mrk 926 (both in Hα region).

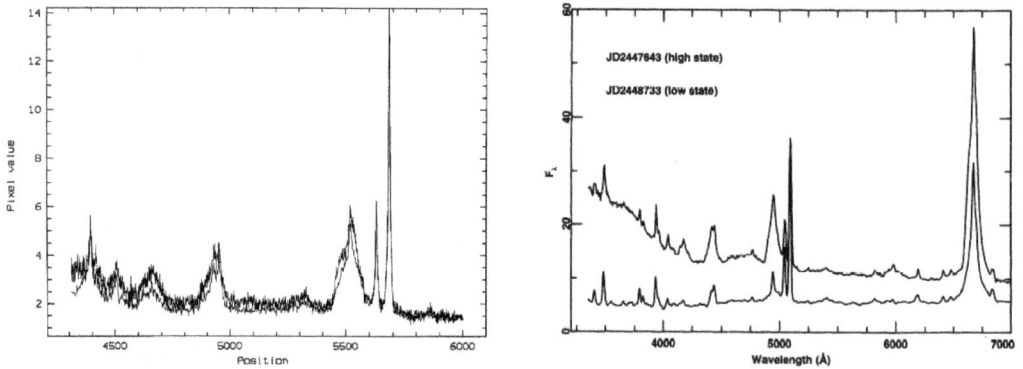

Figure 5: Examples of variable AGN spectra superposed on each other to show the changes in both continuum and emission lines. Left: Kaz 102 (Kazarian & Mickaelian 2007), right: J02447643.

Basic line parameters (of a given profile) are the **Position** of the maximum (to define redshifts), **Intensity** above the continuum level, **Full Width at 0 Intensity (FW0I)** and **Full Width at Half Maximum (FWHM)** and **Equivalent Width (EW)**. All spectral reduction software, including MIDAS, IRAF, etc. can easily measure these parameters. To describe the line width, FWHM is much more important rather than FW0I, as the latter is affected by the inaccurate measurements of the start and end wavelength due to uncertainties and noises in the continuum. FWHM is defined based on the Gaussian built on the observed profile and is measured with high accuracy.

Fine Analysis of Emission Line Spectra helps elimination of host galaxy continuum, accurate line identification, decomposition of composite line profiles and variability studies. We have carried out a number of studies to understand in detail the emission line spectra of active galaxies in collaboration with Véron and Véron-Cetty during 1997-2003. Of course, relatively high resolution spectroscopy is needed to have reasonable results (SDSS spectra can hardly satisfy these requirements; our OHP 1.93m telescope CARELEC 600 /mm grating spectra had 0.9 Å/pix dispersion and 3.2 Åspectral resolution and we obtained two spectra for each galaxy, one around Hβ and the other, around Hα). This study made use of the software SPECTRAI (Véron et al. 1980), in which the user gives start values for each profile: Amplitude, Line center (1+z) and Line width, as well as Continuum value and Slope. Up to 196 free parameters may be used, and we tested 166 spectral lines in the range 3700-7900Å(most important ones are given in Table 1 and may be used for investigation of any AGN spectrum). At the beginning, elliptic galaxy template (stellar continuum) and Fe II template from IZw1 galaxy are subtracted to exclude contamination and have better results. Fine analysis of emission line profiles for fitting of individual Gaussian (or Lorentzian) profiles to the observed summarizing ones was carried out. A comparison between the H and the H ranges observed in frame of the same project was done to accept the best solution. Detailed spectroscopic classification using the derived narrow-line intensities was carried out. Correlations between FWHM of Fe II and Balmer lines, Fe II and [O III] lines were investigated and fine details in spectra of AGN were found, including new lines, often observed very high ionization lines, etc.

Table 1: Most important emission lines in spectra of active galaxies used for fine classification. Spectral lines in the two left columns relate to Hβ range (4200-5300 Å) and those in two right columns relate to Hα range (5600-7300 Å).

Line	λ, Å	Line	λ, Å	Line	λ, Å	Line	λ, Å
Hγ	**4340.47**	[Fe VII]	4942.3	[Fe VII]	5631.1	[Fe X]	6374.51
[O III]	4363.21	**[O III]**	**4958.92**	[Fe VII]	5677.0	[A V]	6435.1
He I	4387.9	[Fe VI]	4967.1	[Fe VII]	5720.9	**[N II]**	**6548.03**
He I	4471.5	[Fe VI]	4972.5	[N II]	5754.8	**Hα**	**6562.79**
Mg I	4571.0	[Fe VII]	4988.9	**He I**	**5875.6**	**[N II]**	**6583.41**
C IV	4658.6	**[O III]**	**5006.84**	Na I	5889.9	[Fe VII]	6599.1
He II	**4685.7**	He I	5015.7	Na I	5895.9	He I	6678.1
[A IV]	4711.3	[Fe VI]	5145.8	He II	5977.0	Ni XV	6701.83
He I	4713.1	[Fe VII]	5158.9	[Ca V]	6086.4	**[S II]**	**6716.47**
[A IV]	4740.2	[Fe VI]	5176.4	**[Fe VII]**	**6086.9**	**[S II]**	**6730.85**
Hβ	**4861.33**	[N I]	5197.9	**[O I]**	**6300.30**	[A V]	7005.7
[Fe VII]	4893.9	[N I]	5200.3	Si II	6347.1	He I	7065.2
He I	4921.9	[Fe VII]	5276.4	**[O I]**	**6363.88**	[A III]	7135.8
[Ca VII]	4940.3			Si II	6371.4	[A IV]	7170.6
				[Fe X]	6372.9	[A IV]	7237.5

4   Diagnostic Diagrams for AGN Classifications

In this section we describe **AGN Optical Line Diagnostics** based on study of spectra in optical range, which allows to distinguish Seyfert galaxies, LINERs and Starburst (or HII regions). First diagrams were introduced by Baldwin, Phillips & Terlevich in 1981 (BPT diagrams; Baldwin et al. 1981). They used emission line intensities ratios ([OIII]5007/Hβ, [NII]6584/Hα, [OIII]5007/[OII]3727 and [OIII]5007/[OI]6300) to distinguish Seyferts against LINERs and Starbursts. Veilleux and Osterbrock improved this technique by modifying line rations to [OIII]5007/Hβ, [NII]6583/Hα, [OI]6300/Hα and [SII]6716+6731/Hα (Veilleux & Osterbrock 1987), as some BPT ratios need reddening correction while the Veilleux & Osterbrock ratios do not (being close in λ). In general Seyferts have strong [OIII]5007/Hβ, strong [NII]6583/Hα and strong [OI]6300/Hα ratios, i. e. a wide range of ionization degree. LINERs have lower ionization degree: weaker [OIII]5007/Hβ but strong [NII]6583/Hα and strong [OI]6300/Hα. HII regions have weaker [OIII]5007/Hβ, much weaker [OI]6300/Hα and somewhat weaker [NII]6583/Hα (Fig. 6).

At present very often these diagrams are called **BPT diagrams**, as later a number of new ones also appeared. However, there are some comments that should be taken into account when classifying galaxies with these diagrams:

- HII regions lie along a sequence in metallicity (Z): low Z have higher ionization. The sequence can reach [OIII]/Hβ $\sim$ 10, so high [OIII]/Hβ is not unique for Seyferts.

- AGN are different because their ionizing spectra are broad, with a high energy component. Behind the fully ionized region lie large partially ionized regions kept hot by X-ray heating. It is in these regions that [OI], [NII] and [SII] lines are generated efficiently.

- Seyfert/LINER group may be considered as a single sequence in radiation parameter, U. Seyferts have intense hard radiation field and LINERs have weak hard radiation

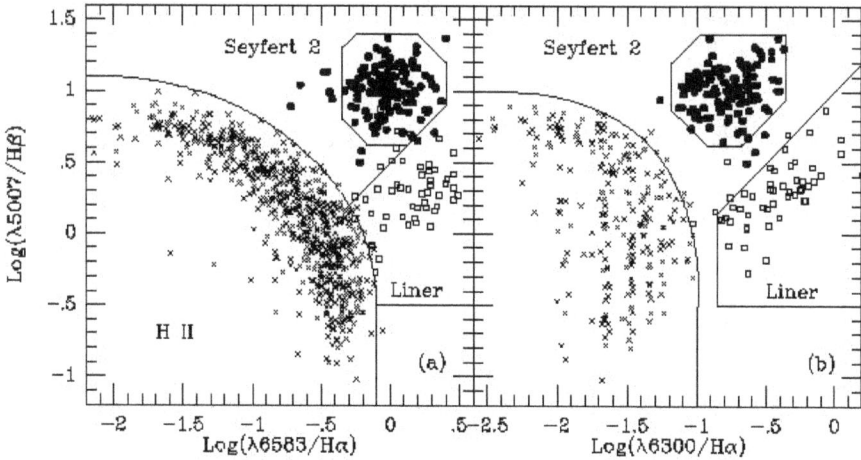

Figure 6: BPT diagnostic diagrams for classical Seyfert 2 galaxies (filled squares), LINERs (open squares) and HII galaxies (crosses). These diagrams allow to define exact regions for these three types in addition to the main lines separating them.

field.

- Very high ionization lines can be unambiguous indicators of AGN: e. g. [NeV]3426.

- Some non-AGN emission regions can show LINER spectra: e. g. cooling flows; shocks in starburst driven winds; bulge inter-stellar medium (ISM) ionized by post-AGB stars.

- Fast (500 km/s) shocks can also show Seyfert spectra, though no such clear cases are known.

- Before and in 1980s, [OIII] line width was used in defining AGN (when FWHM > 300 km/s). However, now this is no more in use, as FWHM tracks bulge mass; many AGN now are known with FWHM < 300 km/s.

Later on, a number of other diagnostic diagrams were suggested, both with the use of line ratios and other parameters, like colours and stellar masses, equivalent widths, etc. E. g. Lamareille and others (Lamareille et al. 2004; Lamareille 2010) have suggested similar diagram for [OIII]5007/Hβ vs. [OII]3727/Hβ ratios (so-called **Blue Diagram**).

In Fig 7 we give three diagnostic diagrams introduced by Veilleux & Osterbrock (1987) and used more frequently based on [NII]6583/Hα vs. [OIII]5007/Hβ (DD I), [SII]6716+6731/Hα vs. [OIII]5007/Hβ (DD II) and [OI]6300/Hα vs. [OIII]5007/Hβ (DD III) line ratios with distribution of Seyferts, LINERs, HII regions and so-called transition objects, which appear as intermediate objects between Seyferts and HIIs, LINERs and HIIs or Seyferts and LINERs. Such objects show features of two or three types and introduce confusion for the classifications. Later on, Véron et al. (1997) called them Composite spectrum objects and showed that these are the same spectra as Sy2, LINER and HII overlapped on each other. However, there have been a number of papers arguing the correctness of the **divisor lines between different types of AGN** (Kewley et al. 2001; 2006; Stasinska et al. 2006, etc.).

Some other diagnostic diagrams in addition to line intensity ratios (excitations) use colours, equivalent widths, stellar masses, etc. Stasinska et al. (2006) have used 4000-Åbreak index, $D_n(4000)$ vs. max (EW[OII]3727, EW[NeIII]3869) equivalent widths to distinguish

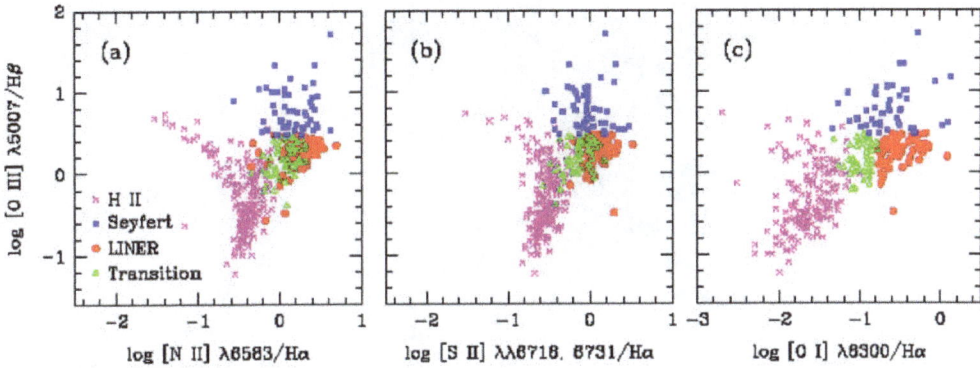

Figure 7: Three diagnostic diagrams (from left to right, DD I, DD II, DD III) with distribution of Seyferts, LINERs, HII regions and so-called transition objects.

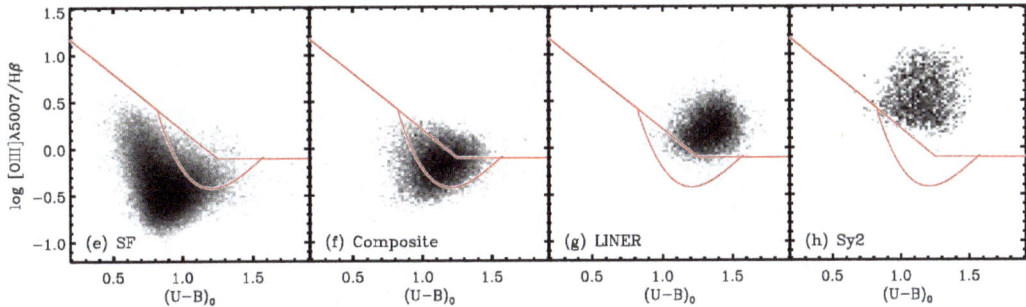

Figure 8: Diagnostic diagrams based on [OIII]5007/Hβ ratios and rest-frame U-B colour separately given for Star-Forming galaxies (SF), Composites, LINERs and Seyferts.

different type of objects (so-called **DEW Diagram**). They showed that none of BPT diagrams is efficient in detecting AGN in metal-poor galaxies, should such cases exist. Their diagram can be used with optical spectra for galaxies with redshifts up to z = 1.3, meaning an important progress over classifications proposed up to now. Since the DEW diagram requires only a small range in wavelength, it can also be used at even larger redshifts in suitable atmospheric windows. It also has the advantage of not requiring stellar synthesis analysis to subtract the stars and of allowing one to see all the galaxies in the same diagram, including passive galaxies.

Fig. 8 gives one of such alternative diagrams based on [OIII]5007/β ratio and rest-frame U-B colour (**Colour-Excitation or CEx Diagram**) separately for Star-Forming galaxies (SF), Composite objects, LINERs and Sy2s introduced by Yan et al. (2011). There is a strict separation for SF and AGN and the authors consider this diagram much more efficient for distinguishing different types.

Trouille et al. (2011) introduced a diagnostic diagram based on the rest-frame SDSS g-z colour vs. [NeIII]3869 / [OII]3726,3729 lines ratios, so-called **TBT Diagram** (after the authors names: Trouille, Barger, Tremonti). Cid Fernandes et al. (2011) introduced diagnostic diagrams based on EW of Hα (WHα) vs. [NII]/Hα (so-called **WHAN Diagram**).

Juneau et al. (2011; 2014) consider [OIII]5007/H line ratios against stellar masses to

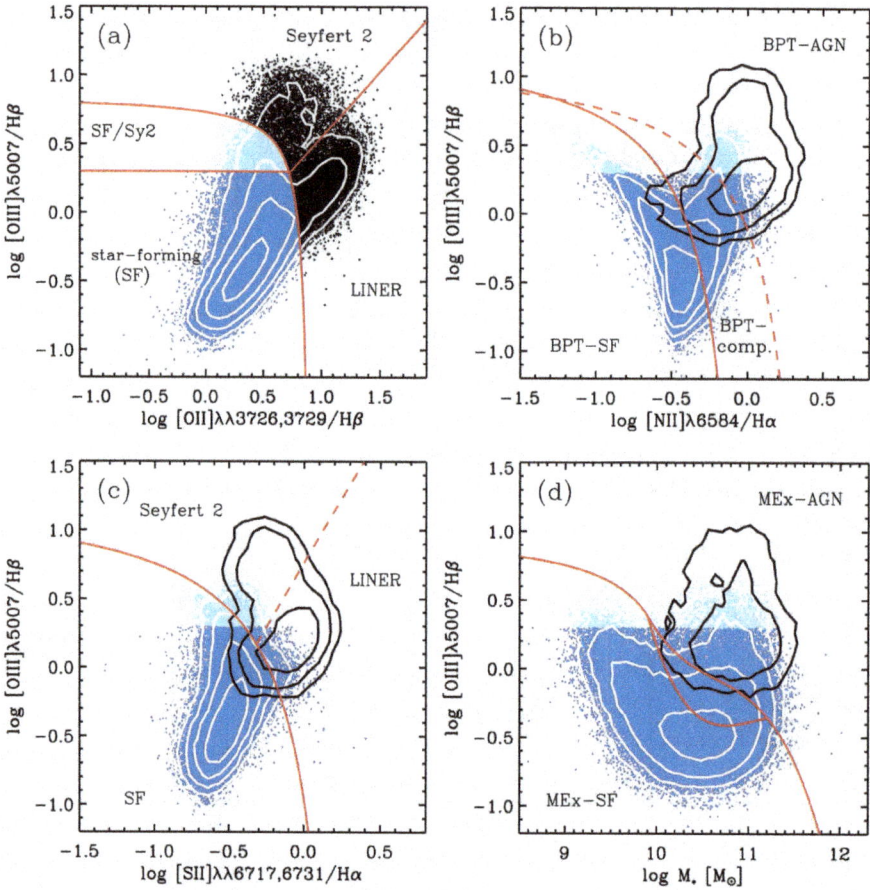

Figure 9: Mass-Excitation (MEx) diagrams by Juneau et al. (2011) demonstrating that combining [OIII]5007/Hβ and stellar mass successfully distinguishes between star formation and AGN.

distinguish Seyfert 2s, LINERs, and star-forming (SF) galaxies (Fig. 9). They are called **Mass-Excitation (MEx) Diagrams**. They are useful to identify AGN in galaxies at intermediate redshift. In the absence of near-infrared spectroscopy, necessary for using traditional nebular line diagrams at z > 0.4, they demonstrate that combining [OIII]5007/Hβ and stellar mass successfully distinguishes between star formation and AGN emission.

AGN Colour-Colour Diagnostics developed by Smolcic et al. (2006, 2008) is based on synthetic colours from a modified set of Stromgren uvby narrow (∼200 Angstroms) bands. They adopt the filters: uz, vz, bz, yz, and show a sequence on the bz − yz vs. vz − yz colour-colour plane. P1 is defined as the location along the sequence, defined by their equations. The adopted dividing line is defined as P1 > 0.15.

Other optical diagnostic diagrams have been developed by Kewley et al. (2001; 2006), Kauffmann et al. (2003), Stern et al. (2005), Hickox et al. (2009), Cid Fernandes et al. (2011), Del Moro et al. (2013), Teimoorinia & Ellison (2014), Vogt et al. (2014) and others. The latter used **ZQE diagrams**, which are a specific set of 3D diagrams that separate the oxygen abundance and the ionization parameter of HII region-like spectra and also enable observers to probe the excitation mechanism of the gas. Stephanie Juneau has developed a com-

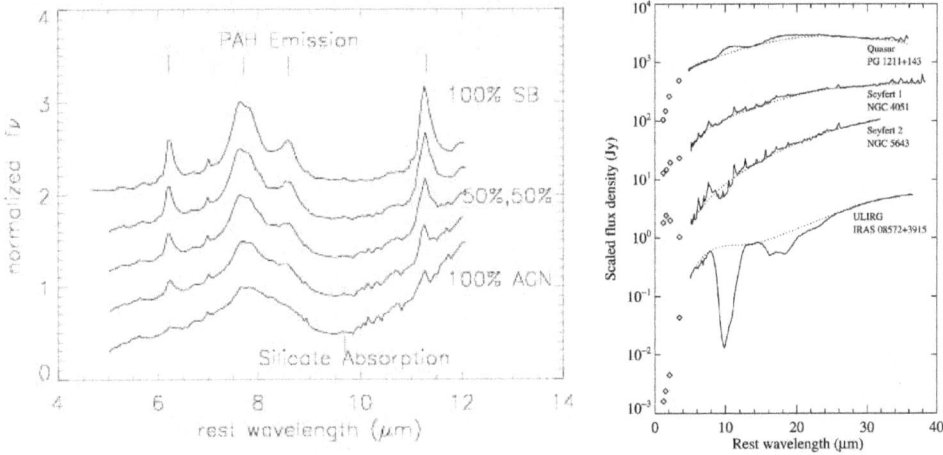

Figure 10: Mid-IR spectra of active galaxies and typical features to distinguish AGN against Starbursts: PAH emission and Silicate absorption (Houck & Weedman 2009; Levenson et al. 2014).

prehensive **AGN diagnostics website** at https://sites.google.com/site/agndiagnostics/, where both optical and other wavelengths are considered.

5   Classifications Based on other Wavelength Ranges

AGN Mid-IR Diagnostics. Mid-IR (MIR) Colour-Colour Diagrams and **MIR Emission Line Diagrams** have been used based on Spitzer Space Telescope IRAC/MIPS colours and IRS spectra (Donley et al. 2012; Weedman & Houck 2009). PAH emission is strong for Starbursts and is also present for objects with intermediate activity type given as 50% AGN and 50% SB. Silicate absorption is deeper for SB as well. These IR features give understanding on fractions of contribution of AGN and SB and may serve as even better classification than precise optical types, because very many nuclei show evidence of mixture of different types, classified as Composite (S2/HII, LINER/HII or other). Levenson et al. (2014) described larger range IR spectrum (5-35 μm) and compared AGN IR spectra to ULIRG one (Fig 10). AGN types are plotted separately as well (Quasar, S1 and S2) showing the change of the same features in the same direction, i. e. from stronger ones for SB to weaker ones for stronger AGN.

Other spectral ranges include **AGN Radio Diagnostics** based on FIR-Radio Correlation and **AGN X-ray Diagnostics**. Compared to optical diagnostic diagrams they are rather limited, however many AGN show no activity in optical wavelengths (obscured or hidden AGN) and one needs alternative means to identify them. Another limitation comes from smaller amount of observations (especially spectral ones) in these wavelengths, namely X-ray, UV, IR, sub-mm and radio. We will show in next section that only MW SEDs and much better, MW spectra can provide full understanding on the nature of the objects, including detection of differences between classes in various wavelength ranges and hence understanding of the processes putting constraints on AGN and/or Starburst emissions in optical range to be identified as classical AGN or SB.

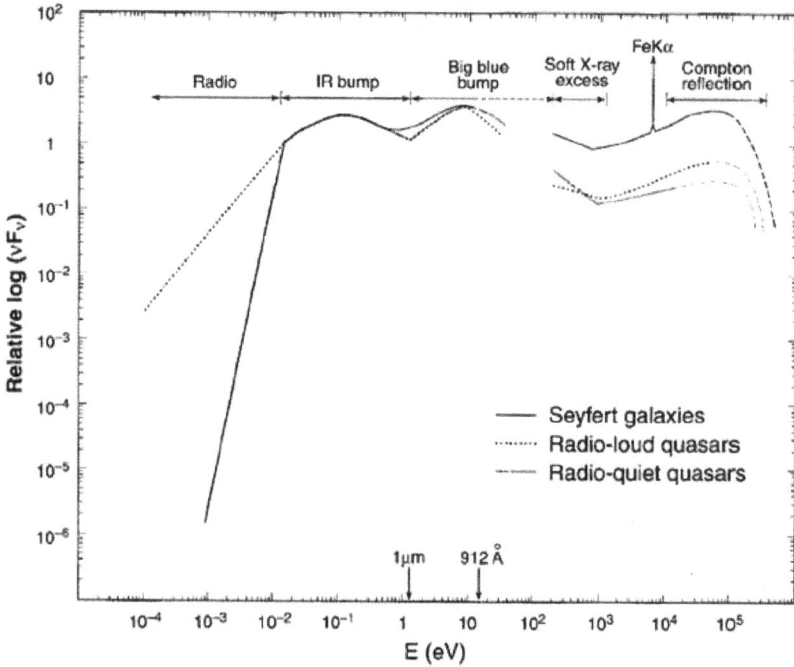

Figure 11: QSOs and Seyfert galaxies SEDs showing big difference between radio-loud and radio-quiet quasars and similarity for radio-quiet quasars and Seyferts.

6   Multiwavelength SEDs of active galaxies

Existence of active galaxies obscured in optical wavelengths and absence of radio and/or X-ray emission at most of the AGN means that multiwavelength spectrum may be the only clue to true classification. **Radio-loud and radio-quiet QSOs** have been discovered since 1960s (Sandage 1965), which already meant a significant difference in their SEDs. Sandage showed that radio-quiet QSOs were 10 times more than radio-loud ones. This meant that no prediction about radio emission could be made based on only optical spectrum. It is sometimes necessary to try to distinguish carefully between radio-loud and radio-quiet QSOs. A useful criterion (Kellermann et al. 1989) appears to be the radio-optical ratio of specific fluxes at 6 cm (5 GHz) and 4400 Å(680 THz); for radio-loud objects it is generally in the range 10-1000, and most radio-quiet objects fall in the range 0.1-1. **AGN MW SEDs** have been built since their first studies in other than optical wavelengths (Fig. 11; Elvis et al. 1994).

To understand the difference in radiation in various wavelength ranges, one should consider the **physical explanation of AGN spectral components**, i. e. mechanisms of radiation at different wavelengths (Fig. 12, Fig. 13). In near-UV (NUV) – optical range, the Big Blue Bump (0.1–1 μm) is emitted by the thermal Compton disc. Sometimes it appears as Small Blue Bump (Fig. 14) and there is also an IR bump coming from dust torus; 1 m inflection and mm break ($\sim$100μm); most of IR is being emitted by dust according to Lν = ν^{-3} at 0.1–3 mm wavelengths. Radio emission is conditioned by synchrotron jet. Soft X-ray excess (at $\sim$0.1 keV), Fe Kα line and Compton reflection in γ rays are conditioned by thermal Compton hot corona (Compton Hump, $\sim$10–30 keV) and inverse Compton jet. Fig. 12 shows differences in SEDs between various AGN types. Fig. 13 shows big differences in

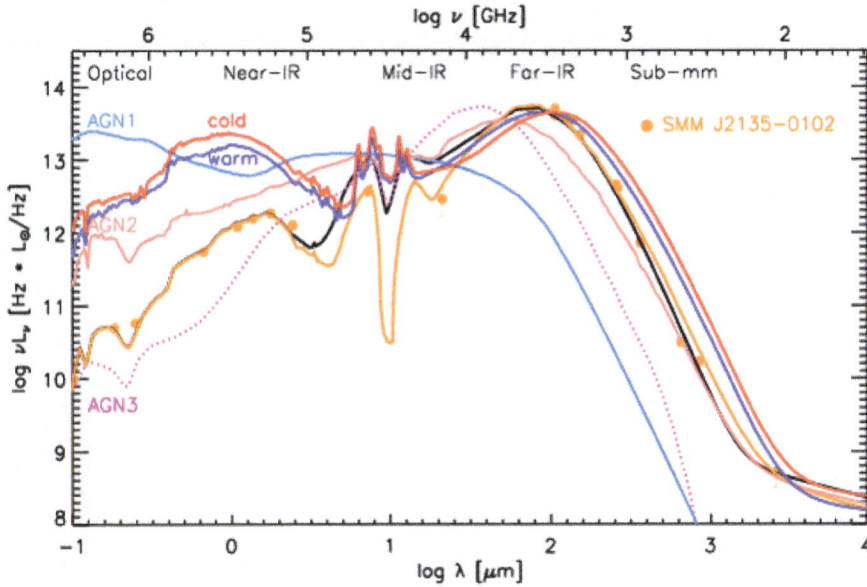

Figure 12: AGN multi-wavelength SEDs from optical to sub-mm wavelength ranges. Differences in SEDs between various AGN types are seen.

distributions for various radiation mechanisms present at different AGN types.

Science Data Center of the Italian Space Agency (ASDC, ASI Science Data Center) gives comprehensive SEDs for any object built by the selected catalogs from the list (www.asdc.asi.it/). Using this facility and other available data, we have carried out MW studies of Markarian galaxies (Mickaelian et al. 2013) and many other bright active galaxies, including IR ones (Sargsyan et al. 2001; Hovhannisyan et al. 2011; Mickaelian et al. 2012). Fig. 15 shows two remarkable Markarian galaxies, Mrk 231 (closest known ULIRG) and Mrk 421 (blazar, one of the rare TeV sources) broad-band spectra (MW SEDs) built by means of ASDC website. To have full understanding on AGN types and activity processes, one should compare all known optical (historical) classifications with MW SEDs and have enough statistics to obtain correspondence and derive relations between radiations in various wavelength ranges.

7 Classifications as a Multi-Parametric Problem

We described the AGN classification principles based on mostly optical, as well as other wavelength ranges. Most of optical diagnostic diagrams are based on emission line ratios. However, AGN classification is a multi-parametric problem. Emission lines (excitation mechanisms) cannot always explain all differences. Here we list those parameters that may somehow be related to AGN types and help the classification:

- Ratios of (narrow!) emission line intensities (diagnostic diagrams)

- Line widths: narrow lines, broad lines (BLS1/NLS1). Through the 1980s [OIII] linewidth was used in defining AGN (FWHM ¿ 300 km/s). At present it is no more used as FWHM tracks bulge mass; many AGN now known with FWHM ¡ 300 km/s. Anyway, linewidth can provide additional hint to classifications

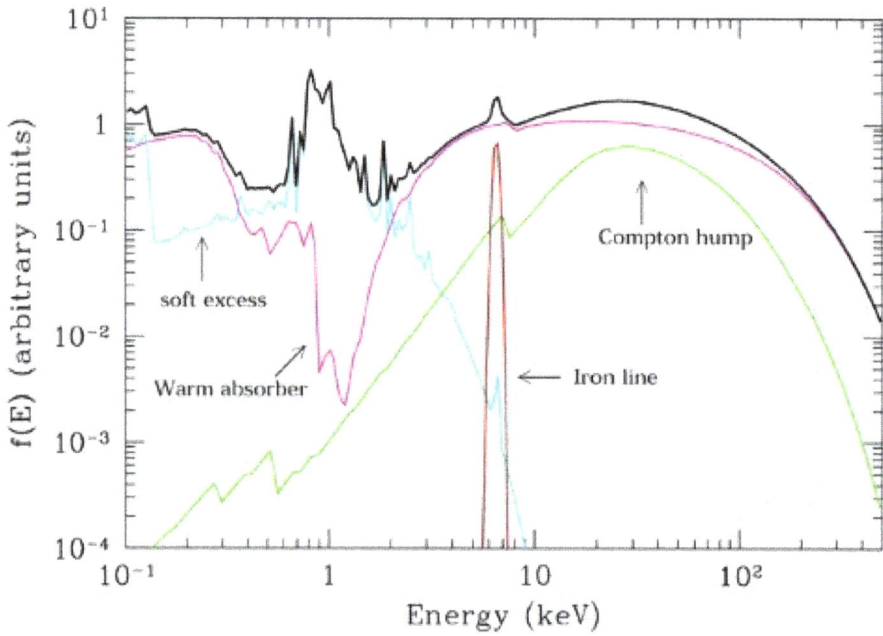

Figure 13: AGN SEDs for X-ray range. Big differences in distributions for various radiation mechanisms present at different AGN types are seen.

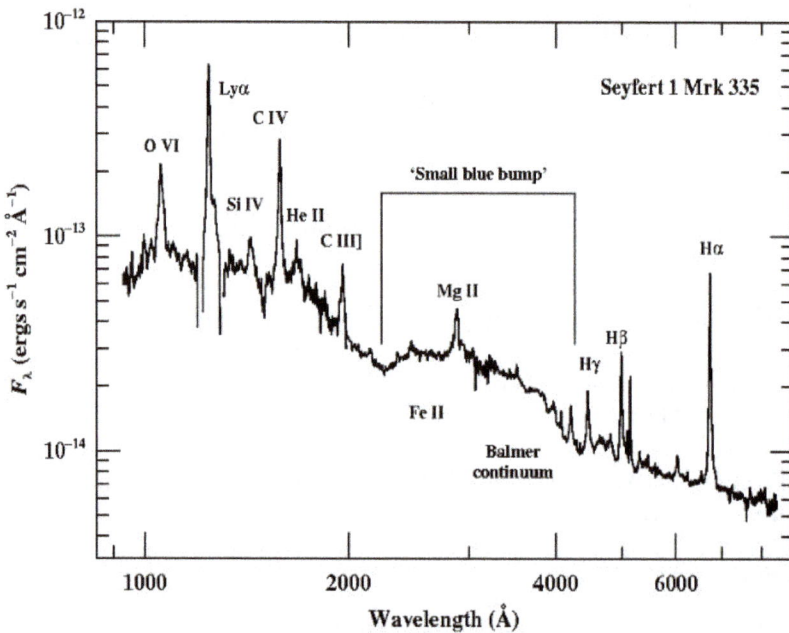

Figure 14: Small Blue Bump in the spectrum of Seyfert 1 galaxy Mrk 335.

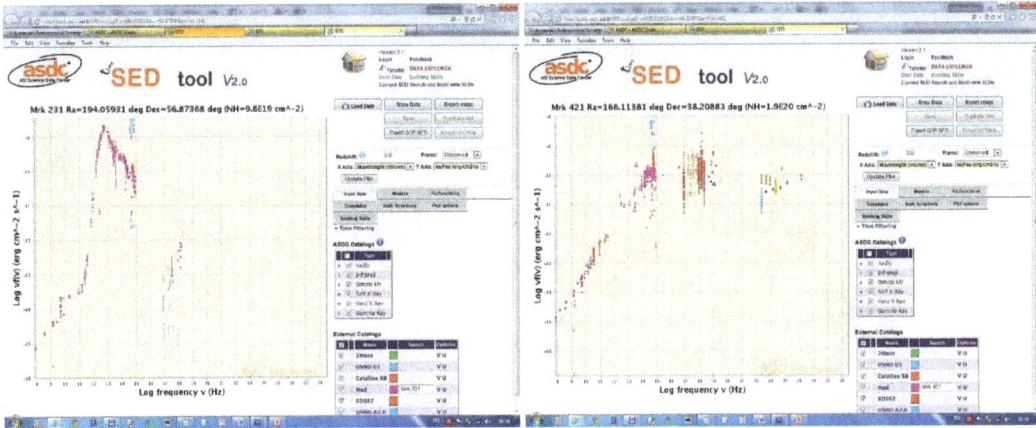

Figure 15: Two remarkable Markarian galaxies, Mrk 231 (ULIRG) and Mrk 421 (Blazar) broad-band spectra (multiwavelength SEDs) from ASDC website.

- Ratio of narrow and broad lines (Osterbrock 1981 subtypes for Seyferts)

- IR diagnostics (estimation of fractions of presence of AGN and SB)

- Sub-mm diagnostics (based on molecular lines)

- Other wavelength ranges (X-ray and radio may serve as criteria for the activity)

- Variability (typically indicates on the activity)

- Polarization (many AGN show highly polarized radiation)

- Host galaxies structure? Jets, lobes, blue components, interactions, merging, etc. may serve as signs of activity

- X-ray / UV / opt / NIR / MIR / FIR / sub-mm / radio flux ratios (typically AGN have higher X-ray/opt and radio/opt flux ratios, subject to be studied statistically in more details)

- Luminosities? AGN are higher luminosity objects, however this may serve as only a statistical criterion

- Other parameters?

8 Types of Active Galaxies

Here we give description of optical emission-line spectrum for known activity types (mainly based on types given in Véron-Cetty & Véron 2010), as well as some other parameters and other than optical wavelengths characteristics. A similar (however less detailed) scheme for activity types was developed by us for HyperLEDA database available at http://leda.univ-lyon1.fr/leda/rawcat/a109.html (Gavrilović et al. 2007). We have developed this classification based on homogeneous SDSS spectra (Ahn et al. 2014). SDSS covers large region of the sky (14,555 deg^2), it has obtained spectra for >4 million objects over 11,600 deg^2, including 3 million galaxies; 200,000 (z<2.3) and 20,000 (z>2.3) QSOs; some 500,000 stars; and other objects.

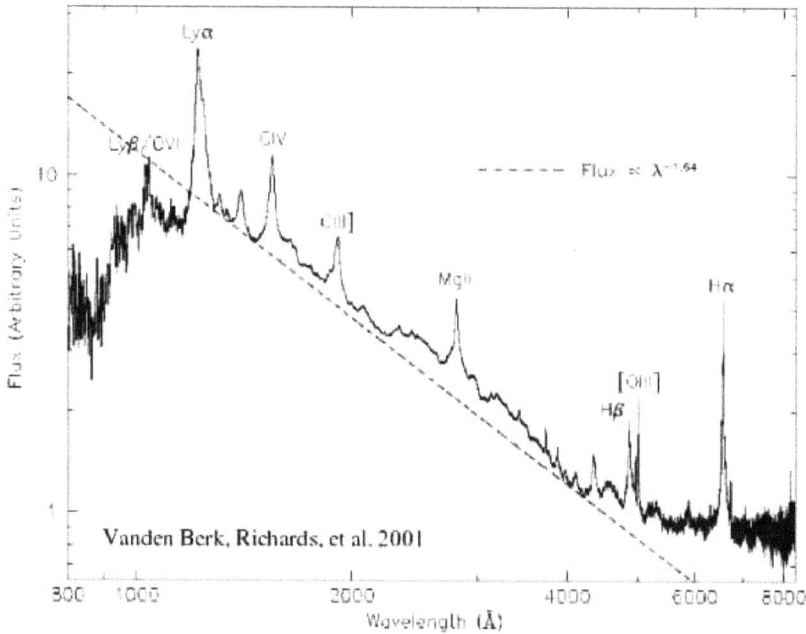

Figure 16: Broad-band synthetic spectrum for QSOs built on the basis of 2dF quasars (Croom et al. 2001; 2004).

Roughly, we distinguish different types of narrow-line active galaxies by following criteria; AGN from HII by [NII]/Hα > 0.6 and [OI]/Hα > 0.1, Seyferts from LINERs by [OIII]/Hβ > 3. Diagnostic diagrams are being used with bordering curves between the 3 types of objects (Seyferts, LINERs and HII) (Kauffmann et al. 2003; Stasinska et al. 2006; Kewley et al. 2006; these recent papers are based on SDSS spectra).

QSO, Quasar – Quasi-Stellar Object, Quasi-Stellar Radio-source (Fig. 16). Have very broad emission lines (FWHM = 5,000–30,000 km/s) with large redshifts. The optical spectra are similar to those of Sy1 nuclei, but the narrow lines are generally weaker. The direct images do not differ from those of the stars on DSS1 and even DSS2, however, objects typically brighter than 17m and/or with redshifts smaller than 0.3 show weak "fuzz", indicating the host galaxy. Have very high luminosities (Mabs>-23). Quasar luminosities are MB < -21.5 + 5 logh$_0$ (Schmidt & Green 1983). The optical spectra are similar to those of S1 nuclei, but the narrow lines are generally weaker. QSO/S1 separation have been conditionally defined by the luminosity limits (MB = -21.5...-24.0), extension (QSOs as star-like and Seyferts as extended objects), and redshift limit (z=0.1; Hewitt & Burbidge 1993), however at present the first one is accepted, though also conditional. Radio-loud QSOs (quasars or **RL QSOs**) and radio-quiet QSOs (or **RQ QSOs**) with a dividing power at P$_{5GHz}$ $\approx$ $10^{24.7}$ W $\cdot$ Hz^{-1}. RL QSOs are 5-10% of the total of QSOs. There is a big gap in radio power between RL and RQ varieties of QSOs. All radio quasars have FR II morphology.

BAL QSO – Broad absorption line QSO (Fig. 17). Besides broad emission lines they show deep blue-shifted very broad (10,000–30,000 km/s) absorption lines with P Cyg type profiles corresponding to resonance lines of CIV, SiIV, NV. All of them are at z $\geq$ 1.5 because the phenomenon is observed in the rest-frame UV. At these redshifts, they are about 10

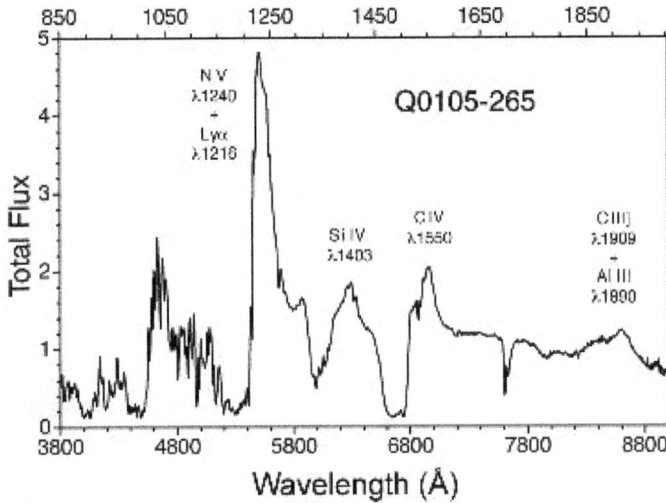

Figure 17: BAL QSO Q0105-265 showing deep blue-shifted very broad absorption lines with P Cyg profiles.

DLA QSO – Damped Ly-alpha QSO. Show unresolved absorption lines even on very high resolution spectra (<1Å), with typical widths of 10–12Å, resulting in a column density of $>10^{23}$, indicating the presence of high density galactic size masses along the line of sight.

OVV QSOs – Optically-Violently Variable QSOs. Similar to BLL but with normal QSO spectrum. They are radio loud.

HPQ, HP – Highly Polarized Quasars. Polarization is typically >3%. Typically are combined with OVV quasars as a single class. The parent population of HPQs is made of FR II radio galaxies.

BL Lac, BLL – BL Lacertae type object (Fig. 18). BL Lac variable "star" is the prototype of this class (Hoffmeister 1929), first such object to be identified as an extragalactic one (Schmitt 1968). This class was proposed by Strittmatter et al. (1972) and BL Lac absorption lines were observed and redshift was measured by Oke & Gunn (1974). Stellar in appearance with variable, intense and highly polarized continuum. Strong featureless continuum; no emission or absorption lines deeper than ∼2% are seen in any part of the optical spectrum, or only extremely week absorption and/or emission lines are observed, as a rule at minimum of their very highly variable phase. The weak lines often just appear in the most quiescent stages. So that their redshifts can only be determined from features in spectra of their host galaxies. Show polarization, and are strong radio sources with flat spectrum (Lawrence 1987; Miller 1978; Miller et al. 1978). The parent population of BLLs is made of FR I radio galaxies.

Blazars – Combination of two most powerful AGN classes; BLL and OVV/HPQ, introduced by E. Spiegel in 1978. Blazars encompass BL Lacs and OVV/HPQ QSOs. These are believed to be objects with a strong relativistically beamed jet in the line of sight. When the angle between the relativistic jet axis and the line of sight is small, the jet is Doppler

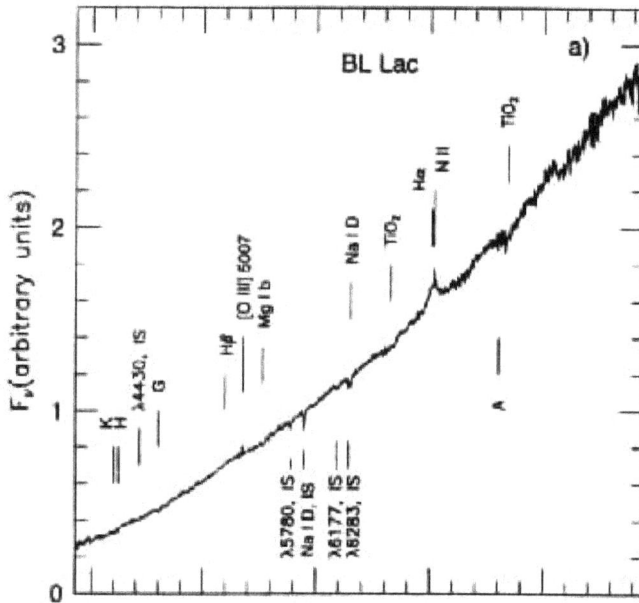

Figure 18: BL Lac optical spectrum showing very weak emission and absorption lines.

boosted by a large factor and the whole spectrum (from radio to γ-ray) is dominated by a compact, highly polarized, highly variable, superluminal, almost featureless continuum, called blazar. As these two types have many common and different physical properties, the question of definition of blazars is still open (Mickaelian et al. 2015). There are many parameters that may be regarded as criteria for definition of blazars, such as high luminosity, radio flat spectrum, X-ray and γ-ray, optical and/or radio variability, polarizations, etc.

S – Seyfert galaxy (no accurate classification if given without a subclass). Relatively low luminosity AGN with MB > -21.5 + $\log_o$. Their host galaxies are clearly detectable. Depending on the width of optical emission lines, Seyfert types (Khachikian & Weedman 1974) and subtypes (Osterbrock 1981) were introduced.

S1, S1.0 or BLS1 – Broad-Line Seyfert 1. Have broad permitted Balmer HI, HeII and other lines (FWHM = 1,000-10,000 km/s; typical is 2,000–6,000 km/s) that originate in a high-density medium ($n_e \leq 10^9$ cm^{-3}), and narrow forbidden lines ([OIII], [NII], [SII], etc. with FWHM = 300-1,000 km/s) that originate in a low-density medium ($n_e \approx 10^3$–10^6 cm^{-3}). Physically are the same objects as QSOs, but having smaller luminosities ($M_{abs i}$-23). They are radio quiet. According to Winkler (1992), Hβ/[OIII]5007 > 5.0. NGC 4151 is the prototype.

NLS1, S1n – Narrow-line Seyfert 1, S1 narrow. Defined by Osterbrock & Pogge (1985) as soft X-ray sources, having narrow permitted lines only slightly broader than the forbidden ones. Many FeI, FeII, FeIII, and often strong [FeVII] and [FeX] emission lines are present, unlike what is seen in Seyfert 2s. The ratio [OIII]5007/Hβ < 3, but exceptions are allowed if there are also strong [FeVII] and [FeX] emission lines present. FWHM(Hβ) < 2000km/s (Goodrich 1989). A spectrum of NLS1 is given in Fig. 19 taken from SDSS (note the wrong

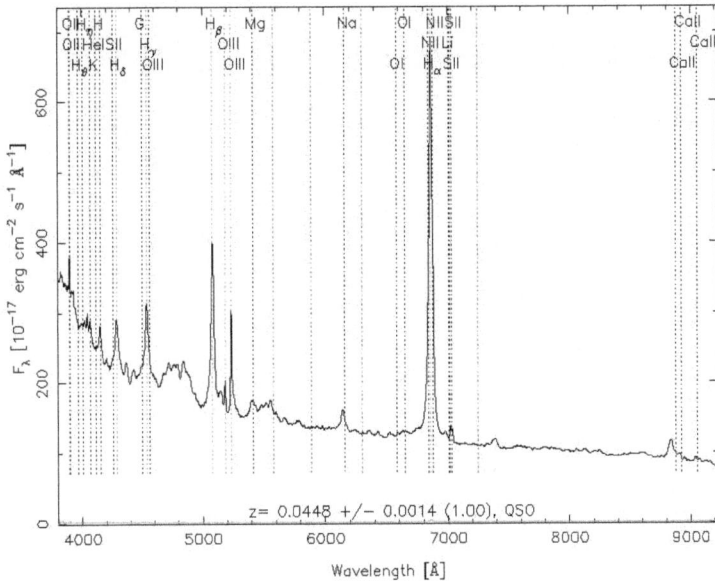

Figure 19: NLS1 galaxy spectrum taken from SDSS. Note relatively narrow broad line components and strong FeII lines on both sides from Hβ.

automatic classification by SDSS as QSO because of FeII features seen as broad emission lines).

S1i – S1 infrared. S1 with a broad Paschen Pa-β line, indicating the presence of a highly reddened BLR (Goodrich et al. 1994). Seyfert 1 with an absorbed BLR visible in NIR.

S1h – S1 hidden. S1 showing S1 like spectra in polarized light (Antonucci & Miller 1985; Miller & Goodrich 1990; Tran et al. 1992). Seyfert 1 with a hidden BLR.

S1.2 – AGN with spectra, which share parameters that are intermediate between those of classical Sy1 and Sy2 galaxies, i. e. both broad and narrow components are present for permitted lines (Osterbrock 1981). According to Winkler (1992), the ratio of the narrow component of Hβ to [OIII]5007 is 2.0 < Hβ/[OIII]5007 < 5.0. Often erroneously related to NLS1s or S1n.

S1.5 – AGN which share parameters that are intermediate between those of classical Sy1 and Sy2 galaxies; have easily discernible narrow HI profile superposed on broad wings (Osterbrock 1981). According to Winkler (1992), the ratio of the narrow component of Hβ to [OIII]5007 is 0.333 < Hβ/[OIII]5007 < 2.0. Fig. 20 shows a typical S1.5 spectrum taken from SDSS (again SDSS automatic classification was incorrect due to the broad line components).

S1.8 – AGN which share parameters that are intermediate between those of classical Sy1 and Sy2 galaxies; have relatively weak broad Hα and Hβ components superposed on strong narrow lines (Osterbrock 1981). According to Winkler (1992), narrow component of Hβ/[OIII]5007 < 0.333.

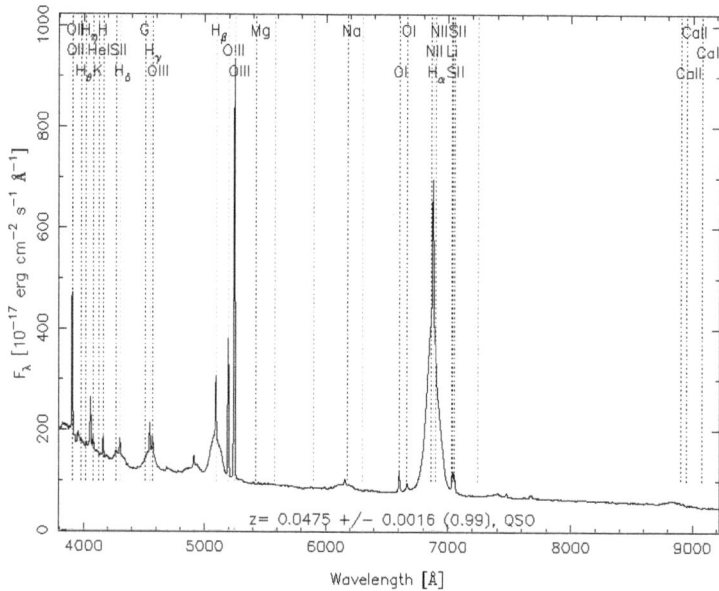

Figure 20: A typical S1.5 spectrum taken from SDSS. Broad line and narrow line components have similar intensities and due to their superposition narrow line peaks are seen higher.

S1.9 – AGN which share parameters that are intermediate between those of classical Sy1 and Sy2 galaxies; have relatively weak broad Hα component superposed on a strong narrow line. The broad component of Hβ is not seen (Osterbrock 1981). According to Winkler (1992), narrow component of Hβ/[OIII]5007 < 0.333.

S2, S2.0 – AGN showing relatively narrow (compared to S1s) emission in both permitted Balmer and forbidden lines (Khachikian & Weedman 1974), with approximately the same FWHM>=300km/s, typically in the range of 300-1000 km/s that originate in a low-density medium ($n_e \approx 10^3$–10^6 cm^{-3}). No broad component is visible. A secondary classification criterion is [OIII]5007/Hβ $\geq$ 3, to distinguish against NLS1s (Veilleux & Osterbrock 1987; Lawrence 1987). NGC 1068 is the prototype.

S3, LINER – Low-Ionization Nuclear Emission-line Region. Introduced by Heckman (1980), they are low activity AGN, the weakest form of AGN activity. They have S2-like spectra with relatively strong low-ionization lines ([OI], [OII]). [OII]3727/[OIII]5007 $\geq$ 1, [OI]6300/[OIII]5007 $\geq$ 1/3. [NII]6584/Hα > 0.6 according to Kauffman et al. (2003). According to Ho et al. (1997), there are 2 classes of LINERs: type 1 shows broad Balmer emission analogous to S1s (weak broad Hα visible), and type 2, without broad Hα analogous to S2s. May be either radio quiet or loud. Most of the nuclei of nearby galaxies are LINERs. However, their emission line spectra are not necessarily caused by active nuclei.

S3b – LINERs with broad Balmer lines, the same as LINER type 1 (Ho, Filippenko & Sargent 1997).

S3h – LINERs with broad Balmer lines seen only in polarized light.

AGN – AGN without a subclass because of spectra with relatively low quality. Show emission line spectra with strong forbidden lines; a few emission lines are observed, mostly Hα with NII lines where NII/Hα ratio indicates an AGN, i.e. either Sy or LINER.

SBN and SBG – Starburst nuclei or Starburst galaxy (Weedman 1977). M82 was the archetype SB galaxy. The major observable feature that distinguishes SB from Sy is their strong narrow emission lines FWHM $\leq$ 300km/s. According to Balzano (1983), SB is a spiral galaxy with a bright, blue nucleus that emits a strong narrow emission line spectrum similar to low-ionization HII region spectra. They have strong, narrow (FWHM $\leq$ 250km/s) low-ionization ([OIII]/Hβ < 3) emission lines; absolute luminosities -17.5 > M > -22.5; conspicuous stellar or semistellar nuclei. SB can occur in disk galaxies, however irregular galaxies often exhibit knots of SB spread throughout the galaxy. SFR is a few M$_\odot$ yr^{-1}, but may reach up to 10^3 M$_\odot$ yr^{-1}. Based on the relative energy output of the SB (LSB) to that of the rest of the galaxy (LG) and the SB age, R. Terlevich classified SB into 3 classes: SB galaxies having LSB $\gg$ LG, Galaxies with SB having LSB $\sim$ LG, and Normal galaxies having LSB $\ll$ LG (Terlevich 1997).

BCDG – Blue Compact Dwarf Galaxy as introduced by Thuan & Martin (1981) and described by Gallego et al. (1996). Subtype of SB. Have HII spectra. Most of them have a high rate of star formation. Dwarf, low-mass, low-metallicity, dust-free objects. The BCDG classification involves spectral-morphological parameters; they are blue objects with M(B)>-17.5 and linear sizes of less than D$\leq$3–4kpc. IZw18 is the most well-known BCDG being the most metal poor one.

WR galaxy – Wolf-Rayet Galaxy (Osterbrock & Cohen 1982; Conti 1991). Subtype of SB having a large portion of bright stars as early-type Wolf-Rayet ones. Because these stars are both very luminous and have very distinctive spectral features, it is possible to identify them in the spectra of the entire galaxies. They show prominent broad emission lines of highly-ionized He and N or C. NGC 6764 and Mrk 309 are WRG prototypes.

HII, H2 – Isolated Extragalactic HII regions, as defined by Sargent & Searle (1970) or HII galaxies as defined by Terlevich et al. (1991). Have spectra similar to SB that is a strong narrow (FWHM $\leq$ 300 km/s) emission line spectrum but with a ratio [OIII]/Hb $\geq$ 3 and [NII]6584/Ha < 0.6, coupled with a blue continuum (Veilleux & Osterbrock 1987).

NSF – normal star-forming galaxies. This classification is often used to distinguish against both AGN and SB.

Composite spectrum objects – galaxies with presence of features of two or more activity types, as a rule, a combination of Seyfert, LINER and/or HII types (Véron et al. 1997; Fig. 21). Before they were regarded as transition objects due to their location in transition regions of diagnostic diagrams. Often they are classified differently on different diagrams. They may be S2/LINER, S2/HII, LINER/HII or even a combination of S1 subtypes (S1.8, S1.9) and a LINER or HII. S2/HII and LINER/HII are considered to be a superposition of S2 or LINER nucleus with circumnuclear HII regions.

ELG, Em – Emission Line Galaxy. Show one or several emission lines (Ha, [NII]6584/6548, and [OII]3727). Absorption lines are very weak or absent. No accurate classification because of relatively low quality of spectra.

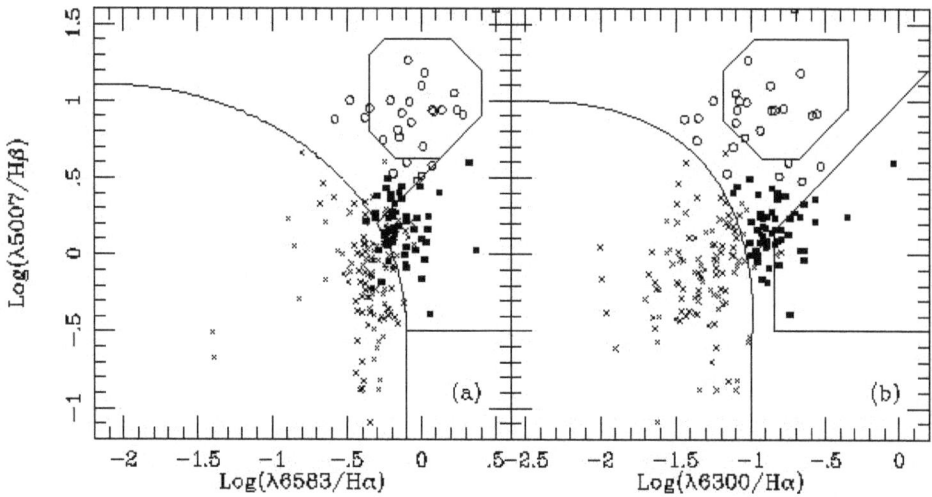

Figure 21: Diagnostic diagrams showing many Composite spectrum objects between the regions of Seyferts, LINERs and HIIs.

Abs – Absorption line galaxy. Only absorption lines are seen. No signs for any (nuclear or starburst) activity. Also called normal galaxies, though the latter ones show some weak emission lines as well.

UVX – Ultra-Violet Excess galaxies, those that are not classified as one of the known activity types or starbursts. Can have both emission and/or absorption lines. Mainly appeared from Markarian Survey (Markarian et al. 1989), Second Byurakan Survey (SBS; Stepanian et al. 2005), Kazarian galaxies (Kazarian et al. 2010) and some other low-dispersion or colorimetric surveys.
Other classifications use other than optical spectrum criteria, including radio, IR and X-ray wavelengths.

Radio wavelengths

RG – Radio Galaxy (Fig. 22). Related objects to radio-loud QSOs and Blazars, they are very luminous in radio wavelengths, with luminosities up to 1039 W between 10 MHz and 100 GHz. The host galaxies are almost exclusively large elliptical galaxies. Optical spectra are similar to Seyfert galaxies. The radio emission is due to the synchrotron processes and the observed radio structure is determined by the interaction between twin jets and the external medium, modified by the effects of relativistic beaming. The lobes of radio galaxies are powered by relativistic jets.

BLRG – Broad-Line Radio Galaxy. Their optical spectra are similar to S1 but they are radio loud.

NLRG – Narrow-Line Radio Galaxy. Their optical spectra are similar to S2 but they are radio loud.

N galaxies – Radio galaxies with compact nucleus.

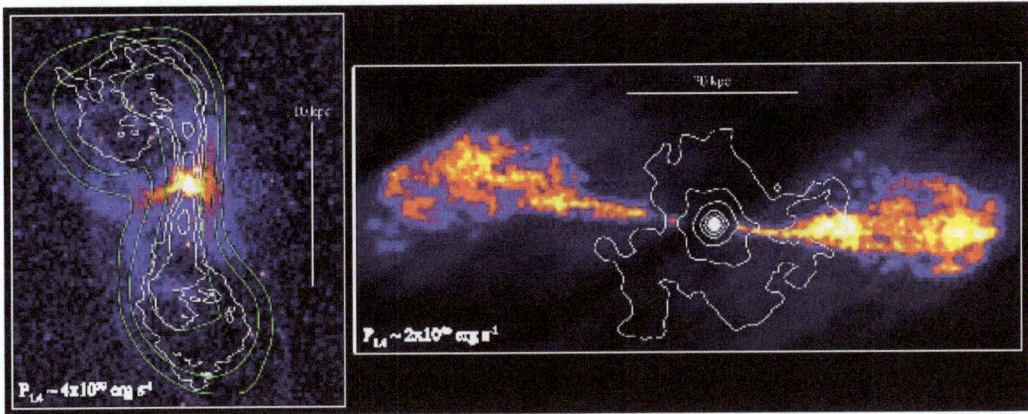

Figure 22: Radio structure of typical radio galaxies. They have central emission, as well as powerful radio lobes.

D galaxies – Giant radio galaxies.

cD galaxies – Giant radio galaxies dominant in their clusters.

FR I – Fanaroff-Riley class I radiogalaxy (Fanaroff & Riley 1974). Sources whose luminosity decreases as the distance from the central galaxy or quasar host increase. These sources are called also edge-darkened. Show radio luminosity in relation to the hosting environment and is measured by the ratio R of the distance between the two brightest spots and the overall size of the radio image; FR I have R < 0.5. These are lower luminosity sources. They make the parent population of BLLs. FR Is have a weak low excitation emission line spectrum, similar to LINERs or no detectable emission at all.

FR II – Fanaroff-Riley class II radiogalaxy (Fanaroff & Riley 1974). Sources exhibiting increasing luminosity in the lobes. Show the radio luminosity in relation to the hosting environment and is measured by the ratio R of the distance between the two brightest spots and the overall size of the radio image; FR II have R > 0.5. Most radio galaxies have double lobe structure; these high radio luminosity sources have edge-brightened lobes. They make the parent population of HPQs. FR II radio galaxies have the nuclear emission line spectrum of Seyferts and are divided into BLRGs and NLRGs based on the width of emission lines. All radio quasars have FR II morphology.

FSR – Flat Spectrum Radio source. Core Dominated Radio Loud Quasar (CD-QSR); optically similar to QSO. Spectral index (α in $F\nu = \nu\alpha$) at $\nu = 1$GHz is flat (with $\alpha = -0.4$ limit).

SSR – Steep Spectrum Radio source. Lobe Dominated Radio Loud Quasar (LD-QSR); usually FR II radio morphology; optically similar to QSO. Spectral index (α in $F\nu = \nu\alpha$) at $\nu = 1$GHz is steep ($\alpha = -0.4$ limit).

CSS – Compact Steep Spectrum radio sources. CSS sources are compact, powerful radio sources with well-defined peaks in their radio spectra (near 100 MHz). They are contained entirely within the host galaxy (≤ 15 kpc). CSS sources at all redshifts exhibit high surface brightness optical light (most likely emission-line gas) that is aligned with the radio axis.

GPS – Gigahertz Peaked Spectrum radio sources. GPS sources are compact, powerful radio sources with well-defined peaks in their radio spectra (near 1 GHz). They are entirely contained within the extent of the NLR ($\leq$ 1 kpc). X-ray observations of high-redshift GPS quasars and a couple of GPS galaxies suggest the presence of significant columns of gas toward the nuclei.

PRG-II – Powerful Radio Galaxy of Fanaroff-Riley class II. Edge brightened, powerful jet; unspecified optical spectrum (could be BLRG/NLRG/LINER).

RG-I – Radio Galaxy of Fanaroff-Riley class I. Similar to PRG-II except lower radio luminosity and FR I class. Edge darkened, lower power jet.

Masers – Extragalactic molecular masers. Conditions for Masers are long low v path through dense gas exposed to strong pumping radiation. These conditions are met in many AGN nuclei; the most famous is NGC 4258. The path lengths will be at their largest for edge-on dense gas disks. Many Sy2s are H2O masers, consistent with expectations for obscuring tori.

IR wavelengths

QSO2, Q2 – Type-2 QSO. Same as QSO but missing broad lines; not many such objects are currently known (some are IRAS QSOs). They are radio quiet. Most of them appeared from NIR surveys (Cutri et al. 2003).

AGN2 – Type-2 AGN. Similar to QSO2 but with lower luminosity.

Obscured AGN – IR revealed AGN not having emissions lines in optical range. Mostly appeared after IRAS mission (e. g. Sargsyan et al. 2008).

LIRG, ULIRG, HLIRG – Luminous, Ultra-Luminous and Hyper-Luminous InfraRed Galaxy (Sanders & Mirabel 1996). They were introduced after IRAS mission and are based on IR (L_{ir}, luminosity in the range 8-1000 μm) and far-IR (L_{fir}, luminosity in the range 40-500 μm) luminosities calculated in $L_\odot$ according to Duc et al. (1997):

$$L_{ir} = 5.6 \cdot 10^5 R^2 (13.56 f_{12} + 5.26 f_{25} + 2.54 f_{60} + f_{100})$$

$$L_{fir} = 5.6 \cdot 10^5 R^2 (2.58 f_{60} + f_{100})$$

where R is the distance of objects in Mpc and f_{12}, f_{25}, f_{60} and f_{100} are IRAS fluxes in the bands at 12, 25, 60, and 100 μm, respectively. LIRGs are galaxies having $L_{ir} > 10^{11}$ $L_\odot$, ULIRGs have 10^{12} $L_\odot < L_{ir} < 10^{13}$ $L_\odot$ and HLIRGs have $L_{ir} > 10^{13}$ $L_\odot$. These are generally extremely dusty objects. The UV radiation produced by the obscured star-formation is absorbed by the dust and reradiated in the IR at around 100μm. It is not known for sure that the UV radiation is produced purely by star-formation, and some authors believe at least part of ULIRGs to be powered by AGN. Well-studied ULIRGs include Arp 220 and Mrk 231. Veilleux (2002) has shown that the fraction of AGN increases among the higher luminosity IR galaxies. Hou et al. (2009) have found 308 ULIRGs among SDSS galaxies. Lee et al. (2011) have classified 115 ULIRGs and showed that they may be either broad-line or narrow-line AGN or show no nuclear activity.

ELF – Extremely Luminous far-IR Galaxy.

X-ray range

Hidden AGN – X-ray revealed AGN not having emissions lines in optical range (Barger et al. 2000; Treister et al. 2005). May be classified as X-ray AGN (XAGN; Paronyan & Mickaelian 2015). Among other AGN detected in optical wavelengths, a small fraction displays X rays (Véron-Cetty et al. 2004).

NELG (NLXG) – Narrow Emission Line Galaxy (Narrow-Line X-ray Galaxy). Most likely obscured Seyfert galaxies.

A homogeneous classification by activity types was carried out for all Markarian galaxies spectroscopically observed in SDSS (Mickaelian et al. 2014; Winkler 2014). Out of 779 objects present in SDSS spectroscopic catalogue (DR7-DR9), we have found 126 AGN or Composites (16.2%). 533 objects are HII, 52 are Em, 65 are Abs, and 3 are stars. Among the Composites, we have LINER/HII, Sy/HII, Sy/LINER or Sy/HII/LINER subtypes. On the other hand, we have a number of NLS1 galaxies with different subtypes, as NLS1.0, so as NLS1.2 and NLS1.5, which we have introduced for more accuracy.

Summarizing, here we described more than 50 activity types (including 30 in optical range), which proves the variety of AGN and other active galaxies present in the Universe. Most probably, there may exist more classes to be discovered in future surveys or more subtypes may be introduced to describe the varieties more accurately, as we did in case of the classification of Markarian galaxies.

9 Summary

AGN may be defined as objects where the total luminosity is radiation not ultimately related to stellar atmospheres. Alternate energy sources should be involved. At present, the most popular explanation for the AGN powerhouse involves accretion of gas onto a most probably spinning SMBH. This is the **Unified Model** or **Unified Scheme**. Probably the first to mention this possibility was Rowan-Robinson (1977). Different regimes of accretion have been invoked to constitute the basis of a unified picture of AGNs. Observed AGN energies are explained through the release of gravitational energy. In fact, the concept of a SMBH surrounded by a viscous disk (of 10^{-3} pc) accreting matter gained popularity in 1960s (Zeldovich & Novikov 1964), and now became the standard model for AGN.

The formation of the torus is crucial to support the unified model of AGN. Further studies strengthened such an understanding of the AGN energy sources. However, there still are many difficulties and the discovery of new objects with new properties encounter challenges in their explanation in frame of the general scheme.

From the point of view of Unified Scheme, we give here the regions and mechanisms that build AGN spectra at different wavelengths. Relativistic Accretion Disk (AD) at 20 AU sizes and 10^{15} cm^{-3} densities and is responsible for Fe Kα line; it also gives variable X-ray emission (minutes-hours). UV accretion disk at 200 AU sizes gives UV radiation. Optical AD (the BLR) at 2000 AU has 10^{10} cm^{-3} densities and gives broad optical emission lines (few 1000 km s^{-1}). Compact, flat-spectrum radio core is the outer BLR at 0.1 pc scale and the jet is detectable with VLBI. Inner NLR (inner bulge) has 10 pc sizes. The main

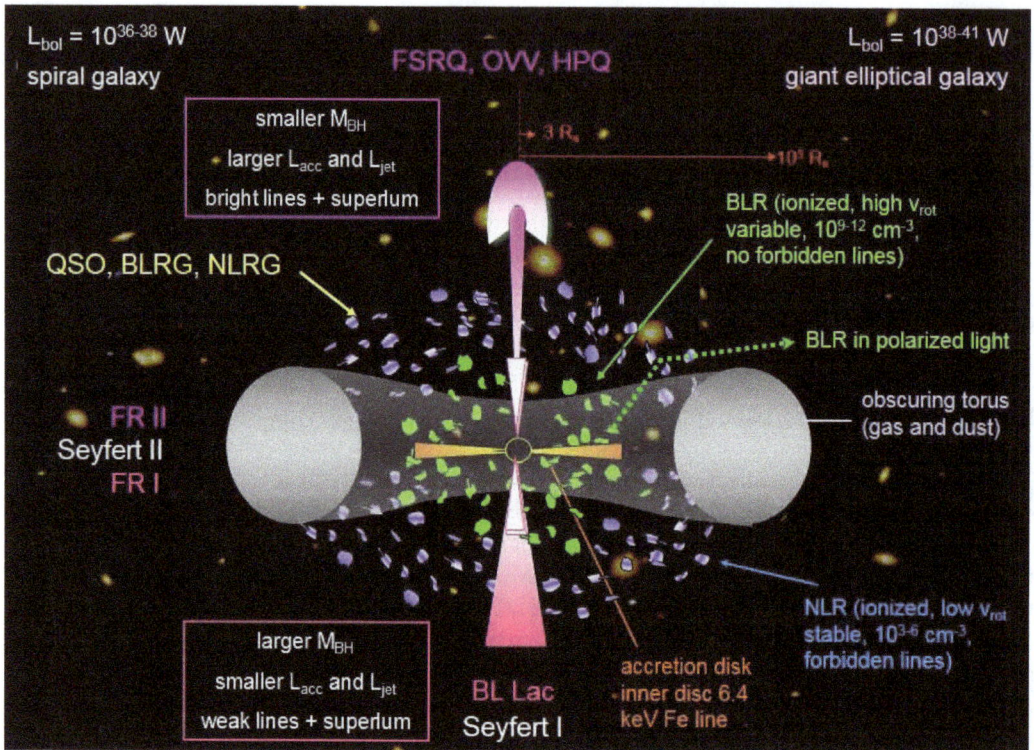

Figure 23: Schematic view of the Unified Model. Different regions emitting with different mechanisms and in different wavelength ranges are given.

bulge (also NLR) at 100 pc size with 10^3–10^6 cm^{-3} densities gives forbidden emission lines (few 100 km s^{-1}) and the radio jet is detectable with VLA. We give in Fig 23 a schematic view of the Unified Model, where different regions emitting with different mechanisms and in different wavelength ranges are given and explained.

However, observationally it is not easy to discover accretion disks or SMBHs. The diameter of an accretion disc around a 108 M BH is at Solar system scale seen at 100 Mpc has an angular resolution of $\sim$ 1 as, which is not yet detectable. The same seen at the distance of Andromeda galaxy (M31) is still 0.1 mas.

There still are a number of **open questions** related to AGN and their classifications. Here we list most important of them:

- Definition and taxonomy of AGN (the problem of historical types and their relevance to present understanding)

- Discontinuity of classes; classification as science trigger or break?

- Non-optical wavelength data; understanding the true fraction of heavily obscured AGN to determine the true luminosity function and its variation with z

- Have all AGN types been discovered?

- Blazars: their duality (BLL and HPQ), luminosities, etc.

- Understanding possible evolutionary and/or physical connection between different classes of AGN, i. e. their consistency with the unification model; AGN vs. ULIRGs and other high-L objects

- AGN growth, structure versus cosmic time

- AGN spectral energy distributions (SEDs) versus cosmic time

- The importance of AGN feedback, relation of AGN to host galaxies

- AGN seeds in the Early Universe

- Future AGN surveys, data mining and overall use of MW data for full understanding of AGN structure, physical processes behind the observed phenomena and energy sources

At the end, we recommend some review papers and books on active galaxies for further reading and study (full bibliography is given in the list of references):

Alloin, Johnson & Lira 2006, *Physics of Active Galactic Nuclei at all Scales*
Aretxaga, Mújica & Kunth (Eds.) 2000, *Advanced Lectures on the Starburst-AGN Connection*
Beckmann & Shrader 2012, *Active Galactic Nuclei*
Harutyunian & Mickaelian 2010, *V.A. Ambartsumian and the Activity of Galactic Nuclei*
Kembhavi & Narlikar 1999, *Quasars and Active Galactic Nuclei*
Krolik 1999, *Active Galactic Nuclei: From the Central Black Hole to the Galactic Environment*
McNamara & Nulsen 2007, *Heating Hot Atmospheres with Active Galactic Nuclei*
Mickaelian & Sanders (Eds.) 2014, *Multiwavelength AGN Surveys and Studies*
Osterbrock 1989, *Astrophysics of gaseous nebulae and active galactic nuclei*
Peterson 1997, *An Introduction to Active Galactic Nuclei*
Robson 1996, *Active Galactic Nuclei*
Sanders & Mirabel 1996, *Luminous Infrared Galaxies*
Véron-Cetty & Véron 2000, *The emission line spectrum of active galactic nuclei and the unifying scheme*

NASA/IPAC Extragalactic Database (NED) has Level 5 "A Knowledgebase for Extragalactic Astronomy and Cosmology" supported by B. F. Madore et al. (https://ned.ipac.caltech.edu/level which we strongly recommend both for AGN and other studies related to galaxies and cosmology.

References

[1] Ahn, C. P.; Alexandroff, R.; Allende Prieto, C.; et al. 2014, ApJS 211, 17

[2] Alloin, D.; Johnson, R.; Lira, P. (Eds.) 2006, Lecture Notes in Physics, Vol. 693

[3] Ambartsumian, V. A. 1955, Some Remarks on Multiple Galaxies, Yerevan

[4] Ambartsumian, V. A. 1956, Izv. Acad. Sci. ArmSSR 9, No. 1, 23

[5] Ambartsumian, V. A. 1958, Proc. 11th Solvay Conf. on Physics: Structure and Evolution of the Universe. Univ. of Brussels, Ed. R. Stoops, Brussels, p. 241

[6] Ambartsumian, V. A. 1961, AJ 66, 536

[7] Antonucci, R. R. J. 1993, ARAA 31, 473

[8] Antonucci, R. R. J.; Miller, J. S. 1985, ApJ 297, 621

[9] Arakelian, M. A. 1975, Com. BAO 47, 3

[10] Aretxaga, I.; Mújica, R.; Kunth, D. (Eds.) 2000, Advanced Lectures on the Starburst-AGN Connection, World Scientific, 372 p.

[11] Arp, H. 1966, Atlas of Peculiar Galaxies, ApJS 14, 1

[12] Baade, W.; Minkowski, R. 1954, ApJ 119, 206

[13] Baldwin, J. A.; Phillips, M. M.; Terlevich, R. 1981, PASP 93, 5

[14] Balzano, V. A. 1983, ApJ 268, 602

[15] Barger, A. J.; Cowie, L. L.; Richards, E. A. 2000, AJ 119, 2092

[16] Beckmann, V.; Shrader, C. R. 2012, Active Galactic Nuclei, Wiley-VCH Verlag GmbH, 350 p.

[17] Bolton, J. G., Stanley, G. J., Slee, O. B. 1949, Nature 164, 101

[18] Cambell, W. W.; Moore, J. H. 1918, Publ. Lick Obs. 13, 75

[19] Cid Fernandes, R.; Stasinska, G.; Mateus, A., Vale Asari, N. 2011, MNRAS, 413, 1687

[20] Collinge, M. J.; Strauss, M. A.; Hall, P. B.; Ivezić, Ž.; Munn, J. A.; Schlegel, D. J.; Zakamska, N. L.; Anderson, S. F.; et al. 2005, 2005, AJ 129, 2542

[21] Conti, P. S. 1991, ApJ, 377, 115

[22] Croom, S. M., Smith, R. J., Boyle, B. J., et al. 2001, MNRAS, 322, L29

[23] Croom, S. M., Smith, R. J., Boyle, B. J., et al. 2004, MNRAS, 349, 1397

[24] Cutri, R. M.; Skrutskie, M. F.; Van Dyk, S.; et al. 2003, The 2MASS All-Sky Catalog, Final Release, University of Massachusetts and IPAC/California Institute of Technology

[25] Del Moro, A. et al 2013, A&A, 549, 59

[26] Donley, J. L.; Koekemoer, A. M.; Brusa, M.; et al. 2012, ApJ 748, 142

[27] Duc, P. A.; Mirabel, I. F.; Maza, J. 1997, A&AS 124, 533

[28] Edelson, R.; Alexander, T.; Crenshaw, D. M.; Kaspi, S.; Malkan, M.; Peterson, B.; Warwick, R. 1998, AdSpR 21, 77

[29] Edge, D. O.; Shakeshaft, J. R.; McAdam, W. B.; Baldwin, J. E.; Archer, S. 1959, Mem. RAS 68, 37

[30] Elvis et al. 1994, ApJS 95, 1

[31] Fanaroff, B. L.; Riley, J. M. 1974, MNRAS 167, 31

[32] Fath, E. A. 1908, Lick Obs. Bull. 5, 71

[33] Gallego, J.; Zamorano, J.; Rego, M.; Alonso, O.; Vitores, A. G. 1996, A&AS 120, 323

[34] Gavrilović, N.; Mickaelian, A. M.; Petit, C.; et al. 2007, Proc. IAU Symp. #238: Black Holes from Stars to Galaxies - Across the Range of Masses, 21-25 Aug 2006, Prague, Czech Rep., Eds. V. Karas & G. Matt, CUP, p. 371

[35] Gigoyan, K. S.; Mickaelian, A. M. 2012, MNRAS 419, 3346

[36] Goodrich, R. W. 1989, ApJ 342, 224

[37] Goodrich, R. W. 1995, ApJ, 440, 141

[38] Goodrich, R. W.; Veilleux, S.; Hill, G. J. 1994, ApJ, 422, 521

[39] Greenstein, J. L.; Matthews, T. A. 1963, AJ 68S, 279

[40] Hao, L.; Strauss, M. A.; Tremonti, C. A.; Schlegel, D. J.; Heckman, T. M.; Kauffmann, G.; Blanton, M. R.; Fan, X.; et al. 2005, AJ 129, 1783

[41] Haro, G. 1956, AJ 61, 178

[42] Harutyunian, H. A.; Mickaelian, A. M. 2010, Proc. Conf. dedicated to Viktor Ambartsumian's 100th anniversary: Evolution of Cosmic Objects through their Physical Activity, 15-18 Sep 2008, Byurakan, Armenia, Eds.: H.A. Harutyunian, A.M. Mickaelian & Y. Terzian, Yerevan, NAS RA "Gitutyun" Publishing House, p. 134

[43] Heckman, T. M. 1980, A&A 87, 152

[44] Hewitt, A.; Burbidge, G. 1993, ApJS, 87, 451

[45] Hey, J. S.; Parsons, S. J.; Phillips, J. W. 1946, Nature 158, 234

[46] Hickox, R. et al 2009, ApJ, 696, 891

[47] Ho, L. C.; Filippenko, A. V.; Sargent, W. L. W. 1997, Proc. IAU Colloquium No. 159

[48] Hoffmeister, C. 1929, AN 236, 233

[49] Hou, L. G.; Wu, Xue-Bing; Han, J. L. 2009, ApJ 704, 789

[50] Hovhannisyan, A.; Sargsyan, L. A.; Mickaelian, A. M.; Weedman, D. W. 2011, Ap 54, 147

[51] Hubble, E. P. 1926, ApJ 64, 328

[52] Humason, M. L. 1932, PASP 44, 267

[53] Juneau, S., Bournaud, F., Charlot, S., Daddi, E. et al, 2014, ApJ, 788, 88

[54] Juneau, S., Dickinson, M., Alexander, D. M., Salim, S. 2011, ApJ, 736, 104

[55] Kauffmann, G.; Heckman, T. M.; Tremonti, C.; et al. 2003, MNRAS 346, 1055

[56] Kazarian, M. A.; Adibekyan, V. Zh.; McLean, B.; Allen, R. J.; Petrosian, A. R. 2010, Ap 53, 57

[57] Kazarian, M. A.; Mickaelian, A. M. 2007, Ap 50, 127

[58] Kellermann, K. I.; Sramek, R.; Schmidt, M.; Shaffer, D. B.; Green, R. 1989, AJ 98, 1195

[59] Kembhavi, A.; Narlikar, J. 1999, Quasars and Active Galactic Nuclei, Cambridge Univ. Press

[60] Kewley, L. J.; Dopita, M. A.; Sutherland, R. S.; Heisler, C. A.; Trevena, J. 2001, ApJ 556, 121

[61] Kewley, L. J.; Groves, B.; Kauffmann, G.; Heckman, T. 2006, MNRAS 372, 961

[62] Khachikian, E. E.; Weedman, D. W. 1971, Ap 7, 231

[63] Khachikian, E. E.; Weedman, D. W. 1974, ApJ, 192, 581

[64] Krolik, J. K. 1999, Active Galactic Nuclei: From the Central Black Hole to the Galactic Environment, Princeton Univ. Press, 632 p.

[65] Lamareille, F. 2010, A&A, 509, A53

[66] Lamareille, F., Mouhcine, M., Contini, T., Lewis, I., Maddox,S. 2004, MNRAS, 350, 396

[67] Lawrence, A. 1987, PASP 99, 309

[68] Lee, J. C.; Hwang, H. S.; Lee, M. G.; Kim, M.; Kim, S. C. 2011, MNRAS, 414, 702

[69] Markarian, B. E. 1963, Com. BAO 34, 3

[70] Markarian, B. E. 1967, Ap 3, 24

[71] Markarian, B. E.; Lipovetski, V. A.; Stepanian, J. A. 1983, Ap 19, 14

[72] Markarian, B. E.; Lipovetski, V. A.; Stepanian, J. A.; et al. 1989, Com. SAO 62, 5

[73] Markarian, B. E.; Lipovetski, V. A.; Stepanian, J. A.; et al. 1997, Vizier Catalogue VII/172

[74] Massaro, F.; Giroletti, M.; D'Abrusco, R.; Masetti, N.; Paggi, A.; Cowperthwaite, P. S.; Tosti, G.; Funk, S. 2014, ApJS 213, 3

[75] Massaro, E.; Maselli, A.; Leto, C.; Marchegiani, P.; Perri, M.; Giommi, P.; Piranomonte, S. 2015, arXiv:1502.07755v1

[76] Mayall, N. U. 1934, PASP 46, 134

[77] Mazzarella, J. M., Balzano, V. A. 1986, ApJS 62, 751

[78] McNamara, B. R.; Nulsen, P. E. J. 2007, ARAA 45, 117

[79] Mickaelian, A. M. 2008, AJ 136, 946. VizieR On-line Data Catalog: III/258

[80] Mickaelian, A. M. 2014, Proc. IAU Symp. #304: Multiwavelength AGN Surveys and Studies, Eds. A. M. Mickaelian & D. B. Sanders, Cambridge University Press, p. 1

[81] Mickaelian, A. M.; Abrahamyan, H. V.; Harutyunyan, G. S.; Paronyan, G. M. 2014, Proc. IAU Symp. #304: Multiwavelength AGN Surveys and Studies, Eds. A. M. Mickaelian & D. B. Sanders, Cambridge University Press, p. 41

[82] Mickaelian, A. M.; Abrahamyan, H. V.; Paronyan, G. M.; Harutyunyan, G. S. 2012, Proc. IAU Symp. #284: The Spectral Energy Distribution of Galaxies, 5-9 Sep 2011, Preston, UK, Eds.: C.C. Popescu & R.J. Tuffs, Cambridge, UK: Cambridge University Press, p. 237

[83] Mickaelian, A. M.; Abrahamyan, H. V.; Paronyan, G. M.; Harutyunyan, G. S. 2013, AN 334, 887

[84] Mickaelian, A. M.; Abrahamyan, H. V.; Paronyan, G. M. 2015, in preparation

[85] Mickaelian, A. M.; Gigoyan, K. S. 2006, A&A 455, 765

[86] Mickaelian, A. M.; Sanders, D. B. 2014, Proc. IAU Symp. #304: Multiwavelength AGN Surveys and Studies, Cambridge Univ. Press., 437 p.

[87] Mickaelian, A. M.; Sargsyan, L. A. 2004, Ap 47, 213

[88] Miller, J. S. 1978, ComAp 7, 175

[89] Miller, J. S.; French, H. B.; Hawley, S. A. 1978, In: Proc. Pittsburgh Conf. BL Lac Objects, Pittsburgh, Pa., Apr 24-26, 1978, University of Pittsburgh, p. 176

[90] Miller, J. S.; Goodrich, R. W. 1990, ApJ, 355, 456

[91] Oke, J. B.; Gunn, J. E. 1974, ApJL 189, L5

[92] Osterbrock, D. E. 1981, ApJ 249, 462

[93] Osterbrock, D. E. 1989, Astrophysics of gaseous nebulae and active galactic nuclei, Univ. Science Books, CA, 422 p.

[94] Osterbrock, D. E.; Cohen, R. D. 1982, ApJ 261, 64

[95] Osterbrock, D. E.; Pogge, R. W. 1985, ApJ 297, 166

[96] Pâris, I.; Petitjean, P.; Aubourg, É.; Bailey, S.; Ross, N. P.; Myers, A. D.; Strauss, M. A.; Anderson, S. F.; et al. 2012, A&A 548, 66

[97] Paronyan, G. M.; Mickaelian, A. M. 2015, ApSS, in press

[98] Peterson, B. P. 1997, An Introduction to Active Galactic Nuclei, Cambridge Univ. Press, 238 p.

[99] Petrosian, A., McLean, B., Allen, R. J., MacKenty, J. W. 2007, ApJS 170, 33

[100] Reber, G. 1940, ApJ 91, 621

[101] Robson, I. E. I. 1996, Active Galactic Nuclei, Wiley, 350 p.

[102] Rowan-Robinson, M. 1977, ApJ 213, 635

[103] Sandage, A. 1965, ApJ 141, 1560

[104] Sanders, D. B.; Mazzarella, J. M.; Kim, D.-C.; Surace, J. A.; Soifer, B. T. 2003, AJ 126, 1607

[105] Sanders, D. B.; Mirabel, I. F. 1996, Ann. Rev. Astron. Astrophys. 34, 749

[106] Sargent, W. L. W.; Searle, L. 1970, ApJ 162, L155

[107] Sargsyan, L.; Mickaelian, A.; Weedman, D.; Houck, J. 2008, ApJ 683, 114

[108] Sargsyan, L.; Weedman, D.; Lebouteiller, V.; Houck, J.; Barry, D.; Hovhannisyan, A.;
       Mickaelian, A. 2011, ApJ 730, 19

[109] Schmidt, M. 1963, Nature 197, 1040

[110] Schmidt, M.; Green, R. F. 1983, ApJ 269, 352

[111] Schmitt, J. L. 1968, Nature 218, 663

[112] Schneider, D. P., Richards, G. T., Hall, P. B., et al. 2010, AJ, 139, 2360

[113] Seyfert, C. K. 1943, ApJ 97, 28

[114] Slipher, V. M. 1917, Pop. Ast 25, 36; Proc. Amer. Phil. Soc. 56, 403

[115] Smolcic, V. et al. 2006, MNRAS, 371, 121

[116] Smolcic, V. et al. 2008, ApJS, 177, 14

[117] Soifer, B. T.; Boehmer, L.; Neugebauer, G.; Sanders, D. B. 1989, AJ 98, 766

[118] Souchay J., Andrei A.H., Barache C., Bouquillon S., Suchet D., Taris F., Peralta R.
       2012, A&A 537, A99

[119] Stasinska, G.; Cid Fernandes, R.; Mateus, A.; Sodré, L.; Asari, N. V. 2006, MNRAS
       371, 972

[120] Stepanian, J. A. 2005, RMxAA 41, 155

[121] Stern, D., et al. 2005, ApJ, 631, 163

[122] Strittmatter, P. A.; Serkowski, K.; Carswell, R.; Stein, W. A.; Merrill, K. M.; Bur-
       bidge, E. M. 1972, ApJ 175, L7

[123] Teimoorinia, H.; Ellison, S, 2014, MNRAS 439, 3526

[124] Terlevich, R. 1997, RMAA 6, 1

[125] Terlevich, R. 2000, In: Advanced Lectures on the Starburst-AGN Connection, Eds.
       Aretxaga, I.; Mújica, R.; Kunth, D., World Scientific, p. 279

[126] Terlevich, R.; Melnick, J.; Masegosa, J.; Moles, M.; Copetti, M. V. F. 1991, A&AS
       91, 285

[127] Thuan, T. X.; Martin, G. E. 1981, ApJ 247, 823

[128] Tran, H. D.; Osterbrock, D. E.; Martel, A. 1992, AJ, 104, 2072

[129] Treister, E., Castander, F. J., Maccarone, T. J., et al. 2005, ApJ, 621, 104

[130] Trouille, L., Barger, A., Tremonti, C. A. 2011, ApJ, 742, 46

[131] Urry, C. M.; Padovani, P. 1995, PASP 107, 803

[132] Veilleux, S. 2002, ASP Conf. Ser. 284, 111

[133] Veilleux, S.; Osterbrock, D. E. 1987, ApJS 63, 295

[134] Véron, P.; Hawkins, M. R. S. 1995, A&A 296, 665

[135] Véron P., Lindblad P.O., Zuiderwijk E.J., Véron M.-P., Adam G. 1980, A&A 87, 245

[136] Véron, P.; Gonçalves, A. C.; Véron-Cetty, M.-P. 1997, A&A 319, 52

[137] Véron-Cetty, M.-P.; Balayan, S.K.; Mickaelian, A.M.; et al. 2004, A&A 414, 487

[138] Véron-Cetty, M.-P.; Véron, P. 2000, A&ARev 10, 81

[139] Véron-Cetty, M.-P.; Véron, P. 2010, A&A 518, A10

[140] Vogt, F. P. A., Dopita, M. A., Kewley, L. J. et al. 2014, ApJ 793, article id. 127

[141] Weedman, D. W. 1977, Vistas in Astronomy 21, 55

[142] Weedman, D. W.; Houck, J. R. 2009, ApJ 693, 370

[143] Weedman, D. W., Khachikian, E. Ye. 1968, Ap 4, 243

[144] Winkler, H. 1992, MNRAS 257, 677

[145] Winkler, H. 2014, Proc. IAU Symp. #304: Multiwavelength AGN Surveys and Studies, Eds. A. M. Mickaelian & D. B. Sanders, CUP, p. 28

[146] Woltjer, L. 1959, ApJ 130, 38

[147] Yan, R., Ho, L. C., Newman et al. 2011, ApJ, 728, 38

[148] Zakamska, N. L.; Strauss, M. A.; Krolik, J. H.; Collinge, M. J.; Hall, P. B.; Hao, L.; Heckman, T. M.; Ivezić, Ž.; et al. 2003, AJ 126, 2125

[149] Zel'dovich, Ya. B.; Novikov, I. D. 1964, SPhD 9, 246

[150] Zwicky, F. 1956, Ergebnisse d. exakt Naturwissenschaften, 29, 344

19

Geomagnetic Field Effect over Azimuth Anisotropy of Cosmic Rays via Study of Primary Particles

Mehdi Khakian Ghomi[1] · Maedeh Fazlalizadeh[1] · Mahmoud Bahmanabadi[2] · Hadi Hedayati Kh.[3]

[1] Department of Physics, Amirkabir University of technology, P.O.Box 15875-4413, Tehran, Iran; email: mehdi.khakian@yahoo.com
[2] Department of Physics, Sharif University of technology, P.O.Box 11155-9161, Tehran, Iran
[3] Department of Physics, Khajenasir University of technology, Tehran, Iran

Abstract. Geomagnetic field is a one of the candidates for creation of anisotropy in azimuth distribution of extensive air showers over the entered cosmic rays to the atmosphere. Here we present the question: *Is there any azimuth anisotropy flux in the upper level of the atmosphere due to the geomagnetic field over the entered cosmic-rays?* The obtained answer is: *yes.* This investigation showed an agreement with a similar functionality to the experimental results, but its amplitude is smaller. We calculated the effect over the primary protons and alpha particles with consideration of their abundances, and found that this effect by itself is not as large as the expected value of experimental results. Therefore we need to investigate the secondary particles for the next step. But for a real simulation of extensive air showers, it is necessary to be applied this anisotropy flux as a seed for the secondary particle in the next step.

Keywords: Azimuth anisotropy, Cosmic Ray, Gamma Ray, Extensive Air Shower, Geomagnetic field

1 Introduction

We receive Cosmic-Rays(CRs) and Gamma-Rays(GRs) from celestial masses in addition to their light. In the energy range of a few TeV and more, CRs and GRs create Extensive Air Showers(EASs) that their secondary particles are detectable on the ground. Gamma ray sources are observable by the GR based EASs, the procedure which is used by Cherenkov Telescope Arrays like CANGAROO, H.E.S.S. , MAGIC, MILAGRO and VERITAS. But GR based EASs are only less than 1% of the EASs, so their contribution cannot explain the anisotropy distribution of the detected EASs in azimuth directions. Since the CR-distribution in TeV energy range inside the Galaxy is homogeneous, we expect an isotropic distribution in azimuth angles. But in the EAS detections; it is observed a slight North-South anisotropy flux of the CRs in the energy range, so it is expected a local effect over the EASs. There are so much attempts to understand the nature of the observed anisotropy in the EAS arrays [1, 2, 3]. Some analysis and calculations have been presented for explanations of it, like slope of the ground, time variations of the geomagnetic field, or dimension and direction of the field in the location of CR observations. But still non of the methods could not present a complete explanation for the anisotropy.

For a complete investigation it is needed to categorize the problem to smaller subproblems, like the geomagnetic field effect over *i)* primaries, *ii)* different secondaries separately and *iii)* investigation of group behavior of the secondaries as an EASs. Meanwhile there are many different works over shadow of the moon and the sun by EAS arrays. It has been well seen that the sun's shadow is weaker than moon's [4]. It means that the effect over primaries is not negligible. The anisotropy is usually compared with a simple harmonic function with the first two harmonics:

$$A = N_0(1 + A_I \cos(x - \phi_I) + A_{II} \cos(2x - \phi_{II})) \tag{1}$$

In each point on the earth, we have two components of the geomagnetic field, one is to the nadir (zenith) in the north (south) hemisphere and the other is horizontal to the north (from the south). So we expect a complete period of 2π anisotropy in azimuth distribution everywhere. Also around the geomagnetic equator we have a symmetry in value of the field in North and South, therefore we expect a half period of π anisotropy in azimuth distribution around the geomagnetic equator. It is interesting that A_I (A_{II}) in equation 1 increases (decreases) with increase of observation latitude, which shows the North-South anisotropy [1, 3, 5]. At Tehran (35°.67 N, 51°.33 E, $\theta_H = 38°$) both A_I and A_{II} are important but always A_I is larger than A_{II}. In Tibet array (30°.11 N, 90°.53 E, $\theta_H = 45°$) both A_I and A_{II} are equally prevailing. At Yakutsk (62° N, 130° E, $\theta_H = 14°$) A_I and in Chakaltaya (16°.35 S, 68°.20 W, $\theta_H = 16°$) A_{II} are dominant [3]. Observations in different points show that the anisotropy amplitude is more in higher zenith angle events [1], and also is more in the results of higher latitude observatories. These results guides us to investigate the geomagnetic field as a good candidate for this phenomena [6]. Since the size of the geomagnetic field is about 50,000 nano-Tesla and its variations in a year is only about a few nano-Tesla (less than 50 nT/yr)(http://ngdc.noaa.gov), it seems that it is more wisely to investigate at first the effect of the geomagnetic field itself respect to its variations. This effect has been studied in the lower energies by A.M. Hillas [7] and J. Clay [8] for solar winds and investigations of van-allen belts and forbidden regions around the earth.

The geomagnetic field is approximately a dipole magnet which is located inside the Earth very near to the center, so its field has been expanded in a vast region around the Earth. Therefore the geomagnetic field at first; affects over primary CRs out of the atmosphere and then on secondary particles into the atmosphere. Energy of the primaries is very larger than the related secondaries, but the affecting time over the primaries is very larger than the secondaries. Dominant contribution of the North-South anisotropy can be due to *i)* deviation of primary CRs through the expanded geomagnetic field out of the atmosphere, *ii)* group deviation of created secondaries of an EAS inside the atmosphere or *iii)* both.

In this work we investigated the effect over primaries. In the second section we discuss about nature of the geomagnetic field and its effect over CRs. In the third section we present our calculation details of the effect over primaries. In the fourth section we show our results, and finally in the fifth section we present our discussion, concluding remarks and our agenda for the following works.

2 Geomagnetic Field

Geomagnetic field is due to a good approximate dipole magnet inside the earth which has an angle of 11°.5 with the earth rotation axis. Its magnetic moment is 8.1×10^{25} Gs.cm^3 and shifted by 342 km relative to the earth center [9]. Magnetic north and south poles are at (83° N, 105° W) and (72° S, 110° W) respectively.

To check our magnetic dipole we used the calculator of "National Geography Data Center (NGDC: http://ngdc.noaa.gov)" for heights above Tehran (35°.67 N, 51°.33 E). For verification of the magnetic field at different heights, we considered Tehran location and obtained the field at different heights until 1000 km (the recommended region in NGDC for calculation of the magnetic field) by steps of 50 km. Figure 1 shows a very good agreement between

NGDC data and the dipole magnetic field. The results of the curve is used in the next steps of our calculations.

Since EASs secondary particles are creating inside the atmosphere with a very smaller

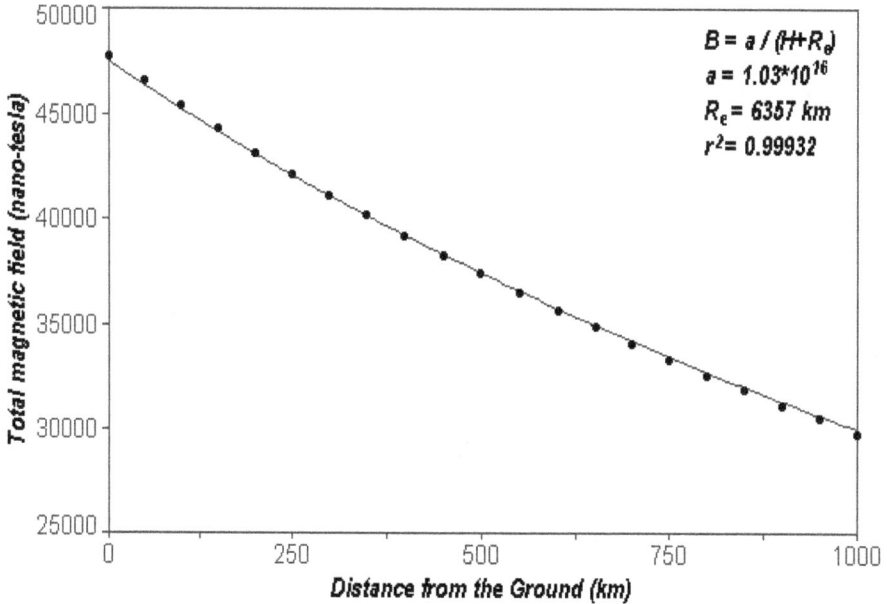

Figure 1: Magnetic field of the Earth, obtained from NGDC. It shows a good fit with dipole magnetic moment field ($B \propto 1/r^3$)[10] in different heights.

thickness relative to the dimensions of the earth magnetic atmosphere, therefore direction and size of the magnetic field is constant for secondaries [11], but this approximation is not true for primary particles. The field zone for the primaries is too vast to consider the geomagnetic factors constant, so they change with height, latitude and longitude which must be consider in our calculations.

3 Calculation Details of The Geomagnetic Field Effect over Primary Particles

In this stage we need a magnetic atmospheric zone. Since the geomagnetic field is a dipole magnet so $B \propto 1/r^3$ and in $r = 20,000$ km from earth center $B = (6,557/20,000)^3 B_0 \simeq 0.03 B_0$. Therefore we define our magnetic atmosphere, a sphere around the earth with $r = 20,000$ km to the ground which most of the primary deviations are in the zone. Details of the deviation calculation of the primaries is presented as follows :

3.1 Calculation Details

Our reference coordinate system (O) is the equatorial coordinates but with longitude and complementary angle of latitude as spherical angles (β, $\pi/2 - \lambda$). Tehran's local coordinate system (O') is; azimuth and zenith angles (ϕ, θ). Direction of the primary CRs in the local coordinates is $\mathrm{P}(\phi, \theta) = |\mathrm{P}| \sin\theta \cos\phi \hat{i}' + |\mathrm{P}| \sin\theta \sin\phi \hat{j}' + |\mathrm{P}| \cos\theta \hat{k}'$ where $\hat{i}', \hat{j}'$ and $\hat{k}'$ are unit vectors in O' which are equivalent to $-\hat{\theta}$, $-\hat{\phi}$ and $\hat{r}$ in O respectively (Figure 2). So $P(\phi, \theta)$ in the equatorial coordinates is $P(\beta, \lambda)$:

$$P(\beta, \lambda) = -|P| \sin \theta \cos \phi \hat{\theta} - |P| \sin \theta \sin \phi \hat{\phi} + |P| \cos \theta \hat{r} \qquad (2)$$

For determination of the inclination, we used *Lorentz Force* over the primary CRs from the starting point (x_{max}) to the upper level of atmosphere (8.4 km effectively) [12] and integrated over all of the path.

Maximum zenith angle in O' coordinates is $\theta_{max} = 60°$ [5], which is observable by our EAS array in Tehran (http://observatory.sharif.ir). Figure 2 shows the geometrical configuration of the two coordinates. An event with zenith angle θ in the local coordinates, has latitude $\lambda = (35 + \omega)°$ where ω is central angle between PO and OO' in Figure 2.

With maximum amount of $x = x_{max} = 20,000$ km and $\theta = \theta_{max} = 60°$ the highest latitude

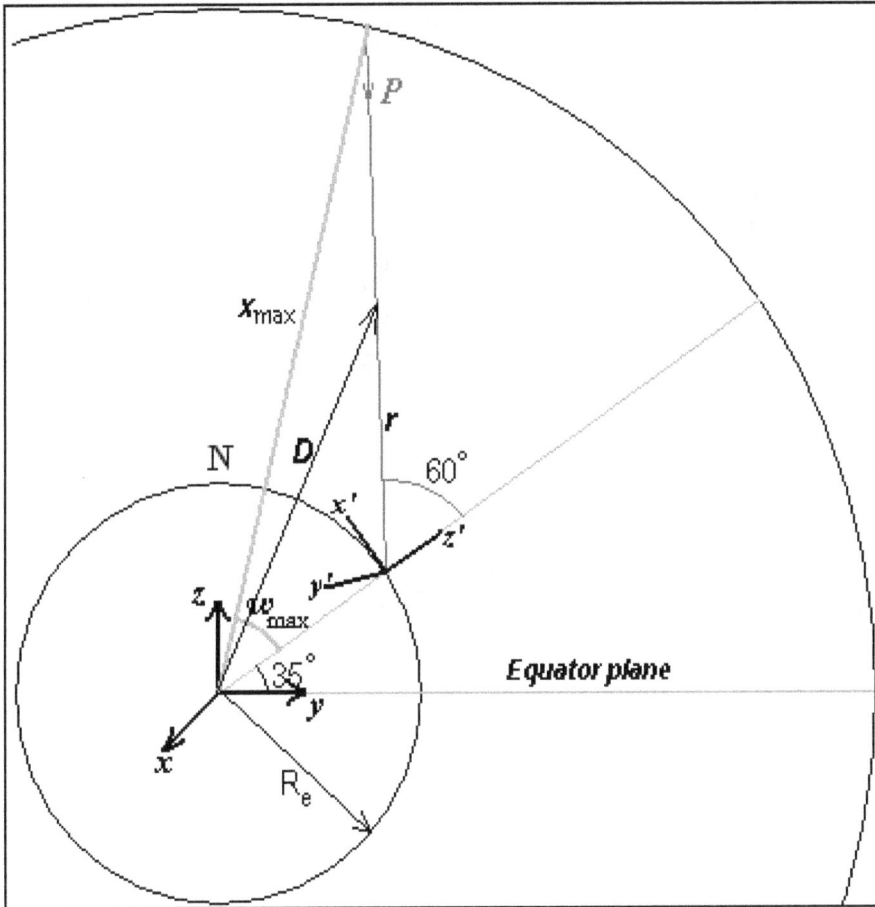

Figure 2: Geometry of the particles around the Earth. The particle comes to the magnetic atmosphere of the earth (x ≤ 20,000 km) and is affected by the magnetic field.

of the CRs is $\lambda_{max} = (35 + \omega_{max})°$, which $\omega_{max} = \cos^{-1}((R_e + r_{max} \cos \theta_{max})/x_{max}) = 44°$ where R_e is the Earth radius. Therefore we need the magnetic field components in an area around Tehran with an angular radius of $44°$.

We obtained $B_{x'}$, $B_{y'}$ and $B_{z'}$ in the intervals of $\lambda \subset [79, -9]$ and $\beta \subset [7, 95]$ by accuracy of $1°$. Therefore we used integration differential $\delta r = r/44$ in the particle's path. In each step of the integration we subtracted δr from distance of the particle to the upper level of atmosphere ($|\mathbf{r}| \longrightarrow |\mathbf{r}| - \delta r$), so λ, β and x change and in follow $B_{x'}$, $B_{y'}$ and $B_{z'}$ change too. Therefore P obtains :

$$P(\phi, \theta; \beta, \lambda) = |\mathbf{P}| \sin\theta \cos\phi (\cos\lambda \cos\beta \hat{i} + \cos\lambda \sin\beta \hat{j} + \sin\lambda \hat{k})$$
$$+ |\mathbf{P}| \sin\theta \sin\phi (\sin\lambda \cos\beta \hat{i} + \sin\lambda \sin\beta \hat{j} - \cos\lambda \hat{k})$$
$$+ |\mathbf{P}| \cos\theta (-\sin\beta \hat{i} + \cos\beta \hat{j})$$

To find location of the particle in equatorial coordinates, in each step we obtained the vector $\mathbf{x} = \mathbf{r} + \mathbf{R_e}$ which $\mathbf{x}$ is the related vector of each step of the particle respect to the earth

$$\tan\beta = \rho_y/\rho_x \qquad , \qquad \tan\lambda = \sqrt{\rho_x^2 + \rho_y^2}/\rho_z \qquad (3)$$

where ρ_x, ρ_y, ρ_z and $x = (r^2 + R_e^2 + 2rR_e \cos\theta)^{1/2}$ are the components and size of the vector $\mathbf{x}$ in the equatorial coordinates. Consequently we will be able to calculate the size of $\mathbf{B}$ components $B(\beta, \lambda, x) = B(\beta, \lambda)(R_e/x)^3$ in each step. At the end of each step, for obtaining final momentum due to Lorentz force $\mathbf{F} = q(\mathbf{P} \times \mathbf{B})/\gamma m_0$, we calculated momentum differential $\Delta\mathbf{P} = \int \mathbf{F} dt$ where $dt = \delta r/c$. In the energy range it is clear that in each step angle between the particle's momentum and the geomagnetic field is constant, so we write $\Delta\mathbf{P} = \mathbf{F}\delta r/c$ and in the next step we have $\mathbf{P}(t + \Delta t) = \mathbf{P}(t) + \Delta\mathbf{P}$, and obtains a new direction. This process will continue until x becomes smaller than the x of Tehran's atmosphere ($R_e + 8.4$ km), and at the end, final angles (ϕ_f, θ_f) are given by $\mathbf{P}_f$ and the deflection angle obtains by comparison of $\mathbf{P}_i$ and $\mathbf{P}_f$.

4   Results

By using a *uniform* random generator for azimuth angles (ϕ_0) of primaries and a "$\sin\theta_0$" random generator (Due to solid angle effect) for zenith directions (θ_0). In our experiment the energy threshold ia about 40 TeV [5], so we used the energies from 1 to 100 TeV, and generated 5 sets of 1.5×10^8 events with energies of 1, 3, 10, 30 and 100 TeV. In each energy, we categorized the events with zenith angles in $0 - 10°$, $10° - 20°$, $\cdots$, $50° - 60°$ bins and azimuth angles in $0 - 30°$, $30° - 60°$, $\cdots$ $330° - 360°$ bins. The obtained results in different energies, zenith angles and azimuth angles distributions are presented respectively as follows :

4.1   Energy Dependence of the Anisotropy

For investigation of flux anisotropy in different azimuth angles, it is used the five large data sets and it is obtained azimuth deflection angle distribution for the data sets. In the investigation at first, we separated azimuth angle of the events in 12 of $30°$ bins, and then fitted the anisotropy function (equation 1) over the five energy sets of the simulated data. The results of the fittings has been presented in Table 1. In Figure 3 it is seen that the amplitude of the first harmonic is decreasing with increase of the primary energy. This means that the anisotropy amplitude in the first harmonic, decreases with increase of energy. But the difference is so small, it means that this effect is essentially very weak but it shows the right physical effect of the geomagnetic field over higher and lower energies. Since the fluctuations in the obtained ensemble ($\Delta\phi = \phi_f - \phi_i$) is high, we see large error bars, but all of the events in the data sets averagely show a very fine right physical effect.

Table 1: Anisotropy coefficients of equation 1 in different energies for all zenith angles, E is the primary particle energy

$E\ (TeV)$	$N_0(\times 10^7)$	$A_I(\times 10^{-4})$	$\phi_I(°)$	$A_{II}(\times 10^{-4})$	$\phi_{II}(°)$
1	1.24962	1.89511	190.19	2.88129	95.04
3	1.24962	1.86892	187.74	2.98770	94.38
10	1.24962	1.85640	188.71	2.92697	91.87
30	1.24962	1.85222	190.86	2.98020	94.10
100	1.24962	1.83365	188.81	3.06506	96.72

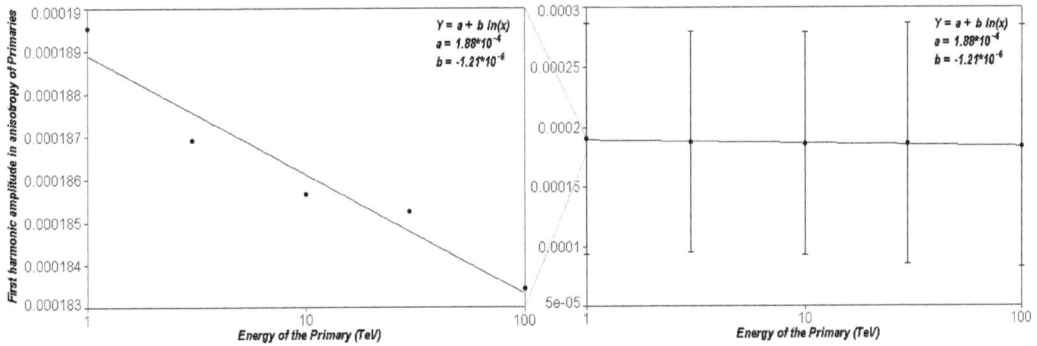

Figure 3: Amplitude of the first harmonic function over azimuth distribution of the simulated events, for 5 differerent energies: 1, 3, 10, 30 and 100 TeV.

4.2   Deflection Angle vs. Primary Zenith Angle

Deflection angle of the particles are obtained by comparing direction of the primary particle before and after passing through the geomagnetic field. As we expected the deflection angle γ ($\cos\gamma = \cos\theta_0 \cos\theta_f + \sin\theta_0 \sin\theta_f \cos(\phi_f - \phi_0)$) increases with increase of θ_0. The best fit for the deflection angle vs. zenith angle of the events obtained by function "$\gamma = \mathbf{a} + \mathbf{b}\theta^{\mathbf{c}}$" over the obtained data. It is seen that in different energies, the variations of $\mathbf{a}$, $\mathbf{b}$ and $\mathbf{c}$ is very small (Table 2), which is in agreement with the results of section 4.1. These results have been averaged over the simulated events for each angle interval. As an example, Figure 4 shows the distribution for 100 TeV events. The results in the *total* row of the Table 2 show that this effect is actually weak-sensitive respect to the energy of the primaries, which is in agreement with the first result.

4.3   North-South Anisotropy

Figure 5 shows a North-South anisotropy respect to the geomagnetic north ($11°.5\ N$) in azimuth (ϕ) angles of the primary particles. The anisotropy is obtained for the five data sets and obtained a very small variation in $\Delta\phi = \phi_f - \phi_i$. In the experiments we have a deficit in North part and an excess in South [1, 2, 3]. If we compare this result with

Table 2: It is seen a similar behavior of deflection angle of the primaries in different energies in the geomagnetic field with increase of zenith angle. a, b and c are fitting constants of the function $\gamma = a + b\theta^c$.

E (TeV)	a	b	c	r^2
1	19.19	0.235	1.232	0.99369
3	19.09	0.253	1.214	0.99283
10	19.54	0.206	1.264	0.99277
30	19.45	0.221	1.247	0.99189
100	19.19	0.235	1.232	0.99253
total	$19.29(1 \pm 0.01)$	$0.230(1 + 0.07)$	$1.238(1 \pm 0.01)$	

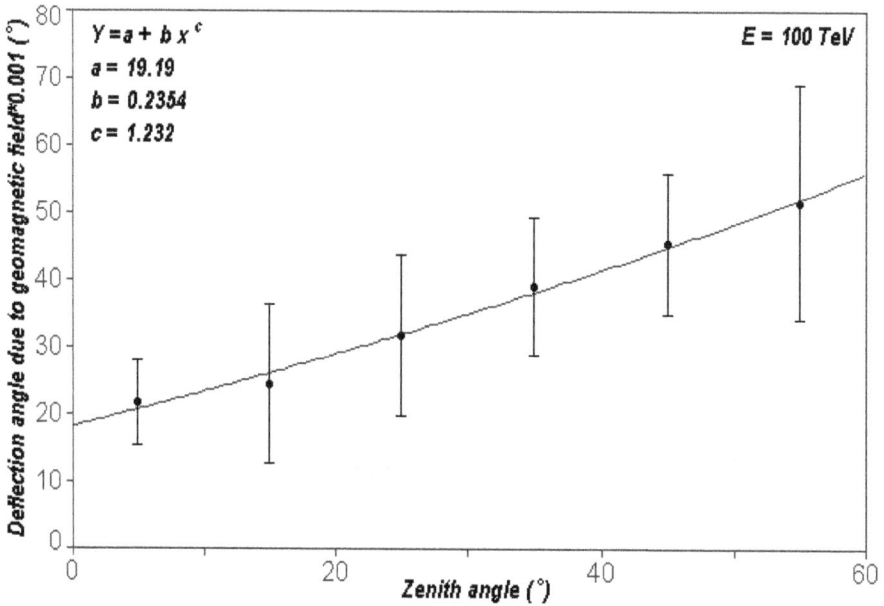

Figure 4: Deviation of the primary particles after passing through the geomagnetic field vs. zenith angle of the events for $E = 100$ TeV. Black points and error bars are average and standard deviations in each angle interval.

the experimental results, it is seen that there is a π phase difference between them. The result is quite right because we drew the deflection angle and in north part we see maximum deflection which shows maximum deficit, and in south part we se minimum deflection which shows a pile up due to deflection from border events in the neighbor bins.

5   Discussion and Concluding Remarks

It is seen that in the experiments results, the observed anisotropy (amplitude of the first or second harmonics) is about few percent, but in the investigation the anisotropy due to the geomagnetic field over primary particles is very smaller than the amount that we expect. It is about %0.01 about hundred times smaller. If it is payed attention to the title of y-axis of

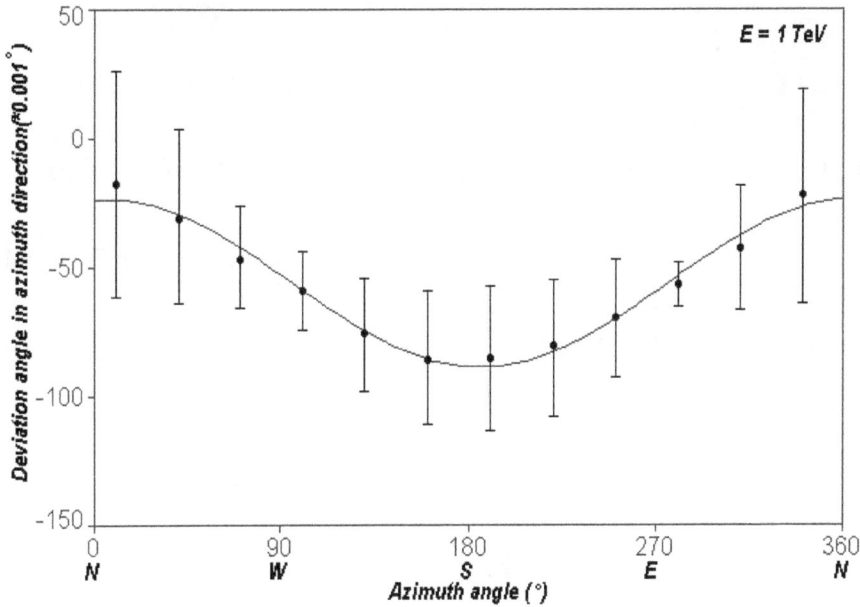

Figure 5: Average azimuth deviation in different azimuth directions which shows a North-South anisotropy.

figure 4 and 5 we have a coefficient of 0.001°. This shows a very small angular deflection, consequently the anisotropy difference in different energies is small too. From Table 1 it is seen that the relative amplitude difference in different energies (ΔA_i, $i = I$, II) to the amplitudes (A_i) is about 0.01, so observation of any variations in different energies from $1\ TeV$ to $100\ TeV$ is not so easy. In this investigation we see that the anisotropy due to the primary particles shows expectable physical behavior (functionalities) but is smaller than our expected experimental results.

Usually for the estimation of angular resolution of EAS arrays the moon and sun shadow are investigated [4]. In most of the works it is observed a weaker and broader shadow of the sun respect to the moon. Usually presented a reason for the phenomena which is the magnetic field of the sun (which affects over primaries). Usually it is seen that the angular size of the sun's shadow is about few degrees. If we compare the $x_{max} \sim 20,000$ km respect to the sun-earth distance$\sim 150 \times 10^6$ km it is about 7 to 8 thousand times. Therefore we can expect that the deflection is about the order of thousandth of degrees.

Hence for the next agenda we investigate the geomagnetic field effect over the secondary particles of EASs and their group behavior with consideration of the obtained anisotropy as a seed for their primaries.

Acknowledgment

This research was supported by a grant from office of vice president for science, research and technology of Amirkabir University of Technology (Tehran Polytechnic).
I thank Dr. Paolo Bernardini for his so much constructive advises for understanding the geomagnetic field problem over EASs.
Also I thank Prof. G. Schutz for his constructive comments.

References

[1] Ivanov A. A., Egorova V. P., Kolosov V. A., Krasilnikov A. D., Pravdin M. I., Sleptsov H. Y., 1999, JETP letters, 69(4), 288

[2] He H. H., Bernardini P., Calabrese Melcarne A. K., Chen S. Z., 2007, Astropart. Phys., 27, 528

[3] Bahmanabadi M., Anvari A., Khakian Ghomi M., Pourmohammad D., Samimi J., Lamehi Rachti M., 2002, Experimental Astronomy, 13, 39

[4] Oshima A., Gupta S. K., Hayashi K., et al., 2008, Proc. 30th ICRC, Mexico, 3, (OG part 2), 1515-1518

[5] Khakian Ghomi M., Bahmanabadi M., Samimi J., 2005, A&A, 434, 459

[6] Khakian Ghomi M., Bahmanabadi M., Samimi J., Shadkam A.H., Sheydaei F., Anvari A., 2007, Proc. 30th ICRC, Yukatan Mexico, HE11-0263

[7] Hillas A. M., 1972, Cosmic Rays. Pergamon Press, p. 22

[8] Clay J., 1972, Proc. Roy. Acad., Amsterdam, Vol. 30, p. 1115

[9] Dorman L. I., 2008, Cosmic Rays in Magnetospheres of the Earth and other Planets, Spriger, P. 9

[10] Reitz J. R., Milford F. J., Christy R. W., 1979, Foundations of Electromagnetic Theory, 3rd edition, Adison Wesley

[11] He H. H., Bernardini P., Calabrese Melcarne A. K., Chen S. Z., 2005, Proc. 29th ICRC, Pune India, Vol. 6, P. 5

[12] Gaisser T.K., 1990, Cosmic Rays and particle Physics, cambridge University Press New York, p. 35

PERMISSIONS

LIST OF CONTRIBUTORS

Habib G. Khosroshahi
School of Astronomy, Institute for Research in Fundamental Sciences (IPM), Tahran, Iran

Louisa A. Nolan
School of Physics and Astronomy, The University of Birmingham, Birmingham B15 2TT, UK

Mahdi Yousefzadeh
Department of Physics, Institute for Advanced Studies in Basic Sciences (IASBS), Zanjan, Iran

Mohsen Javaherian and Hossein Safari
Department of Physics, University of Zanjan, Zanjan, Iran

Vahid Abbasvand, Hossein Ebadi and Zahra Fazel
Astrophysics Departartment, Physics Faculty, University of Tabriz, Tabriz, Iran

Maryam Ghasemnezhad
Faculty of physics, Shahid Bahonar University of Kerman, Kerman, Iran

Akram Gheidi Shahran and Taghi Mirtorabi
Physics Department, Alzahra University, Vanak, 1993891176, Tehran, Iran

S. Y. Rokni, H. Razmi and M. R. Bordbar
Department of Physics, University of Qom, Qom, I. R. Iran

Azam Rafiei, Kurosh Javidan and Mohammad Ebrahim Zomorrodian
Department of Physics, Ferdowsi University of Mashhad, 91775-1436, Mashhad, Iran

S. Karbasi
Department of Physics, University of Qom, Qom, I. R. Iran

Mohsen Bigdeli, Nariman Roohi and Mina Zamani
Department of Physics, University of Zanjan, Zanjan, Iran

Neda Amjadi, Vahid Abbasvand and D.M Jassur
Astrophysics Department, Physics Faculty, University of Tabriz, Tabriz, Iran

Ali Koohpaee
Department of Energy Engineering and Physics, Amirkabir University of Technology, Tehran, Iran

Mehdi Khakian Ghomi
Department of Energy Engineering and Physics, Amirkabir University of Technology, Tehran, Iran

Amir Asaiyan
Department of Physics, Faculty of Basic Sciences, University of Mazandaran, Babolsar, Iran

A. Aghamohammadi
Sanandaj Branch, Islamic Azad University, Sanandaj, Iran

Elham Saremi and Abbas Abedi
Department of Physics, Faculty of Science, University of Birjand, Birjand

Atefeh Javadi and Habib Khosroshahi
School of Astronomy, Institute for Research in Fundamental Sciences (IPM), Tehran, Iran

Jacco van Loon
Lennard-Jones Laboratories, Keele University, ST5 5BG, UK

Seyede Tahere Kash and Shahram Abbassi
Department of Physics, Ferdowsi University of Mashhad, Mashhad, Iran

Elham Bazyar
Research Institute for Astron. & Astrophys. of Maragha (RIAAM), 55134-441 Maraghe, Iran

Ali Ajabshirizadeh
Research Institute for Astron. & Astrophys. of Maragha (RIAAM), 55134-441 Maraghe, Iran

Jean-Pierre Rozelot
Nice   university,   OCA-CNRS-Lagrange
Depart., Av. Copernic, 06130 Grasse, France

Zahra Fazel
Astrophysics   Depart.,   Physics   Faculty,
University of Tabriz, Tabriz, Iran

Areg M. Mickaelian
Byurakan Astrophysical Observatory (BAO),
Byurakan 0213, Aragatzotn Province, Armenia

**Mehdi   Khakian   Ghomi   and   Maedeh
Fazlalizadeh**
Department of Physics, Amirkabir University
of technology, Tehran, Iran

Mahmoud Bahmanabadi
Department of Physics, Sharif University of
technology, Tehran, Iran

Hadi Hedayati Kh.
Department of Physics, Khajenasir University
of technology, Tehran, Iran

Index